H.F. Hameka
University of Pennsylvania

Quantum Theory of the Chemical Bond

HAFNER PRESS
A Division of Macmillan Publishing Co., Inc.
New York
Collier Macmillan Publishers
London

Copyright © 1975 by Hafner Press, A Division of
Macmillan Publishing Co., Inc.

All rights reserved. No part of this book may be reproduced
or transmitted in any form or by any means, electronic or
mechanical, including photocopying, recording or by any information
storage or retrieval system, without the permission in writing
from the Publisher.

Hafner Press
A Division of Macmillan Publishing Co., Inc.
866 Third Avenue, New York, N.Y. 10022

Collier-Macmillan Canada Ltd.

Library of Congress Cataloging in Publication Data
Hameka, Hendrik F
   Quantum theory of the chemical bond.

Includes bibliographies.
    1. Quantum chemistry. 2. Chemical bonds.
    I. Title
QD462.H28   541'.28   74-14742
ISBN 0-02-845-5660-4

Printed in the United States of America

TO CHARLOTTE
AND TO MY PARENTS

PREFACE

This book is written for a one-semester course at the undergraduate or first-year-graduate level in quantum chemistry. It discusses the properties of the electronic structure of molecules from a theoretical point of view and gives the theoretical background of the customary experimental methods and for explaining molecular structure, such as spectroscopy and magnetic resonance.

The mathematical level is kept as low as possible, but it is necessary for the reader to be familiar with calculus to gain an adequate understanding of the theory. My approach to each topic has been, first, to give a general discussion from a chemical or physical point of view, next to state the mathematical expressions that describe the phenomenon, and, finally, to prove the validity of these mathematical expressions. In a course for graduate students, the instructor will want to cover these more sophisticated derivations. Therefore, I have placed most mathematical proofs at the end of each section or in a separate section by themselves.

Most chemistry departments offer a course in quantum chemistry variously called "Valence," "Molecular Structure," "Advanced Physical Chemistry," or "Quantum Chemistry," designed for junior/senior undergraduates or for first-year graduate students. The first semester of a two-semester course is usually devoted to a thorough discussion of the principles of quantum mechanics; applications of quantum theory to chemical systems are discussed during the second semester. In the one-semester course the general theory of quantum mechanics must be formulated briefly to leave enough time for treating the chemical aspects of the subject.

While my book is written for the one-semester course, it purposely contains more material than one can comfortably cover in one semester. This will enable instructors, few of whom agree on which topics are essential, to structure the course according to their own tastes. Therefore, I hope this book will be found suitable for a variety of courses at various levels.

CONTENTS

1   The principles of quantum mechanics

1. Introduction ... 1
2. The wave function ... 5
3. The Bohr atom ... 11
4. Stationary states and the particle in a box ... 14
5. The color of conjugated organic molecules ... 19
6. Operators and their properties ... 22
7. Approximate methods in quantum mechanics ... 28

   Problems ... 34
   Bibliography ... 35

2   Exact solutions of the Schrödinger equation

1. Introduction ... 37
2. The harmonic oscillator ... 38
3. Vibration of diatomic molecules, an example ... 45
4. The rigid rotor ... 48
5. Angular momentum ... 55
6. The hydrogen atom ... 63
7. Properties of the eigenvalues and eigenfunctions of the hydrogen atom ... 72

   Problems ... 83
   Bibliography ... 85

3 Atomic structure

1. Introduction — 87
2. Calculations on the He atom — 96
3. The electron spin — 102
4. Exclusion principle and spin — 109
5. Atomic orbitals — 114

   Problems — 120
   Bibliography — 121

4 Light and spectroscopy

1. The nature of light — 123
2. Transition probabilities — 131
3. Transition probabilities and quantum theory — 135
4. The maser and the laser — 139

   Problems — 143
   Bibliography — 144

5 The spectra of diatomic molecules

1. Experimental information — 145
2. The Born-Oppenheimer approximation — 150
3. The vibrational motion of a diatomic molecule — 158
4. Molecular symmetry — 166
5. Selection rules and spectral intensities in electronic bands — 169
6. Intensities in the infrared and microwave regions — 182

   Problems — 183
   Bibliography — 185

6 The chemical bond

1. Introduction — 187
2. The hydrogen molecular ion — 189
3. The hydrogen molecule — 196
4. Diatomic molecules — 202
5. Electronegativity — 212
6. Hybridization — 216

7. Unsaturated molecules		227
8. Conjugated and aromatic molecules		232
Appendix		244
Problems		246
Bibliography		248

7 The solid state

1. Crystal structures	249
2. Ionic crystals	252
3. Metals and semiconductors	255
4. Molecular crystals	265
Problems	270
Bibliography	271

8 Magnetic resonance

1. Introduction	273
2. Relaxation phenomena	279
3. Chemical shifts	286
4. Spin-spin coupling	290
5. Electron spin resonance	300
Problems	306
Bibliography	307

Index   309

CHAPTER 1

THE PRINCIPLES OF QUANTUM MECHANICS

1. Introduction

The theory of valence is concerned with the problem of predicting the properties of a molecule by calculating the motion of its nuclei and its electrons. It is well known that the motion of macroscopic systems may be described by means of classical or Newtonian mechanics. The term "macroscopic system" means here the objects with which we deal in everyday life, such as cars, trains, ships, baseballs, and so on, or much larger objects such as stars or planets. However, if we study the motion of an electron in a molecule, then we deal with a particle that has a mass of the order of 10^{-27} g and that moves around in an orbit of the order of 10^{-8} cm. For such a small system classical mechanics is no longer valid. Instead we use a new type of mechanics, which was developed in the period between 1900 and 1930 and which is known as quantum mechanics or wave mechanics.

Many people think that quantum mechanics is much more difficult to learn than classical mechanics because the mathematical techniques in quantum mechanics are more difficult. Actually, the mathematical techniques in both types of mechanics are of equal complexity. Classical mechanics seems easier because most people have an intuitive feeling for the principles of classical mechanics from their experience in everyday life. For example, driving a car, hitting a baseball, or throwing a football all involve classical mechanics. Most people are familiar with the physical quantities that play a role in classical mechanics even through they may not be aware of it. It is convenient to begin our study of mechanics with a discussion of these quantities.

2   THE PRINCIPLES OF QUANTUM MECHANICS

The simplest situation that we can consider in classical mechanics is the motion of a point particle in one dimension. An example would be a car moving along a straight highway if we do not worry about the dimensions of the car. The motion of the particle (or of the car) is completely determined if we know its position as a function of time. If we use the symbol x to describe the position, we can write this as $x(t)$.

The main purpose of classical mechanics is to determine how the position x depends on the time t. This is usually achieved by solving the equation of motion, which is a differential equation. In order to find the solution of this equation, it is necessary to know both the position and the velocity of the particle at a given time, say $t = 0$. In this procedure it is assumed that $x(0)$ and $v(0)$, the velocity at $t = 0$, are known exactly. In that case the function $x(t)$ determines exactly where the particle is at any given time t.

If we consider the same problem, namely the motion of a point particle in one dimension, in quantum mechanics, then we must solve a different equation of motion, which is again a differential equation. However, at this stage we should not worry about differences in mathematical techniques, because the difference between quantum mechanics and classical mechanics is much more basic than just these mathematical details. We should realize that the essential concepts in the two kinds of mechanics are fundamentally different. We have already mentioned that the main purpose of classical mechanics is to determine the position x of the particle as a function of the time t, that is, to derive the function $x(t)$.

In principle, classical mechanics leads to exact predictions about the function $x(t)$, which is called the orbit of the particle. In practice, however, this is usually not true. Even if we consider the simple situation we discussed above, we have to measure $x(0)$ and $v(0)$, the position x and the velocity v of the particle at time $t = 0$, before we can calculate $x(t)$. The observations of $x(0)$ and $v(0)$ are necessarily subject to the possible inaccuracies of our measurements, and our prediction of $x(t)$ should also allow for these possible deviations. One of the basic differences between classical mechanics and quantum mechanics is in the approach to the treatment of these possible experimental errors.

In classical mechanics it is assumed that there are no limits to the accuracy of our observations and that it is in principle possible to improve our measuring techniques to the extent where experimental errors become arbitrarily small. In quantum mechanics, on the other hand, we have to recognize that there is a finite limit to the accuracy of our measurements.

Let us illustrate this difference in viewpoint by considering a simple and familiar situation, namely the motion of an ocean liner. If it is a calm day in mid-ocean and the weather forecast predicts no change, then it may be assumed that the ship will move with a constant velocity v. A measurement of v shows that this velocity is 16 nautical miles per hour. On most ocean liners the purser runs a betting pool every day, where every passenger can predict the distance that the ship covers from noon on one day until noon the next day. The mileage is rounded off at the nearest integer, and the passenger who predicts the right number (or comes closest to it) wins the pot. Obviously, the calculation of the ship's daily course is a useful application of classical mechanics to anyone who happens to be a passenger on the ship.

Actually, we do not need to know much classical mechanics to calculate the ship's course over a 24-hr period. Since the velocity v is 16 miles/hr, the result is simple $24 \times 16 = 384$ nautical miles. According to classical mechanics this result should be rigorous; this means that if we give this number to the purser we should have a 100% chance of winning the pool. We know that in real life we are usually not that lucky. There are many unforeseeable factors that might cause the ship to speed up or slow down during the next 24 hr, the measurement of the velocity v is subject to possible errors, the weather might change and, as a consequence, it is easily possible that the ship's course might be different from 384 miles. An experienced bettor does not believe that the prediction of 384 miles is 100% accurate, and he will attempt instead to make a guess about the accuracy of his number. Now, in the situation that we have described there is not enough information available to make such a guess, but for the sake of the argument, we assume that we know the probability pattern for the prediction and that the probability pattern is given under A in Table 1.1. We should imagine that this probability pattern contains all the uncertainties in the weather, the measurements, possible engine trouble, and so on. We may then conclude from probability pattern A that there is only a 22%

TABLE 1.1 ● Probability patterns A and B for the prediction of the ship's course.

Course	A Probability (%)	Course	B Probability (%)
<380	1		
380	2		
381	6	<382	1
382	12	382	4
383	18	383	25
384	22	384	40
385	18	385	25
386	12	386	4
387	6	>386	1
388	2		
>388	1		

chance that the prediction of 384 miles will turn out to be exact, that there is a 36% chance that the prediction is 1 mile off, and so on. We may also imagine a different situation with fewer uncertainties, which is described by probability pattern B, where the chance for a correct prediction is 40%, the chance for a 1-mile error is 50%, and so on.

We see that classical mechanics is actually an idealization of everyday life situations, because we assume that in classical mechanics everything can be measured with 100% accuracy and that there are no unforeseeable changes in our system. The result of these assumptions is that in classical mechanics the predictions about the future behavior of our system should be rigorously accurate. We have seen in the above example that in real life these assumptions are not always satisfied, and that predictions about the future positions of our system involve probability patterns rather than exact predictions. Such probability patterns play a prominent role in the formalism of quantum mechanics.

The above example is perhaps somewhat misleading, because we know that for macroscopic systems it is in principle possible to improve our measuring techniques to any extent that we desire. However, if we now consider the motion of an electron in a molecule, then the situation becomes completely different. Here it is not only impossible in practice to measure the position and velocity of the electron with unlimited accuracy, but it is not even possible in principle to perform such measurements. The only way to obtain information about the electron is by letting it interact with radiation or with other particles. Such interactions necessarily affect the motion of the electron. Consequently, we have to choose between two unattractive alternatives in planning experiments on the motion of the electrons. In the first alternative we use as little radiation as possible—then the motion of the electron is not seriously affected by our measurement, but on the other hand we get only a very crude idea of the electron's position. In the second experiment we use enough radiation to obtain an accurate result for the electron's position—but then the electron will be subject to a drastic change in velocity as a result of our measurement. It follows that we cannot measure both the position and the velocity of the electron with unlimited accuracy. As soon as we try to make the result for one of the two quantities more accurate, the other becomes less accurate. In order to put this limitation into mathematical form, we introduce Δx as the possible error in our determination of the electron's position and Δv as the possible error for the velocity. It follows that there is a lower limit for the product of the uncertainties, and we can write this as

$$\Delta x \cdot \Delta v > A \qquad (1)$$

This result was first obtained by Heisenberg in 1927. We should point out that Heisenberg arrived at this conclusion by following a different line of reasoning than we used and also that his argument was based on the momentum p of the particle rather than the velocity v. The momentum p is simply defined as the velocity v multiplied by the mass m of the particle,

$$p = mv \tag{2}$$

Heisenberg derived the condition

$$\Delta x \cdot \Delta p > h \tag{3}$$

where Δp is the uncertainty in the observed momentum and h is Planck's constant (it has the value $h = 6.62 \times 10^{-27}$ erg sec). Equation (2) is known as Heisenberg's uncertainty principle. In three dimensions the position of the particle is given by the three Cartesian coordinates x, y, and z, and the momentum also has three components, p_x, p_y, and p_z. The Heisenberg uncertainty relations refer to each pair of coordinates and momentum components that fall along the same axis,

$$\begin{aligned} \Delta x \cdot \Delta p_x &> h \\ \Delta y \cdot \Delta p_y &> h \\ \Delta z \cdot \Delta p_z &> h \end{aligned} \tag{4}$$

According to Heisenberg's uncertainty relations, the product of the uncertainties Δx and Δp_x must be larger than h but the product of, for example Δx and Δp_z, can be arbitrarily small.

It follows from the Heisenberg uncertainty relations that we cannot describe the motion of an electron in terms of the functions $x(t)$, $y(t)$, and $z(t)$ and that classical mechanics is not applicable to electronic motion. Instead we should recognize the uncertainties of our observations as an essential part of the theory, and we should describe the motion of the particle by means of probability patterns, like the patterns we discussed in Table 1.1. In the following section we shall discuss the mathematical description of such probability patterns in quantum mechanics.

2. The Wave Function

Let us begin by trying to give a more precise mathematical description of the probability patterns that we introduced above as examples. For instance, consider pattern A of Table 1.1. We should realize that the numbers we have listed are all rounded off to the nearest integer, so that there is a 22% probability that x lies between 383.5 and 384.5, an 18% possibility that x lies between 384.5 and 385.5, and so on. We can give a graphical representation of this probability pattern by means of a step function, as

6 THE PRINCIPLES OF QUANTUM MECHANICS

shown in Fig. 1.1. We have divided the possible values of x into 1-mile intervals, and we are interested only in the probability distribution over these relatively large intervals.

It is obvious that we would have obtained a more precise probability pattern if we had taken smaller intervals. For example, if we take 0.2-mile intervals, we get a finer-grained step function and, consequently, a more accurate probability pattern. Clearly, if we let the intervals get smaller and smaller, then we should get a continuous function at the limit. This is the probability distribution function $\rho(x)$ that we have drawn in Fig. 1.1.

By definition, the probability of finding x between the two values x_1 and x_2 is given by

$$P(x_1, x_2) = \int_{x_1}^{x_2} \rho(x)\, dx \qquad (5)$$

It follows that

$$\int_{382.5}^{383.5} \rho(x)\, dx = \tfrac{18}{100}$$
$$\int_{383.5}^{384.5} \rho(x)\, dx = \tfrac{22}{100} \qquad (6)$$

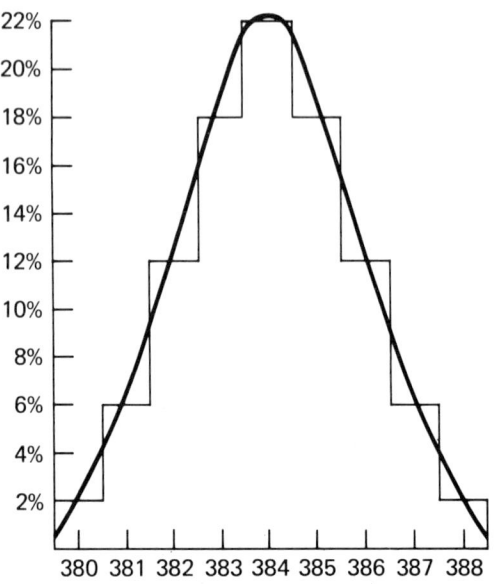

FIGURE 1.1 ● Continuous and discontinuous probability functions.

and so on. If we take the interval between x_1 and x_2 very small, for example,

$$x_1 = x_0 - \tfrac{1}{2} \Delta x \\ x_2 = x_0 + \tfrac{1}{2} \Delta x \tag{7}$$

then the probability becomes

$$P(x_0, \Delta x) = \int_{x_0 - \frac{1}{2}\Delta x}^{x_0 + \frac{1}{2}\Delta x} \rho(x)\, dx \tag{8}$$

If Δx becomes small enough, this may be written as

$$P(x_0, \Delta x) = \rho(x_0) \cdot \Delta x \tag{9}$$

Since the sum of all probabilities must be equal to unity, we have

$$\int_{-\infty}^{\infty} \rho(x)\, dx = 1 \tag{10}$$

In general, we may expect that our predictions for x also depend on the time that has elapsed, so that our forecasts for x are represented by a probability function $\rho(x, t)$, which also depends on the time. However, for the sake of simplicity we have disregarded this time dependence in the above discussion.

In order to avoid confusion, it may be useful to discuss the certainties and uncertainties of probability theory. Let us consider some examples from everyday life. If we flip pennies, we know that there is a 50% chance of getting heads and a 50% chance of getting tails. These probabilities are exact, and this can be verified by flipping a large number of pennies. On the other hand, if we flip a penny only once, we have absolutely no idea what will turn up; the only thing we know is that the chances of heads or tails are exactly equal. Another example is the life expectancy tables that life insurance companies use to determine their premiums. Nobody knows exactly when an individual death will occur, but the life expectancy tables are quite accurate if we consider a large group of people; if this were not the case, the life insurance companies would soon be bankrupt. Similarly, if we predict the future motion of a particle by means of a probability function $\rho(x, t)$, then we have no exact knowledge about the positions of the particle and we can predict only the relative probabilities. On the other hand, the probability function is quite rigorous and exact. If we could consider a large number of identical systems and take the average over all systems, then we would find that these averages agree exactly with the predictions from the probability density function.

Let us now return to quantum mechanics. One of the basic concepts of quantum mechanics is that the motion of a particle, that is, its position as a function of space and time, is exactly described by means of a probability density function $\rho(x, y, z, t)$. By definition, the probability of finding the particle at a time t_0 in a volume element dv, surrounding the point (x_0, y_0, z_0), is given by

$$P = \rho(x_0, y_0, z_0; t_0)\, dv \tag{11}$$

In quantum mechanics the probability density function ρ is derived from another function ψ, which is known as the wave function. The relation between ρ and ψ is given by

$$\rho(x, y, z; t) = \psi(x, y, z; t)\psi^*(x, y, z; t) \tag{12}$$

This definition allows for the possibility that the wave function ψ is a complex function, which may contain a real part ψ_1 and an imaginary part ψ_2,

$$\psi(x, y, z; t) = \psi_1(x, y, z; t) + i\psi_2(x, y, z; t) \tag{13}$$

The symbol ψ^* stands for the complex conjugate of ψ, which is defined as

$$\psi^*(x, y, z; t) = \psi_1(x, y, z; t) - i\psi_2(x, y, z; t) \tag{14}$$

if ψ is given by Eq. (13). It follows that

$$\psi\psi^* = (\psi_1 + i\psi_2)(\psi_1 - i\psi_2) = \psi_1^2 - i^2\psi_2^2 = \psi_1^2 + \psi_2^2 \tag{15}$$

because $i^2 = -1$.

We should point out that it is always possible to write a molecular or atomic wave function as a real function. In that case the probability density function ρ is given simply by

$$\rho(x, y, z; t) = [\psi(x, y, z; t)]^2 \tag{16}$$

and there is no need to consider the complications that result from ψ being complex.

The wave function is obtained as the solution of a differential equation known as the Schrödinger equation. For a particle of mass m, which moves in three dimensions, the Schrödinger equation has the form

$$-\frac{\hbar^2}{2m}\left(\frac{\partial^2\psi}{\partial x^2} + \frac{\partial^2\psi}{\partial y^2} + \frac{\partial^2\psi}{\partial z^2}\right) + V(x, y, z)\psi = E\psi \tag{17}$$

where $\hbar$ is the Dirac constant, which will be defined in Eq. (23). Let us now discuss the various aspects of this equation. In order to do this, it is useful to return to classical mechanics.

In classical mechanics the motion of a particle is determined by the forces that act on it. In three-dimensional space the total force that acts on a particle can be written as a sum of three components, namely the forces in the X, Y, and Z directions, F_x, F_y, and F_z, respectively. In general, these force components are functions of the coordinates x, y, and z of the particle and of the time. However, we shall consider only situations where the forces are not directly dependent on the time and where they can all be represented as the derivative of a function $V(x, y, z)$,

$$F_x = -\frac{\partial V}{\partial x} \qquad F_y = -\frac{\partial V}{\partial y} \qquad F_z = -\frac{\partial V}{\partial z} \tag{18}$$

The function $V(x, y, z)$ is known as the potential energy of the particle.

The total energy E of the particle is the sum of the potential energy V and its kinetic energy T,

$$E = T + V \tag{19}$$

The kinetic energy is given by

$$T = \tfrac{1}{2}mv^2 \tag{20}$$

or by

$$T = \frac{p^2}{2m} \tag{21}$$

if we use the momentum p instead of the velocity v.

It may be interesting to give the classical equations of motion for the particle. They are

$$m\frac{d^2x}{dt^2} = -\frac{\partial}{\partial x} V(x, y, z)$$
$$m\frac{d^2y}{dt^2} = -\frac{\partial}{\partial y} V(x, y, z) \tag{22}$$
$$m\frac{d^2z}{dt^2} = -\frac{\partial}{\partial z} V(x, y, z)$$

If we are able to solve them, we obtain x, y, and z as functions of time.

Because V depends on the coordinates, the potential energy depends implicitly on the time and the kinetic energy T is also time-dependent.

However, it can be shown that the total energy E is time-independent in the situation described above. It follows, then, that E remains constant during the motion of the particle and that the total energy is also independent of the position coordinates. Hence, the energy is a number, and it is called a constant of motion.

In both classical mechanics and quantum mechanics we are interested mainly in stationary states, that is, states where the energy remains constant during the motion. This means that in the Schrödinger equation (17), the symbol E may be considered a number.

Let us now return to the Schrödinger equation (17). It contains the mass m of the particle and the Dirac constant $\hbar$, which is defined as

$$\hbar = \frac{h}{2\pi} = 1.0544 \times 10^{-27} \text{ erg sec} \tag{23}$$

Here, h is known as Planck's constant. In quantum mechanics we often find Planck's constant divided by 2π, and as a result a separate symbol $\hbar$ was introduced.

In atoms and molecules the forces between the particles are all electrostatic, apart from small perturbations that we shall neglect for the time being. The electrostatic interaction between two particles with charges Z_1 and Z_2 is given by

$$V = \frac{Z_1 Z_2}{r} \tag{24}$$

if the charges have the same sign and by

$$V = \frac{-Z_1 Z_2}{r} \tag{25}$$

if they have different signs. Here r is the distance between the particles.

To give an example, the Schrödinger equation for the hydrogen atom is

$$-\frac{\hbar^2}{2m}\left(\frac{\partial^2}{\partial x^2} + \frac{\partial^2}{\partial y^2} + \frac{\partial^2}{\partial z^2}\right)\psi(x, y, z) - \frac{e^2\psi(x, y, z)}{r} = E\psi(x, y, z) \tag{26}$$

if we take the position of the nucleus fixed, or

$$-\frac{\hbar^2}{2m}\left(\frac{\partial^2}{\partial x^2} + \frac{\partial^2}{\partial y^2} + \frac{\partial^2}{\partial z^2}\right)\psi(x, y, z; X, Y, Z)$$

$$-\frac{\hbar^2}{2M}\left(\frac{\partial^2}{\partial X^2} + \frac{\partial^2}{\partial Y^2} + \frac{\partial^2}{\partial Z^2}\right)\psi(x, y, z; X, Y, Z)$$

$$-\frac{e^2\psi(x, y, z; X, Y, Z)}{r} = E\psi(x, y, z; X, Y, Z) \tag{27}$$

if we are more precise. Here m is the mass of the electron, M is the mass of the proton, (x, y, z) are the coordinates of the electron, (X, Y, Z) are the coordinates of the proton, and r is the distance between the electron and the proton. We shall discuss how this equation is solved in the following chapter.

It should be noted that the Schrödinger equation (17) is suitable only for deriving the wave functions of stationary states, which satisfy the condition that the energy of the system is a constant. The equation is known as the time-independent Schrödinger equation, and its solutions are also time-independent. For the time being we are interested only in stationary states, but later in this book we shall wish to consider the quantum theoretical aspects of spectroscopy and magnetic resonance. At that stage we shall have to use the time-dependent Schrödinger equation (IV-43), which is suitable for nonstationary states.

3. The Bohr Atom

The quantum mechanics of bound particles is in many respects dependent on the Bohr theory of the atom. We shall give a brief outline of the experimental results of atomic spectroscopy, followed by a discussion of the Bohr theory.

In studying the spectroscopic properties of a given sample, we measure either the absorption spectrum or the emission spectrum. Atomic spectra are usually measured in emission, where we subject the sample to a very high temperature by placing it in a high-voltage discharge and where we measure the spectroscopic distribution of the light that is emitted. The result of the experiment is often a photographic plate from which we derive the intensity of the emitted light as a function of the wavelength. It has been known for many years that atomic spectra are usually line spectra; a typical example may be seen in Fig. 1.2. Here light is emitted only in very narrow wavelength ranges; in a typical case the range might be of the order of 0.001 Å for a spectral line with a wavelength of about 4,000 Å. The wavelength of a spectral line in an atomic spectra is thus measured with an accuracy of $1:10^7$ or even better.

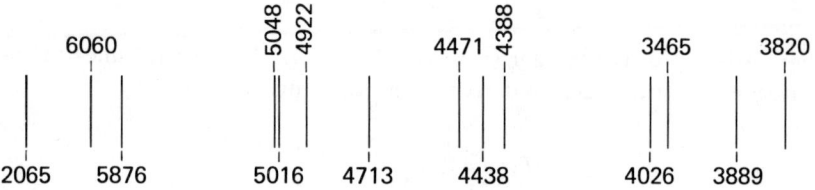

FIGURE 1.2 ● Line spectrum of He atom. Wavelengths are in terms of angstroms.

In reporting spectroscopic measurements, various sets of units may be used. Wavelengths are usually reported in terms of angstrom units (1 Å = 10^{-8} cm). It is possible also to describe a spectrum in terms of the frequency of the light. The frequency and the wavelength are related by

$$c = \lambda \nu \qquad (28)$$

where c is the velocity of light, which is approximately equal to 3×10^{10} cm/sec. Obviously a wavelength of 5,000 Å corresponds to a frequency of 6.10^{14} sec^{-1}. It is often advantageous to use the wave number σ instead of the wavelength ν for the description of a spectrum. The wave number σ is defined simply as the reciprocal of the wavelength,

$$\sigma = \frac{1}{\lambda} \qquad (29)$$

and it is expressed in terms of reciprocal centimeters (cm^{-1}). We have seen that the accuracy of the results for spectroscopic lines is sometimes better than the accuracy with which we know the velocity of light, and it is therefore better to express spectroscopic results in terms of σ than in terms of ν.

The final result of an atomic spectrum measurement consists of a very large collection of numbers, each of which represents a spectroscopic line. Around the turn of the century, some semiempirical rules were discovered that led to a more concise description of these numbers. The most important of these rules was discovered by Ritz in 1908. Ritz described the atomic spectra in terms of wave numbers instead of wavelengths, and he noted that it is then possible to construct a set of terms T_i so that each spectral line is the difference of a pair of terms,

$$T_{ij} = |T_j - T_i| \qquad (30)$$

The spectrum of the hydrogen atom can be represented in a very simple form, because in this spectrum the terms are given by

$$T_n = \frac{R_H}{n^2} \qquad (31)$$

where n is a positive integer ($n = 1, 2, 3, 4, \ldots$). All spectral lines in the hydrogen spectrum may be derived from the equation

$$T_{n,m} = R_H \left(\frac{1}{n^2} - \frac{1}{m^2} \right)$$
$$m, n = 1, 2, 3, \ldots \qquad n < m \qquad (32)$$

The constant R_H is known as the Rydberg constant for the hydrogen atom, and its value is $R_H = 109{,}677.501$ cm^{-1}. It should be noted that Eq. (32) represents exactly the value of each hydrogen line, to within the accuracy that these lines are measured.

Some attempts were made to explain the above results in terms of classical mechanics, but these attempts met with very serious difficulties. If we accept the Rutherford model of the hydrogen atom, in which the electron moves in a closed orbit around the proton, then it is very difficult to understand even why an atomic spectrum consists of discrete lines.

According to classical mechanics, we may represent the motion of the electron as an oscillating charge, or an oscillating electric dipole, and we may derive from electromagnetic theory that such a dipole emits electromagnetic radiation and, consequently, produces light. However, such a system should emit light of a broad range of frequencies, so that the classical theory predicts a band spectrum instead of a line spectrum. Furthermore, it follows from the principle of conservation of energy that the emission of light would cause a decrease in the energy of the atom. As more and more light is emitted, the atom would lose more and more energy, and the electron would orbit in a smaller orbit, closer to the nucleus. Eventually the electron would fall down on the nucleus, and the atom would collapse. It follows that classical theory predicts that a hydrogen atom should emit a short burst of light with a continuous spectroscopic distribution and subsequently collapse. It is easy to see that this prediction cannot be true, because it does not agree with the experimental facts.

In order to avoid the above difficulties, Bohr introduced a set of new fundamental assumptions to be used in the theoretical description of atomic structure. This quantum theory of atomic structure was published in a series of papers around the beginning of World War I.

The first Bohr assumption states that an atom (or molecule) can only be in certain stable (or stationary) states with a fixed and well-defined energy. The atom does not absorb or emit radiation as long as it remains in one of these stationary states.

The second Bohr assumption states that an atomic spectral line corresponds to a transition from one stationary atomic state to another. Bohr made use of the ideas that had been advanced earlier by Planck and Einstein, and he proposed that a spectral transition between two stationary states with energies E_n and E_m has a frequency given by

$$h\nu_{nm} = |E_n - E_m| \qquad (33)$$

We see that the two Bohr assumptions account for those features of atomic spectra that cannot be explained classically: The first assumption prevents the atomic collapse, and the second assumption leads to a discrete

14 THE PRINCIPLES OF QUANTUM MECHANICS

line spectrum. Bohr and Sommerfield proposed a third assumption, which enabled them to make theoretical predictions of the energies belonging to the stationary states of the hydrogen atom. In this way some useful theoretical results were derived.

On the other hand, the Bohr quantum theory by itself does not offer a really fundamental explanation of the structure of an atom even though it was one of the major scientific breakthroughs of this century. We know that Bohr's assumptions are basically correct, but we want to know why an atom has stationary states and how we can explain these assumptions from a more fundamental theoretical description. In the following section we shall show, therefore, how the Bohr theory can be derived in a simple way from the Schrödinger equation.

4. Stationary States and the Particle in a Box

It follows from the Bohr model of atomic structure that in an atomic system only specific, discrete values of energy are permissible. This cannot be explained from classical mechanics. We wish to show now how the existence of discrete energy states is derived from quantum theory.

Let us again review the procedure that we must follow in order to derive the quantum mechanical description of the behavior of a particle moving in three-dimensional space. First, we must solve the Schrödinger equation,

$$\frac{-\hbar^2}{2m}\left(\frac{\partial^2 \psi}{\partial x^2} + \frac{\partial^2 \psi}{\partial y^2} + \frac{\partial^2 \psi}{\partial z^2}\right) + V\psi = E\psi \tag{34}$$

Next we derive the probability density function ρ from the solution ψ of Eq. (34),

$$\rho = \psi^2 \tag{35}$$

Finally, we use the function ρ for the physical interpretation of the motion of the particle, following the concepts of probability theory that we outlined in the first section of this chapter.

The first stage of this procedure is purely mathematical. Let us imagine that we are able to solve Eq. (34). Then the solution is dependent on the value of E, and for every value of E we obtain a different solution. In other words, the solution may be written as $\psi(x, y, z; E)$, and it contains E as a parameter. If we then proceed to the next stage, we find in many cases (the hydrogen atom is one of them) that the general solution is not suitable for the purpose of physical interpretation. We then have to impose the condition that our solution must make sense from a physical point of view, and in most cases this condition leads to restrictions for the energy values that may be used.

It is difficult to put these conditions into an exact mathematical form that has general validity, because the conditions that we impose depend on the circumstances and these vary from one problem to the next. But in general they are determined from simple common sense, and for a particular problem it is usually quite easy to see what should be done. Let us consider specific examples.

It is easily seen that the probability density function should satisfy the condition,

$$\iiint \rho(x, y, z; E)\, dx\, dy\, dz = 1 \qquad (36)$$

because the integral represents the total probability of finding the particle anywhere in space. The solution of a differential equation may always be multiplied by an arbitrary constant. Therefore, if we find that

$$\iiint \psi^2(x, y, z; E)\, dx\, dy\, dz = C \qquad (37)$$

where C is a finite number, we can make sure that the condition of Eq. (36) is satisfied by multiplying by a suitable constant. However, if the integral of Eq. (37) is infinite, this procedure does not work and we find that the corresponding wave function is not suitable. Those wave functions for which the integral of Eq. (37) is finite are called normalizable. It follows that one of the conditions we have to impose on the wave function is that it can be normalized.

A second condition for the wave function is that it should be continuous at every point. This means that if we consider the value of the wave function at a given point P (given by x_0, y_0, z_0), then the function should change by only a small amount ϵ if we move from the point P by a small distance δ. In the limit where δ approaches zero, ϵ should also approach zero. The wave function should be continuous at every point; if it is not continuous in even one single point, then it becomes difficult to interpret the probability patterns. Also, if the function is not continuous, its derivative is not defined and it cannot satisfy the Schrödinger equation too well.

There may be other conditions that the wave function must satisfy, depending on the nature of the system. All these conditions may be summarized in a simple way: The wave function should give a probability density function that makes sense from a physical point of view.

As an illustration, we shall solve the Schrödinger equation for the simplest possible system in quantum mechanics, namely the one-dimensional particle in a box. So far we have mentioned only the three-dimensional Schrödinger equation, but it is easily seen that if we restrict the motion of the particle to only one dimension, then the Schrödinger equation takes the form

16 THE PRINCIPLES OF QUANTUM MECHANICS

$$\frac{-\hbar^2}{2m}\frac{d^2\psi}{dx^2} + V(x)\psi = E\psi \qquad (38)$$

A one-dimensional box is defined by a potential that is zero for $0 \le x \le a$ and that is infinite elsewhere (see Fig. 1.3). Algebraically, the potential is given by

$$\begin{aligned} V(x) &= \infty & x &< 0 \\ V(x) &= 0 & 0 &\le x \le a \\ V(x) &= \infty & a &< x \end{aligned} \qquad (39)$$

It is clear that the particle cannot move outside the box and we have, therefore,

$$\psi(x) = 0 \qquad x < 0 \text{ and } x > a \qquad (40)$$

Inside the box the Schrödinger equation reduces to

$$\frac{-\hbar^2}{2m}\frac{d^2\psi}{dx^2} = E\psi \qquad (41)$$

where E must be real and positive. We may introduce a new parameter λ by means of

$$\frac{2mE}{\hbar^2} = \lambda^2 \qquad (42)$$

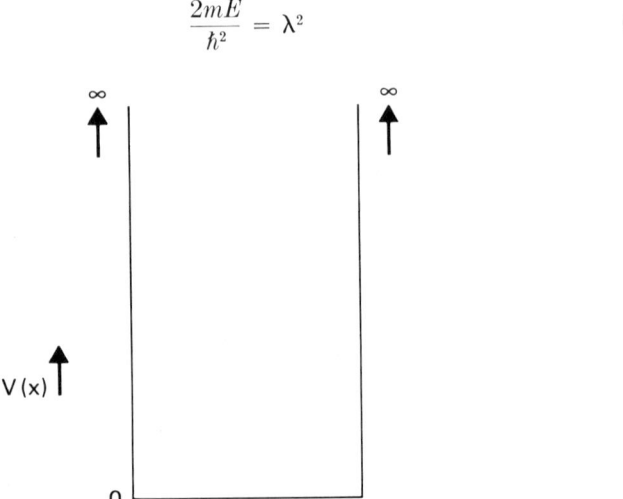

FIGURE 1.3 ● Potential function of particle in a box.

STATIONARY STATES AND THE PARTICLE IN A BOX

The Schrödinger equation then becomes

$$\frac{d^2\psi}{dx^2} = -\lambda^2\psi \qquad (43)$$

It is easily verified that this differential equation has two solutions, namely

$$\psi_1(x) = \sin \lambda x \qquad (44)$$

and

$$\psi_2(x) = \cos \lambda x \qquad (45)$$

The general solution of Eq. (43) is a linear combination of the two solutions of Eq. (44) and (45) and is represented as

$$\psi(x) = A \sin \lambda x + B \cos \lambda x \qquad (46)$$

This expression contains three parameters, namely A, B, and λ, and it satisfies the Schrödinger equation (43) for any values of these parameters.

Let us now consider the various conditions the wave function must satisfy. The function of Eq. (46) is normalizable because it is nonzero only in the finite interval $0 \le x \le a$ and is finite everywhere within the interval. The important condition for the wave function is the continuity condition. If we summarize the results we obtained for the wave function, we have

$$\begin{array}{ll} \psi(x) = 0 & x < 0 \\ \psi(x) = A \sin \lambda x + B \cos \lambda x & 0 \le x \le a \\ \psi(x) = 0 & a < x \end{array} \qquad (47)$$

The above function is continuous within any of the three intervals, but it is not necessarily continuous if we go from one interval to the other, that is, at the points $x = 0$ and $x = a$. We must impose the conditions that the function be continuous at these points, which gives

$$\lim_{x \to 0}(A \sin \lambda x + B \cos \lambda x) = B = 0 \qquad (48)$$

and

$$\lim_{x \to a}(A \sin \lambda x + B \cos \lambda x) = A \sin \lambda a + B \cos \lambda a = 0 \qquad (49)$$

By making use of Eq. (48) we find that Eq. (49) reduces to

$$A \sin \lambda a = 0 \qquad (50)$$

18   THE PRINCIPLES OF QUANTUM MECHANICS

It is easily seen that the parameter A cannot be zero. If it were zero the wave function would be zero everywhere, and the total probability of finding the particle would also be zero; this does not make sense, so we do not allow for it. Consequently, we have the condition

$$\sin \lambda a = 0 \tag{51}$$

This equation has an infinite number of solutions, namely,

$$\lambda a = n\pi \qquad n = 0, \pm 1, \pm 2, \pm 3, \ldots \tag{52}$$

Some of these solutions may be eliminated. If $n = 0$, the wave function would again be zero everywhere, and this is not allowed. Also, if we consider the two solutions $n = \nu$ and $n = -\nu$, we should recognize that it is only one solution because they lead to the same probability density function. Consequently, the possible values of n are

$$n = 1, 2, 3, 4, \ldots \tag{53}$$

We should remember that λ is related to the energy E by means of Eq. (42). If λ must satisfy Eq. (52), then the possible values of E are given by

$$E_n = \frac{n^2 \hbar^2 \pi^2}{2ma^2} \tag{54}$$

where the possible values of n are listed in Eq. (53).

It follows from the above considerations how the use of quantum mechanics leads to the existence of stationary states because only the energy values that are given by Eq. (54) are allowed. All the other energy values are not allowed, because they would lead to results that do not make sense from a physical point of view.

Each of the energy values E_n has a corresponding wave function ψ_n that is obtained by substituting Eq. (52) into Eq. (46),

$$\psi_n = A_n \sin \frac{n\pi x}{a} = \sqrt{\frac{a}{2}} \sin \frac{n\pi x}{a} \tag{55}$$

Here we have used the result $B = 0$, and we have substituted the value for the normalization constant A_n, which is determined from the normalization condition

$$\int_0^a [\psi_n(x)]^2 \, dx = A_n^2 \int_0^a \sin^2 \frac{n\pi x}{a} \, dx = 1 \tag{56}$$

THE COLOR OF CONJUGATED ORGANIC MOLECULES   19

In Fig. 1.4 we show the behavior of some of the eigenfunctions. This type of representation is fairly common. First we draw the potential function as a function of x, just as we did in Fig. 1.3, and we draw horizontal lines at the positions of the eigenvalues. We show only the lowest eigenvalues, E_1, E_2, E_3, and E_4. Then we plot the eigenfunction $\psi_1(x)$ as a function of x with respect to the horizontal line that corresponds to E_1, we plot the eigenfunction $\psi_2(x)$ with respect to the line belonging to E_2, and so on. In this way we can see the behavior of both the eigenvalues and eigenfunctions in a convenient way.

5. The Color of Conjugated Organic Molecules

The particle in a box has served as a starting point for a number of simple theories in chemistry. We think that it may be interesting to discuss one of them here in order to show that even simple calculations in quantum

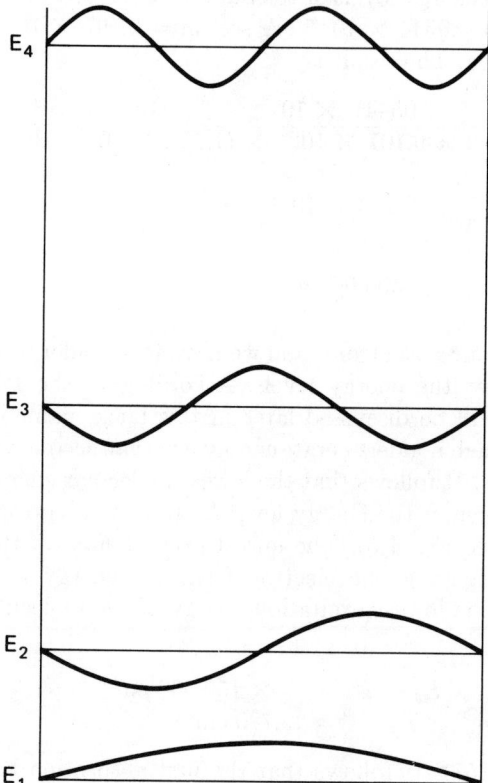

FIGURE 1.4 ● Eigenvalues and eigenfunctions of the particle in a box.

mechanics can have some practical applications. We have to make use of some concepts that we have not yet discussed, but hopefully this will not cause difficulties.

In Kuhn's theory of the color of conjugated organic molecules, it is assumed that the color is due mostly to the π electrons in the molecule. It is further assumed that these electrons move freely along the molecular bonds and that the particle in a box offers a reasonable approximation to their quantum mechanical description.

The energy levels of the π electrons are derived in this model from Eq. (54), since this represents the eigenvalues of the particle in a box. Let us consider the conjugated hydrocarbons, and let us start with butadiene. This molecule has three carbon–carbon bonds; if we take the average length of these bonds as 1.38 Å, then the total length of the carbon–carbon bonds is 5.14 Å. We assume that the π electrons may move slightly beyond the carbon skeleton, and we allow them half a bond length, or 0.69 Å at each end of the molecule. The length of the box is therefore 4×1.38 Å.

We evaluate the energy levels of the butadiene π electrons from Eq. (54). We substitute $\hbar = 1.0544 \times 10^{-27}$ erg sec, $m = 9.107 \times 10^{-28}$ g, and $a = 4 \times 1.38 \times 10^{-8}$ cm. The result is

$$E_n = \frac{(1.0544)^2 \times 10^{-54} \times (3.14159)^2}{2 \times 9.107 \times 10^{-28} \times (1.38)^2 \times 10^{-16}} \frac{n^2}{4^2} \text{ erg}$$

$$= \frac{n^2}{4^2} \times 3.163 \times 10^{-11} \text{ erg} \qquad (57)$$

$$= \frac{n^2}{4^2} \times 159{,}300 \text{ cm}^{-1}$$

Butadiene has four π electrons, and we have to consider how they should be distributed over the energy levels. According to the Pauli exclusion principle, which will be discussed later in this book, namely Chapter III Sections 1 and 4, each nondegenerate energy level can accommodate no more than two electrons. It follows that the lowest molecular energy is obtained if two electrons occupy the energy level E_1 and two electrons occupy the energy level E_2 (see Fig. 1.5). The lowest excited state of the molecule is obtained by transferring one electron from the energy level E_2 to the energy level E_3. The lowest excitation energy of the molecule is therefore given by

$$E_3 - E_2 = \tfrac{5}{16} \times 159{,}300 \text{ cm}^{-1}$$
$$= 49{,}780 \text{ cm}^{-1} \qquad (58)$$

according to Eq. (57). It follows that the first absorption band occurs at about 2,000 Å and, because this is in the ultraviolet, that the molecule is colorless.

THE COLOR OF CONJUGATED ORGANIC MOLECULES

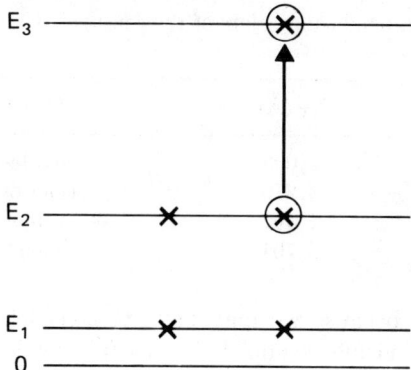

FIGURE 1.5 ● Distribution of electrons over the energy eigenstates in butadiene.

Let us now investigate how many carbon atoms a conjugated hydrocarbon must have in order that the compound be colored. An arbitrary conjugated hydrocarbon has $2N$ carbon atoms and $(2N - 1)$ carbon–carbon bonds. Its energy levels $E_n(N)$ may be derived from Eq. (57) be replacing the number 4 in this equation by $2N$.

$$E_n(N) = \frac{n^2}{4N^2} \times 159{,}300 \text{ cm}^{-1} \tag{59}$$

The molecule has $2N$ π electrons, and its lowest excitation energy is obtained as the difference between E_{N+1} and E_N,

$$\begin{aligned} \Delta E_N &= E_{N+1}(N) - E_N(N) \\ &= \frac{(2N+1)}{4N^2} \times 159{,}300 \text{ cm}^{-1} \end{aligned} \tag{60}$$

The wavelength of the first absorption band, λ_N, is given by the inverse of Eq. (60),

$$\lambda_N = \left(1 - \frac{1}{2N+1}\right) N \times 1.255 \text{ Å} \tag{61}$$

In Table 1.2 we have listed the values of λ_N and the colors for the first four conjugated hydrocarbons. It should be noted that the colors that we see are complementary to the colors represented by λ_N. For instance, in the compound C_8H_{10}, light is absorbed in the blue-violet region and the compound looks yellow to us because yellow is complementary to blue-violet. We have to be careful if we want to extend our predictions beyond

TABLE 1.2 • Absorption bands and colors of the conjugated hydrocarbons $C_{2N}H_{2N+2}$.

N	λ_N (Å)	Color
2	2,008	colorless
3	3,270	colorless
4	4,462	yellow
5	5,704	violet

the compound $C_{10}H_{12}$, because we may have to consider more than one absorption band in the visible region. It is a useful exercise to look up the colors of the compounds that we have listed in Table 1.2 and verify the accuracy of our theoretical predictions.

6. Operators and Their Properties

It is useful to introduce the concept of operators, because it relates the results of quantum mechanics to physical observations. Let us first define an operator. Any procedure that transforms an arbitrary function f into another function g may be written in the form

$$g = \Lambda f \tag{62}$$

where Λ is defined as the operator that represents this procedure. The most common examples of operators are multiplicative operators and differential operators. If g is obtained by multiplying f by another function ϕ, we may write

$$g = \phi f \tag{63}$$

and consider ϕ a multiplicative operator. If g is obtained by differentiating f,

$$g = \frac{df}{dx} = \frac{d}{dx} f \tag{64}$$

then the operator is the differential operator (d/dx). A well-known operator is the Laplace operator, for which we use the symbol Δ. It is defined by

$$\Delta = \frac{\partial^2}{\partial x^2} + \frac{\partial^2}{\partial y^2} + \frac{\partial^2}{\partial z^2} \tag{65}$$

It is also convenient to use operator language to abbreviate the Schrödinger equation (34) for a particle. By using the symbol Δ, we may rewrite the Schrödinger equation (34) in the form

OPERATORS AND THEIR PROPERTIES

$$-\frac{\hbar^2}{2m}\Delta\psi + V\psi = \left(-\frac{\hbar^2}{2m}\Delta + V\right)\psi = E\psi \tag{66}$$

The Hamiltonian operator $\mathcal{H}_{op}$ is defined as

$$\mathcal{H}_{op} = -\frac{\hbar^2}{2m}\Delta + V \tag{67}$$

The Schrödinger equation may thus be written in the simple form

$$\mathcal{H}_{op}\psi = E\psi \tag{68}$$

Let us now rewrite the problem of the particle in a box in terms of operator language. We have a one-dimensional problem,

$$\mathcal{H}_{op}(x)\psi(x) = E\psi(x) \qquad 0 \le x \le a \tag{69}$$

and the solutions must satisfy the conditions

$$\psi(0) = \psi(a) = 0 \tag{70}$$

We have seen in Section 4 that Eqs. (69) and (70) are satisfied for a certain set of energy values E_n ($n = 1, 2, 3, \ldots$). The corresponding solutions are the functions $\psi_n(x)$, and they satisfy the equations

$$\begin{aligned}\mathcal{H}_{op}(x)\psi_n(x) &= E_n\psi_n(x) \\ \psi_n(0) &= \psi_n(a) = 0\end{aligned} \tag{71}$$

The conditions of Eq. (70) are called the boundary conditions of the differential equation (69). The differential equation is defined by the operator $\mathcal{H}$. The set of numbers E_n are called the eigenvalues of the operator $\mathcal{H}$, and the corresponding functions ψ_n are the eigenfunctions of $\mathcal{H}$.

In the example we have discussed, namely the particle in a box, there is exactly one eigenfunction belonging to each eigenvalue. In that case we call the eigenvalue nondegenerate. In general, it is possible that there is more than one eigenfunction belonging to the same eigenvalue. If that is the case we call the eigenvalue degenerate. We speak of a twofold degenerate eigenvalue if it has two different eigenfunctions, a threefold degenerate eigenvalue if there are three different eigenfunctions, and so on.

Let us consider a twofold degenerate eigenvalue λ_n of an operator Λ. We have the two equations

$$\begin{aligned}\Lambda\phi_{n,1} &= \lambda_n\phi_{n,1} \\ \Lambda\phi_{n,2} &= \lambda_n\phi_{n,2}\end{aligned} \tag{72}$$

where $\phi_{n,1}$ and $\phi_{n,2}$ are the two eigenfunctions belonging to λ_n. It follows from Eq. (72) that we also have

$$\Lambda(a\phi_{n,1} + b\phi_{n,2}) = \lambda_n(a\phi_{n,1} + b\phi_{n,2}) \tag{73}$$

where a and b are arbitrary parameters. In other words, if a degenerate eigenvalue has a set of different eigenfunctions, then any linear combination of those eigenfunctions is also an eigenfunction belonging to that eigenvalue.

We have seen that we can make probability predictions about the behavior from the wave function. In particular, we discussed the predictions about the position of the particle. By using probability theory we can more or less predict what the most probable position of the particle is. In fact, the different definitions give somewhat different results, and probability theory is somewhat ambiguous in this respect. If the probability density function is symmetric, then the various definitions of the most probable position all give the same result. But if the probability function is not symmetric, then the result depends on the definition.

It has been found that in quantum mechanics the most probable values of experimental quantities must be defined by means of the expectation value of the operator. According to this definition, we predict the position of a particle in one dimension by means of the integral

$$\bar{x} = \frac{\int \psi^*(x) x \psi(x)\, dx}{\int \psi^*(x) \psi(x)\, dx} \tag{74}$$

where ψ is the wave function of the particle. In three dimensions the expectation value of the position is given by the three integrals

$$\begin{aligned}\bar{x} &= \iiint \psi^*(x, y, z) x \psi(x, y, z)\, dx\, dy\, dz \\ \bar{y} &= \iiint \psi^*(x, y, z) y \psi(x, y, z)\, dx\, dy\, dz \\ \bar{z} &= \iiint \psi^*(x, y, z) z \psi(x, y, z)\, dx\, dy\, dz\end{aligned} \tag{75}$$

Here it is assumed that the wave function ψ is normalized to unity,

$$\int \psi^*(x, y, z) \psi(x, y, z)\, dx\, dy\, dz = 1 \tag{76}$$

By definition, the expectation value is the number that we should measure if we perform an experiment. It follows, then, that the position of the particle should be the vector $\bar{\mathbf{r}}$ with components $\bar{x}$, $\bar{y}$, and $\bar{z}$, defined by Eq. (75). This is the most probable value as predicted by quantum mechanics.

If we wish to find the expectation value for any other physical observable, we should know the operator that is representative for this observable.

OPERATORS AND THEIR PROPERTIES 25

For example, the energy E is represented by the Hamiltonian operator $\mathcal{H}_{op}$, and its expectation value $\bar{E}$ is given by

$$\bar{E} = \iiint \psi^*(x, y, z)\, \mathcal{H}_{op}\psi(x, y, z)\, dx\, dy\, dz \tag{77}$$

In the case where the wave function is an eigenfunction ψ_n of the Hamiltonian operator, we have

$$\bar{E} = \iiint \psi_n^*(x, y, z)\, \mathcal{H}_{op}\psi_n(x, y, z)\, dx\, dy\, dz = E_n \tag{78}$$

The expectation value of the energy is the eigenvalue E_n of the operator.

It is obvious that the result of a physical measurement cannot be an imaginary number. It follows that a physical observable should always be real. Its expectation value should also be real, otherwise the above definitions would not make much sense. If we consider a physical observable L, which is represented by an operator Λ, then the expectation value L is given by

$$\bar{L} = \int \psi^*\, \Lambda \psi\, dv \tag{79}$$

The complex conjugate of Eq. (79) is

$$(\bar{L})^* = \int \psi\, \Lambda^* \psi^*\, dv \tag{80}$$

The condition that L is a real quantity is equivalent to the condition

$$\int \psi^*\, \Lambda \psi\, dv = \int \psi\, \Lambda^* \psi^*\, dv \tag{81}$$

This means in mathematical language that the operator Λ must be Hermitian, because a Hermitian operator is defined as an operator that satisfies the condition

$$\int f\, \Lambda g\, dv = \int g\, \Lambda^* f\, dv \tag{82}$$

for any pair of functions f and g. In summary, we have the condition that any operator that represents a physical observable must be Hermitian.

Hermitian operators have some useful properties. First, the eigenvalues of a Hermitian operator are all real, and second, the eigenfunctions of a Hermitian operator are orthogonal to each other. We shall prove both properties and discuss what they mean.

Let H be a Hermitian operator and ϵ_n and ϕ_n one of its eigenvalues and the corresponding eigenfunction, respectively. We then have

$$H\phi_n = \epsilon_n \phi_n \tag{83}$$

26   THE PRINCIPLES OF QUANTUM MECHANICS

If we multiply both sides of this equation on the left by ϕ_n^* and then integrate, we obtain

$$\int \phi_n^* H \phi_n \, dv = \epsilon_n \int \phi_n^* \phi_n \, dv \tag{84}$$

The complex conjugate of this equation must also be valid; it is

$$\int \phi_n H^* \phi_n^* \, dv = \epsilon_n^* \int \phi_n^* \phi_n \, dv \tag{85}$$

The left-hand side of Eq. (84) is equal to the left-hand side of Eq. (85) because the operator is Hermitian. Therefore, if we subtract the two equations, we obtain

$$(\epsilon_n - \epsilon_n^*) \int \phi_n^* \phi_n \, dv = 0 \tag{86}$$

The integral must be nonzero and positive, so that

$$\epsilon_n = \epsilon_n^* \tag{87}$$

In other words, the eigenvalue ϵ_n is real.

By definition, two functions ψ and χ are orthogonal to each other if they satisfy the condition

$$\int \chi^* \psi \, dv = 0 \tag{88}$$

We wish to prove that two eigenfunctions ϕ_n and ϕ_m, belonging to different eigenvalues ϵ_n and ϵ_m of the Hermitian operator H, are always orthogonal. The function ϕ_m satisfies the equation

$$H \phi_m = \epsilon_m \phi_m \tag{89}$$

If we multiply this equation on the left by ϕ_n^* and integrate, we obtain

$$\int \phi_n^* H \phi_m \, dv = \epsilon_m \int \phi_n^* \phi_m \, dv \tag{90}$$

The function ϕ_n must satisfy the complex conjugate of Eq. (83), which is

$$H^* \phi_n^* = \epsilon_n \phi_n^* \tag{91}$$

We should remember here that ϵ_n is real. We multiply Eq. (91) on the left by ϕ_m and integrate. The result is

$$\int \phi_m H^* \phi_n^* \, dv = \epsilon_n \int \phi_n^* \phi_m \, dv \tag{92}$$

Again, the left-hand side of Eq. (90) is equal to the left-hand side of Eq. (92) because the operator is Hermitian. Consequently, subtraction of the two equations gives

$$(\epsilon_m - \epsilon_n) \int \phi_n^* \phi_m \, dv = 0 \tag{93}$$

We have assumed that the two eigenvalues are different, hence

$$\int \phi_n^* \phi_m \, dv = 0 \tag{94}$$

This means that the two eigenfunctions ϕ_n and ϕ_m are orthogonal to each other. The two theorems about Hermitian operators that we have proved above will be very useful later on.

At this point we wish to mention an alternative notation for integrals of the type of Eqs. (90), (92), and so on. If we have two functions ϕ and ψ and an operator Ω, we define

$$\begin{aligned}\langle \phi \mid \Omega \mid \psi \rangle &= \int \phi^* \Omega \psi \, dv \\ \langle \phi \mid \psi \rangle &= \int \phi^* \psi \, dv\end{aligned} \tag{95}$$

This notation has the advantage that we do not have to list the integration variables specifically, because by definition we must integrate over all pertinent variables. This is obviously a useful abbreviation if we deal with the wave functions of a large molecule, such as naphthalene for example.

It is useful to know that in a molecule the Hamiltonian is always real if there are no exterior electromagnetic fields and that the molecular eigenfunctions may then also be taken as real. If we have a complex eigenfunction ψ_n, then it can be written as

$$\psi_n = \psi_{n,1} + i\psi_{n,2} \tag{96}$$

where $\psi_{n,1}$ and $\psi_{n,2}$ are real functions. It is easily seen that we then have

$$\begin{aligned}\mathcal{H}\psi_{n,1} &= E_n \psi_{n,1} \\ \mathcal{H}\psi_{n,2} &= E_n \psi_{n,2}\end{aligned} \tag{97}$$

because both $\mathcal{H}$ and E_n must be real. If the eigenvalue E_n is nondegenerate, then $\psi_{n,1}$ and $\psi_{n,2}$ must be proportional to one another and we can take either one as the eigenfunction. If E_n is degenerate, then we can take the real functions $\psi_{n,1}$ and $\psi_{n,2}$ as the eigenfunctions rather than the complex function ψ_n. Hence, if we do not like to deal with complex functions, we can choose our eigenfunctions all real.

7. Approximate Methods in Quantum Mechanics

We mentioned that the various properties of an atom or molecule may all be derived from the wave function, which may in principle be obtained by solving the corresponding Schrödinger equation. Unfortunately, the Schrödinger equation becomes quite complicated even for small atoms or molecules, and it can be solved exactly only for a few simple systems. The largest system for which the exact eigenfunctions and eigenvalues have been derived from the Schrödinger equation is the hydrogen molecular ion, H_2^+, and here the solution is so complicated that it is of little practical use. If we want to apply quantum mechanics to problems that are of chemical interest, then we must resort to approximate methods. We shall discuss the two methods of approximation that are most widely used in quantum chemistry, namely the variation principle and perturbation theory.

If we consider a molecule such as benzene or even a simple molecule such as H_2, we know that it is not possible to find the exact eigenfunctions of the system. However, by making use of our intuition and knowledge of chemistry, we may write down approximate expressions for the eigenfunctions. The variation principle enables us to find out how good these approximate functions are. Also, if our approximate wave function contains any parameters, the variation principle may be used to find the most suitable values for these parameters.

The variation principle states that if we have an approximate wave function ψ for a molecule and if we use this wave function to calculate the molecular energy, then the result must always be higher than the true molecular ground state energy, E_0. In mathematical language, if the atom or molecule is represented by a Hamiltonian $\mathcal{H}$, then we have

$$E(\psi) = \frac{\langle \psi | \mathcal{H} | \psi \rangle}{\langle \psi | \psi \rangle} \geq E_0 \qquad (98)$$

where E_0 is the smallest of the set of eigenvalues E_n of $\mathcal{H}$,

$$\mathcal{H}\Phi_n = E_n \Phi_n \qquad E_0 \leq E_1 \leq E_2 \ldots \qquad (99)$$

The expectation value $E(\psi)$ is always larger than the exact eigenvalue E_0 except when ψ is equal to the eigenfunction Φ_0, in which case the expectation value $E(\Phi_0)$ is equal to E_0.

The theorem of Eq. (98) can be proved fairly easily, but before we do that let us see first how we make use of it. The approximate energy $E(\psi)$ is equal to the exact eigenvalue E_0 only when ψ is equal to Φ_0. This means that the lower the energy $E(\psi)$, the closer it is to E_0 and the more the function ψ resembles the function Φ_0. In practice, if we have a number of different approximate functions ψ that contain some parameters, then we should take the values of the parameters that give the lowest energy. Those parameter values then give the best form of the wave function.

As an illustration, let us consider the hydrogen molecule (see Fig. 1.6). We can construct an approximate eigenfunction for its ground state by assuming that the first electron 1 is in the vicinity of nucleus a and the second electron 2 is in the vicinity of nucleus b. The ground-state eigenfunction of a hydrogen atom is known, and we denote it by $s(r)$, where r is the distance to the nucleus. We may thus approximate the above situation by the wave function

$$\psi_I = s_a(1)s_b(2) \tag{100}$$

where s_a is a hydrogen eigenfunction centered on nucleus a and s_b is a hydrogen eigenfunction centered on nucleus b. Obviously, it is just as likely that electron 1 is on nucleus b and electron 2 is on nucleus a; this situation is represented by

$$\psi_{II} = s_a(2)s_b(1) \tag{101}$$

There is also a possibility that the two electrons are centered on the same nucleus, either on a, which is described by

$$\psi_{III} = s_a(1)s_a(2) \tag{102}$$

or on b, where we have

$$\psi_{IV} = s_b(1)s_b(2) \tag{103}$$

We have sketched the four situations in Fig. 1.6. It may be seen that I and II describe an ordinary chemical bond, composed of two electrons, and

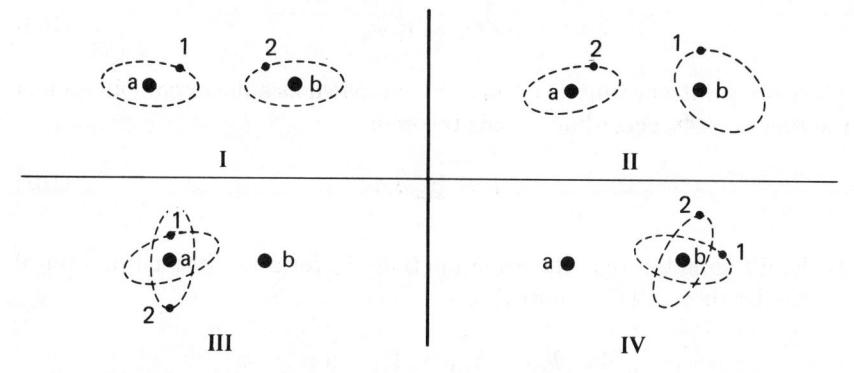

FIGURE 1.6 • Possible electronic structures of the hydrogen molecule.

that III and IV represent ionic structures. The total wave function may now be written as a superposition of the above four functions,

$$\psi = \lambda(\psi_{\text{I}} + \psi_{\text{II}}) + \mu(\psi_{\text{III}} + \psi_{\text{IV}}) \qquad (104)$$

It may be derived from symmetry considerations that ψ_{I} and ψ_{II} must have the same coefficient, and ψ_{III} and ψ_{IV} also.

If we want to make use of the variation principle, we calculate the expectation value $E(\psi)$ from the function (104) and we determine for which values of the parameters λ and μ this expectation value has a minimum. From the result we may then derive the probability for ionic structures in the H_2 molecule. This probability is approximately given by $\mu^2/(\lambda^2 + \mu^2)$.

In the above example we have used the principle of superposition of states, which has been widely applied in the theory of the chemical bond. In order to describe a given molecule, we write down a number of chemical structures of a given molecule. Each of these structures is then represented by a wave function Φ_{I}, Φ_{II}, Φ_{III}, and so on. The total molecular wave function is now written as a linear combination of these separate wave functions,

$$\psi = \sum_n a_N \Phi_N \qquad (105)$$

We determine the values of the coefficients by using the variation principle, that is, by minimizing the energy. The relative probability of each structure is then given by the square of its coefficient.

The principle of superposition is a special case of a mathematical theorem that states that any function ψ may be expanded in terms of a complete set of functions. The theorem also states that the set of eigenfunctions of a Hamiltonian $\mathcal{H}$ forms such a complete set. Let us define the complete set of eigenfunctions of $\mathcal{H}$ by

$$\mathcal{H}\Phi_n = E_n \Phi_n \qquad (106)$$

Then any arbitrary function ψ may be expressed as a linear combination of the functions Φ_n according to this theorem:

$$\psi = \sum_n a_n \Phi_n \qquad (107)$$

It should be noted that the eigenfunctions Φ_n form an orthonormal set of functions; they satisfy the relations

$$\langle \Phi_n \mid \Phi_m \rangle = \delta_{n,m} = 1 \quad \text{if } n = m$$
$$\qquad\qquad\qquad\qquad = 0 \quad \text{if } n \neq m \qquad (108)$$

The variation principle (98) is easily derived from Eqs. (106), (107), and (108). We take E_0 as the smallest eigenvalue of the operator $\mathcal{H}$ and we determine

$$(\mathcal{H} - E_0)\psi = ? \tag{109}$$

where ψ is an arbitrary function. We know that ψ may be expressed in terms of the expansion (107). We do not know the values of the coefficients a_n, but that does not matter; we just substitute Eq. (107) into Eq. (109) and find

$$\begin{aligned}(\mathcal{H} - E_0)\psi = (\mathcal{H} - E_0)\sum_{n=0}^{\infty} a_n \Phi_n &= \sum_{n=0}^{\infty} a_n(E_n - E_0)\Phi_n \\ &= \sum_{n=1}^{\infty} a_n(E_n - E_0)\Phi_n\end{aligned} \tag{110}$$

If we multiply this equation on the left by ψ^* and then integrate, we obtain

$$\begin{aligned}\langle \psi | \mathcal{H} - E_0 | \psi \rangle &= \left\langle \psi \Big| \sum_{n=1}^{\infty} a_n(E_n - E_0)\Phi_n \right\rangle \\ &= \sum_{n=1}^{\infty} a_n(E_n - E_0)\langle \psi | \Phi_n \rangle \\ &= \sum_{n=1}^{\infty} a_n(E_n - E_0)\left\langle \sum_{m=0}^{\infty} a_m \Phi_m \Big| \Phi_n \right\rangle \\ &= \sum_{n=1}^{\infty} a_n a_n^*(E_n - E_0)\end{aligned} \tag{111}$$

In the last step of Eq. (111) we have made use of the orthogonality relation (108).

It is easily seen that the last term of Eq. (198) must be positive, because each term in the sum must be positive. We thus find that

$$\langle \psi | \mathcal{H} - E_0 | \psi \rangle \geq 0 \tag{112}$$

or

$$\langle \psi | \mathcal{H} | \psi \rangle - E_0 \langle \psi | \psi \rangle \geq 0 \tag{113}$$

This may be rewritten as

$$\langle \psi | \mathcal{H} | \psi \rangle \geq E_0 \langle \psi | \psi \rangle \tag{114}$$

or

$$\frac{\langle \psi | \mathcal{H} | \psi \rangle}{\langle \psi | \psi \rangle} \geq E_0 \tag{115}$$

which is the variation principle of Eq. (98).

The second method of approximation in quantum mechanics is perturbation theory. This method is particularly suitable to study the effects of electric or magnetic fields on atoms or molecules, and it plays a prominent role in the theories of magnetic resonance and spectroscopy.

The detailed formulation of perturbation theory is all mathematical, but we shall attempt to outline it in general terms. Again, we consider an atom or molecule that is represented by a Hamiltonian, which we now call $\mathcal{H}_0$. This Hamiltonian has a set of eigenvalues ϵ_n and eigenfunctions Φ_n, which are defined by the relation

$$\mathcal{H}_0 \Phi_n = \epsilon_n \Phi_n \tag{116}$$

If we place this molecule in an electric or magnetic field, then the system is described by a slightly different Hamiltonian $\mathcal{H}$ and the new eigenfunctions and eigenvalues are ψ_n and E_n. They are defined by

$$\mathcal{H} \psi_n = E_n \psi_n \tag{117}$$

It may be seen that if the Hamiltonian $\mathcal{H}$ differs only slightly from $\mathcal{H}_0$,

$$\mathcal{H} \approx \mathcal{H}_0 \tag{118}$$

then the sets of eigenvalues and eigenfunctions must also be similar; that is,

$$E_n \approx \epsilon_n \tag{119}$$

and

$$\psi_n \approx \Phi_n \tag{120}$$

Of course we assume here that the two sets of eigenvalues and eigenfunctions are numbered the same way.

The main problem of perturbation theory is to formulate the expressions (119) and (120) in a more precise mathematical form. This can be done only if we write Eq. (118) in a more precise form first. If $\mathcal{H}$ and $\mathcal{H}_0$ differ only slightly, then we can write

$$\mathcal{H} = \mathcal{H}_0 + \mathcal{H}' \tag{121}$$

with

$$\mathcal{H}' \ll \mathcal{H}_0 \tag{122}$$

However, the precise formulation of perturbation theory makes use of power series expansions, and we write instead

$$\mathcal{H} = \mathcal{H}_0 + \lambda V \tag{123}$$

with

$$\lambda \ll 1 \tag{124}$$

Here λ is a somewhat artificial quantity known as a scaling parameter. Since it is much smaller than unity, it may be assumed that power-series expansions in terms of λ are rapidly convergent.

If we introduce the scaling parameter λ and then expand the eigenvalues and eigenfunctions of $\mathcal{H}$ as power series in λ, they take the form

$$\begin{aligned} E_n &= \epsilon_n + \lambda E'_n + \lambda^2 E''_n + \cdots \\ \psi_n &= \Phi_n + \lambda \psi'_n + \lambda^2 \psi''_n + \cdots \end{aligned} \tag{125}$$

Since λ is supposed to be small, it is assumed that each successive term in these expansions is an order of magnitude smaller than its preceding term, where an order of magnitude is defined here as a factor λ.

It is customary to call the term E'_n the first-order perturbation to the energy, E''_n the second-order perturbation to the energy, and so on. Similarly, ψ'_n is the first-order perturbation to the eigenfunction Φ_n, and so on. The goal of perturbation energy is to express the various perturbation terms $E'_n, E''_n, \psi'_n, \ldots$, in terms of the eigenvalues and eigenfunctions ϵ_n and Φ_n of the unperturbed Hamiltonian $\mathcal{H}_0$.

We shall just mention some of the results of perturbation theory and not give the mathematical derivations because these are fairly elaborate. Also, we limit ourselves to perturbations of nondegenerate energy levels because the perturbation theory of degenerate states is much more involved.

If the state n is nondegenerate, then the first-order perturbation to the energy, E'_n, is given by

$$E'_n = \langle \Phi_n | V | \Phi_n \rangle = V_{n,n} \tag{126}$$

This is simply the expectation value of the perturbation term V. The second-order perturbation E''_n is given by an infinite sum,

$$E''_n = -\sum_{m \neq n} \frac{\langle \Phi_n | V | \Phi_m \rangle \langle \Phi_m | V | \Phi_n \rangle}{\epsilon_m - \epsilon_n} = -\sum_{m \neq n} \frac{V_{n,m} V_{m,n}}{\epsilon_m - \epsilon_n} \tag{127}$$

The perturbation terms of the eigenfunction ψ_n are usually expressed as series expansions in terms of the set of functions Φ_n. The first-order perturbation ψ'_n is given by

34 THE PRINCIPLES OF QUANTUM MECHANICS

$$\psi'_n = -\sum_{m \neq n} \frac{\langle \Phi_m | V | \Phi_n \rangle \Phi_m}{\epsilon_m - \epsilon_n} = -\sum_{m \neq n} \frac{V_{m,n}}{\epsilon_m - \epsilon_n} \Phi_m \qquad (128)$$

In magnetic resonance and in spectroscopy the value of the scaling parameter λ is usually very small, between 10^{-3} and 10^{-7}. The perturbation expansions thus converge very rapidly and there is no need to consider the terms beyond second order in the energy and beyond first order in the wave function. Perturbation theory has also been applied in situations where λ is much larger, say between 0.1 and 0.5, but then the method becomes less satisfactory.

Finally, we should point out that we have oversimplified perturbation theory in our discussion. There is a lot more to it than we have discussed, but we have limited ourselves to the main features of the theory.

Problems

1. Look up in the literature the colors of the four conjugated hydrocarbons C_4H_6 (butadiene), C_6H_8 (hexatriene), C_8H_{10} (octatetraene), and $C_{10}H_{12}$ (decapentaene) and verify the accuracy of the theoretical results of Table 1.2.
2. Calculate the first excitation energy $(E_2 - E_1)$ in terms of electron volts for an electron and for a proton in a one-dimensional box of 1 Å.
3. Write down the Hamiltonian operator and the Schrödinger equation for the electron in the hydrogen molecular ion H_2^+, for the He atom, and for the hydrogen molecule H_2.
4. Calculate the wavelength of the absorption line that corresponds to an energy difference of 4 eV and also for an energy difference of 80 kcal/mole. (Look up the values of the physical constants.)
5. By making use of the Heisenberg uncertainty principle, we can predict a lower limit for Δp, Δv, and for the kinetic energy of a particle in a box because Δx is equal to the length a of the box. Show that these predictions are consistent with the result we obtained for the lowest-energy eigenvalue by solving the Schrödinger equation.
6. Calculate the lower limit Δv, Δx for a car (mass 100 kg), a bullet (mass 10 g), and a hydrogen molecule by making use of the Heisenberg uncertainty principle. Determine Δx for the car if it moves at 60 miles/hr, for the bullet if it moves at 2000 ft/sec, and for the hydrogen molecule if it moves at 20 km/sec.
7. How long should a linear conjugated hydrocarbon molecule be in order to absorb light between 8,000 Å and 9,000 Å in an electronic transition?
8. We know that one throw of a die will give equal possibilities for throwing 1, 2, 3, 4, 5, or 6, so that the probability density function is a straight line. Derive the probability density functions for the average result of two throws of the die and of three throws of the die.
9. Show explicitly that for a particle in a box the eigenfunctions belonging to the eigenvalues E_1 and E_3 are orthogonal.

10. It is easily shown that the multiplicative operator x is Hermitian. What about the operators (d/dx) and (d^2/dx^2)?
11. If we take it that visible light is light with a wavelength between 4,000 Å and 7,000 Å, which spectral transitions of the hydrogen atom, given by Eq. (32), are observed in the visible?
12. If we have an electron in a box 10 Å long, what is the wavelength of the light that is absorbed in a transition from the ground state to the first excited state?
13. Let the probability density function of a particle be given by $(x^2 - 4x + 5)^{-1}$. Determine the expectation value x_0 for this probability density function. Show that x_0 is equal to the point where the probability density function has its maximum. Determine also the probability for finding the particle between the two points $x_0 - 1$ and $x_0 + 1$ and between the two points $x_0 - \frac{1}{2}$ and $x_0 + \frac{1}{2}$.
14. Consider the probability density function

$$P(x) = x^6 e^{-x} \quad x \geq 0$$
$$P(x) = 0 \quad x < 0$$

Determine the position x_m of the maximum of this function and the expectation value x_0 with respect to this function. According to quantum mechanics, where are we supposed to find the particle?

Bibliography

In Chapter I we gave a brief outline of quantum theory; for a more complete treatment of the subject, we refer the reader to any of the quantum mechanics texts listed at the end of Chapter II. We feel that it may be of interest to read the historical development of quantum theory, and we recommend reading the two Bohr biographies that we have listed below (1, 2) and the two histories (A1, A2) listed in Chapter II. In addition, we have listed here a description of the old quantum theory (3) and the original Bohr paper (4).

1. R. Moore. *Niels Bohr: The Man, His Science and the World They Changed.* Alfred A. Knopf, New York (1966).
2. S. Rozental. *Niels Bohr, His Life and Work as Seen by His Friends and Colleagues.* North-Holland, Amsterdam (1967).
3. D. ter Haar. *The Old Quantum Theory.* Pergamon Press, Oxford (1967).
4. N. Bohr. On the quantum theory of line-spectra. *Kgl. Danske Vid. Selsk. Skr., Nat.-Math. Afd.,* 8 Raekke IV, 1, Part I, April 1918.

CHAPTER 2

EXACT SOLUTIONS OF THE SCHRÖDINGER EQUATION

1. Introduction

In 1926 Erwin Schrödinger published the famous paper in which he proposed that the wave functions of the stationary states of a system may be derived from the differential equation that now carries his name. Schrödinger published two additional papers at the same time, in which he showed how the equation is solved for three systems that play an essential role in the theory of atomic and molecular structure. These three systems are the harmonic oscillator, the rigid rotor, and the hydrogen atom. In this chapter we shall discuss the quantum mechanical description of these three systems, because they form the basis of many of our subsequent discussions. It is somewhat ironical that since 1926 hardly any new systems have been solved exactly beyond the three systems that were treated in the original paper. A few years later the Schrödinger equation was solved for the hydrogen molecular ion H_2^+, but this is a very complex mathematical problem and the solutions are so complicated that they are of little practical use.

Apparently Schrödinger discussed practically every system for which his equation can be solved exactly. In a subsequent paper he even developed perturbation theory, which he used to calculate the Stark effect of the hydrogen atom. However, if we want to understand the electronic structure of atoms and molecules and the theoretical description of the chemical bond, we must have some idea about the general behavior of the wave functions of larger molecules even though we cannot solve the Schrödinger equation exactly for these molecules. It is thus necessary to resort to approximate methods. We discuss these approximations in subsequent chapters, and we shall show that many of these approximations begin with the exact solution of the few systems that are available. It is thus helpful to discuss the exact solutions of the Schrödinger equation for these systems—the harmonic oscillator, the rigid rotor, and the hydrogen atom. Our treatment is a compromise between rigorous mathematical derivation and no mathematics at all. We list the exact solutions and we show that they satisfy the Schrödinger equation for each system. At the same time, we introduce the concept of angular momentum and mention some aspects of the classical mechanical descriptions wherever that seems relevant.

2. The Harmonic Oscillator

Around the turn of the century, it was not known how the electrons in an atom move around, but it was fairly obvious that their motion must satisfy two conditions. First, the electrons are not standing still but are moving in some fashion and, second, the electrons stay within the vicinity of the nucleus. A simple one-dimensional model that satisfies both conditions is the harmonic oscillator. This is a particle moving in the X direction that is subject to a force F given by

$$F = -kx \tag{1}$$

It may be seen that the force always tends to move the particle to the origin (see Fig. 2.1). The farther the particle gets away from the origin, the larger the force is that tends to move it back. The potential energy that corresponds to the above force is determined by the equation

$$F = -\frac{dV}{dx} \tag{2}$$

It follows easily that V is given by

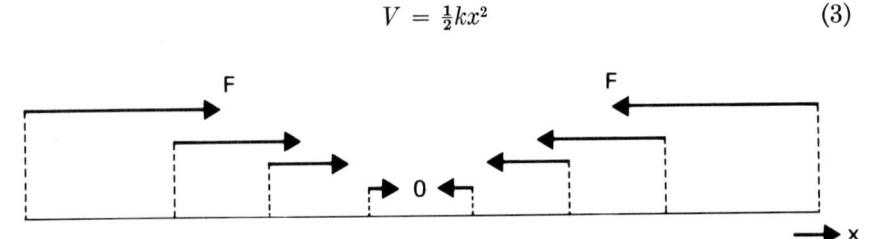

$$V = \tfrac{1}{2}kx^2 \tag{3}$$

FIGURE 2.1 ● The force acting on a harmonic oscillator. The force F tends to move the particle back to the origin.

THE HARMONIC OSCILLATOR

It may be seen that in classical mechanics the particle will oscillate back and forth around the origin. If it moves to the right, the force will tend to slow it down until the particle is turned around, then the force will accelerate it toward the origin. As soon as it has passed the origin, moving to the left, the force will slow it down again, and so on.

It is relatively easy to solve the classical equation of motion for the harmonic oscillator, and it may then be shown that the particle behaves exactly as we expect. We have

$$ma = m\frac{d^2x}{dt^2} = -kx \tag{4}$$

where the acceleration of the particle a multiplied by its mass m must be equal to the force F that acts on it, where F is given by Eq. (1). It is convenient to introduce the angular frequency ω of the harmonic oscillator by defining it as

$$\omega^2 = \frac{k}{m} \tag{5}$$

If we substitute this into Eq. (4), we obtain

$$\frac{d^2x}{dt^2} = -\omega^2 x \tag{6}$$

This equation has two solutions,

$$\begin{aligned} x &= \sin \omega t \\ x &= \cos \omega t \end{aligned} \tag{7}$$

and the general solution is

$$x = A \cos \omega t + B \sin \omega t \tag{8}$$

where A and B are arbitrary parameters.

We can describe the classical behavior of the harmonic oscillator by means of the simple model of Fig. 2.2. Here we consider a particle that moves with a constant velocity along the circumference of a circle with radius R. The position of the particle is determined by the polar angle ϕ, because the radius R is constant. The dependence of the angle ϕ on the time t must be given by the linear relationship,

$$\phi = \omega t \tag{9}$$

40 EXACT SOLUTIONS OF THE SCHRÖDINGER EQUATION

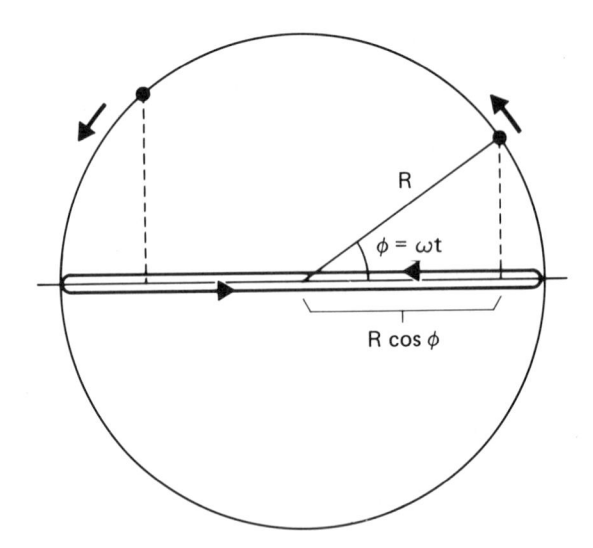

FIGURE 2.2 ● The harmonic oscillator moves as the projection of the motion of the particle around the circle.

because the particle has a constant angular velocity ω. The position of the projection is then given by

$$x(t) = R \cos \phi = R \cos \omega t \tag{10}$$

which is the first of the two solutions of Eq. (8).

Let us now derive an expression for the energy of the harmonic oscillator in classical mechanics. We find the velocity $v(t)$ of the particle by differentiating Eq. (10) with respect to time,

$$v(t) = -R\omega \sin \omega t \tag{11}$$

The total energy is obtained as the sum of the kinetic energy T and the potential energy V,

$$E = T + V = mv^2 + \tfrac{1}{2}kx^2 \tag{12}$$

By substituting Eqs. (10) and (11), we obtain

$$E = \tfrac{1}{2}mR^2\omega^2 \sin^2 \omega t + \tfrac{1}{2}kR^2 \cos^2 \omega t = \tfrac{1}{2}kR^2 = \tfrac{1}{2}m\omega^2 R^2 \tag{13}$$

We consider now the description of the harmonic oscillator in quantum mechanics. We note that the Hamiltonian is given by

$$\mathcal{H} = \frac{p^2}{2m} + \frac{kx^2}{2} \tag{14}$$

so that the Schrödinger equation is

$$-\frac{\hbar^2}{2m}\frac{d^2\psi}{dx^2} + \frac{kx^2}{2}\psi = E\psi \qquad (15)$$

We must solve this differential equation first and then we must determine for which values of the energy parameter E these solutions are acceptable from a physical point of view. These values, E_n, are then eigenvalues of the Schrödinger equation (15), which belong to the stationary states. The corresponding solutions $\psi_n(x)$ are the eigenfunctions, which describe the motion of the particle when it is in the stationary state n.

We shall show that the energy eigenfunctions of the harmonic oscillator are given by

$$E_n = (n + \tfrac{1}{2})\hbar\omega$$
$$n = 0, 1, 2, 3, 4, \ldots \qquad (16)$$

where ω is equal to the classical frequency that we defined in Eq. (5).

In Fig. 2.3 we have given the customary representation of the quantum mechanical eigenstates of the harmonic oscillator. We have plotted the potential energy V of Eq. (3) as a function of x, and we have drawn horizontal lines at the positions of the energy eigenvalues E_n. It is obvious from Eq. (16) that these lines must be equidistant; that is, they are all a distance $\hbar\omega$ apart.

Before we consider the eigenfunctions of the various eigenstates, let us determine first the points of intersection between the horizontal line belonging to the eigenstate E_0 and the parabolic curve of the potential curve $V(x)$. These are the points $\pm x_0$ in Fig. 2.3. These are determined from the condition

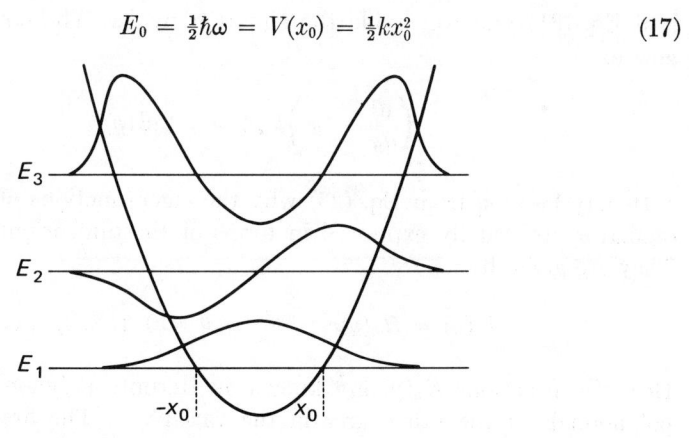

$$E_0 = \tfrac{1}{2}\hbar\omega = V(x_0) = \tfrac{1}{2}kx_0^2 \qquad (17)$$

FIGURE 2.3 ● Eigenvalues and eigenfunctions of the harmonic oscillator.

42 EXACT SOLUTIONS OF THE SCHRÖDINGER EQUATION

It follows easily that

$$x_0^2 = \frac{\hbar\omega}{k} = \frac{\hbar}{\sqrt{km}} \qquad (18)$$

or

$$x_0 = \frac{\sqrt{\hbar}}{(km)^{1/4}} \qquad (19)$$

It seems logical to use the quantity x_0 as the unit of length in the eigenfunctions of the harmonic oscillator. This means that we introduce the new variable y,

$$y = \frac{x}{x_0} \qquad (20)$$

This is a dimensionless quantity that represents the position x of the particle in terms of x_0 as its unit of length.

In order to find the eigenfunctions of the harmonic oscillator, let us first rewrite the Schrödinger equation (15) in terms of the variable y. This means that we take x_0 as the unit of length. We find

$$-\frac{\hbar^2}{2mx_0^2}\frac{d^2\psi}{dy^2} + \frac{kx_0^2 y^2}{2}\psi = E\psi \qquad (21)$$

At the same time we take $\hbar\omega$ as our new unit of energy; thus we substitute

$$E = \hbar\omega\epsilon \qquad (22)$$

into Eq. (21) and we divide the equation by $\hbar\omega$. The equation reduces now to

$$\left(\frac{d^2}{dy^2} - y^2\right)\psi(y) = -2\epsilon\psi(y) \qquad (23)$$

It may be seen from Eq. (23) why the eigenfunctions of the harmonic oscillator are usually expressed in terms of the dimensionless variable y. They are given by

$$\psi_n(y) = H_n(y)e^{-(1/2)y^2} \qquad n = 0, 1, 2, 3, \ldots \qquad (24)$$

Here the functions $H_n(y)$ are known as Hermite polynomials: They are polynomials of the nth degree in the variable y. The first few Hermite polynomials are given by

$$H_0(y) = 1$$
$$H_1(y) = 2y$$
$$H_2(y) = 4y^2 - 2 \tag{25}$$
$$H_3(y) = 8y^3 - 12y$$

and so on. We show the corresponding eigenfunctions $\psi_0(x), \ldots, \psi_3(x)$ in Fig. 2.3.

The Hermite polynomial $H_n(y)$ of the nth order is defined by the expression

$$H_n(y) = e^{y^2}\left(\frac{d^n e^{-y^2}}{dy^n}\right) \tag{26}$$

Obviously the eigenfunctions $\psi_n(y)$ may then be written as

$$\psi_n(y) = e^{(1/2)y^2}\left(\frac{d^n e^{-y^2}}{dy^n}\right) \tag{27}$$

We show that the functions $\psi_n(y)$ satisfy the Schrödinger equation (23) by determining the second derivative. We write the second derivative in the form

$$\frac{d^2\psi_n}{dy^2} = \left(\frac{d^2 e^{(1/2)y^2}}{dy^2}\right)\left(\frac{d^n e^{-y^2}}{dy^2}\right) + 2\left(\frac{de^{(1/2)y^2}}{dy}\right)\left(\frac{d^{n+1} e^{-y^2}}{dy^{n+1}}\right) + e^{(1/2)y^2}\left(\frac{d^{n+2} e^{-y^2}}{dy^{n+2}}\right)$$

$$= \left(\frac{d^2 e^{(1/2)y^2}}{dy^2}\right)\left(\frac{d^n e^{-y^2}}{dy^n}\right) + 2\left(\frac{de^{(1/2)y^2}}{dy}\right)\left(\frac{d^{n+1} e^{-y^2}}{dy^{n+1}}\right) + e^{(1/2)y^2}\left(\frac{d^{n+1}}{dy^{n+1}}\right)\left(\frac{de^{-y^2}}{dy}\right)$$

$$= (y^2 + 1)e^{(1/2)y^2}\left(\frac{d^n e^{-y^2}}{dy^n}\right) + 2ye^{(1/2)y^2}\left(\frac{d^{n+1} e^{-y^2}}{dy^{n+1}}\right)$$

$$+ e^{(1/2)y^2}\left(\frac{d^{n+1}}{dy^{n+1}}\right)(-2ye^{-y^2}) \tag{28}$$

The last term of this equation may be written as

$$\frac{d^{n+1}}{dy^{n+1}}(-2ye^{-y^2}) = -2y\frac{d^{n+1} e^{-y^2}}{dy^{n+1}} - 2(n+1)\frac{d^n e^{-y^2}}{dy^n} \tag{29}$$

Substitution into Eq. (28) gives

$$\frac{d^2\psi_n}{dy^2} = (y^2 + 1)e^{(1/2)y^2}\frac{d^n e^{-y^2}}{dy^n} + 2ye^{(1/2)y^2}\frac{d^{n+1} e^{-y^2}}{dy^{n+1}}$$

$$- 2ye^{(1/2)y^2}\frac{d^{n+1} e^{-y^2}}{dy^{n+1}} - 2(n+1)e^{(1/2)y^2}\frac{d^n e^{-y^2}}{dy^n}$$

$$= (y^2 - 2n - 1)\psi_n \tag{30}$$

Obviously,

$$\left(\frac{d^2}{dy^2} - y^2\right)\psi_n = -(2n+1)\psi_n \tag{31}$$

It thus follows that ψ_n is an eigenfunction of the Schrödinger equation (23) belonging to the eigenvalue

$$\begin{aligned} -2\epsilon_n &= -(2n+1) \\ E_n &= (n + \tfrac{1}{2})\hbar\omega \end{aligned} \tag{32}$$

It may be instructive to consider the probability density of the lowest eigenstate. We have plotted the normalized probability density

$$\rho_0(y) = \frac{1}{\sqrt{\pi}} e^{-y^2} \tag{33}$$

in Fig. 2.4. We notice two things. First, the probability density is largest for $y = 0$. Second, there is a finite probability for $y > y_0$, that is, beyond the two points where the energy line intersects with the potential energy curve. This second feature is quite remarkable. We should remember that the points $y = \pm y_0$ represent the two points where the total energy E is equal to the potential energy V, so that the kinetic energy T is zero. In the outside regions E is smaller than V, and in our classical picture the kinetic energy would be negative. We find, therefore, that there is a finite probability of finding the particle in a region where its kinetic energy is

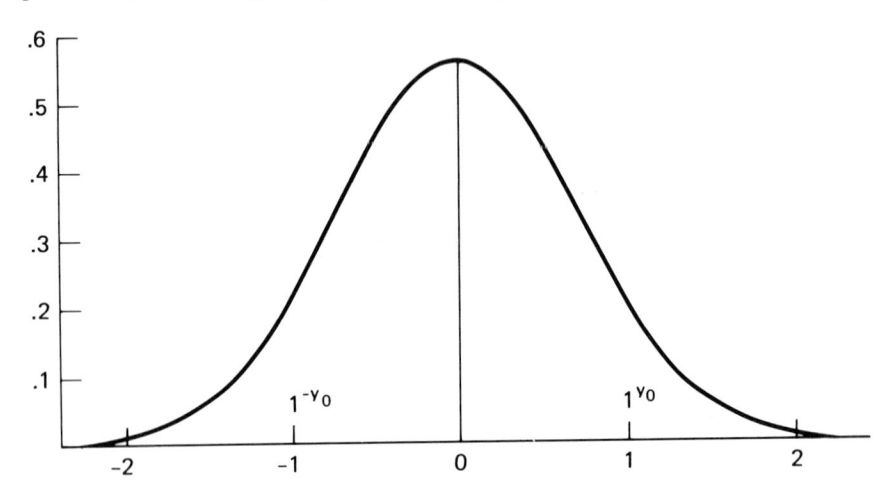

FIGURE 2.4 ● Charge distribution in the ground state of the harmonic oscillator.

negative. This type of behavior is fairly common in quantum mechanics, but it is not possible in classical mechanics where the kinetic energy must always be positive.

We illustrate the practical use of the harmonic oscillator in the next section by calculating the average displacement from equilibrium of the nuclei in a diatomic molecule due to vibrations.

3. Vibration of Diatomic Molecules, An Example

In a diatomic molecule the distance R between the two nuclei is not a fixed quantity. The two nuclei oscillate around the equilibrium position R_0 (see Fig. 2.5), and this oscillation, which is called the molecular vibration, may be represented as a harmonic oscillator with the distance $q = R - R_0$ as its coordinate. We shall try to derive the magnitude of the nuclear motion from the experimental data by making use of the theoretical results of the harmonic oscillator.

The vibrational motion is described by the Hamiltonian

$$\mathcal{H}_{vib} = -\frac{\hbar^2}{2M}\frac{d^2}{dq^2} + \tfrac{1}{2}kq^2 \tag{34}$$

where M is the reduced mass of the nuclei with masses M_a and M_b,

$$\frac{1}{M} = \frac{1}{M_a} + \frac{1}{M_b} \tag{35}$$

and k is the force constant of the harmonic oscillator.

We wish to find out how much the nuclei move away from their equilibrium position due to the vibrational motion when the molecule is in its vibrational ground state. In quantum mechanical terms this means that we wish to calculate the expectation value of the quantity q^2. We can define the extent of the vibrational motion in terms of a quantity Δq, which is related to this expectation value by

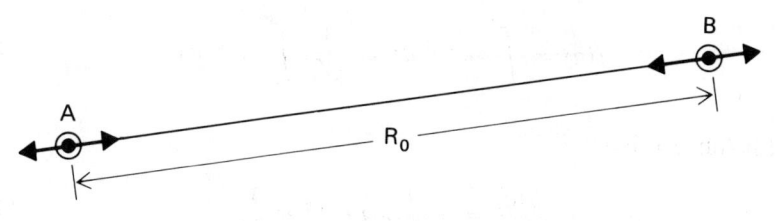

FIGURE 2.5 ● Vibrational motion of the nuclei in a diatomic molecule AB.

$$(\Delta q)^2 = \langle \psi_0 | (R - R_0)^2 | \psi_0 \rangle = \langle \psi_0 | q^2 | \psi_0 \rangle \tag{36}$$

Here ψ_0 is the normalized eigenfunction of the ground state of the harmonic oscillator.

We have seen that the ground-state eigenfunction ψ_0 of the harmonic oscillator is given by Eqs. (24) and (25) as

$$\psi_0(y) = \exp(-\tfrac{1}{2}y^2) \tag{37}$$

where the variable y is given by Eq. (20),

$$y = \frac{q}{q_0} \tag{38}$$

In our problem the variable is q, and q_0 is our unit of length as given by Eq. (19). It should be noted that the function (37) is not normalized; if we wish to use it for the evaluation of $(\Delta q)^2$ we must use the expression

$$(\Delta q)^2 = \frac{\int q^2 \exp(-q^2/q_0^2)\, dq}{\int \exp(-q^2/q_0^2)\, dq} \tag{39}$$

The calculation of Eq. (39) is not too difficult because we do not have to calculate the two integrals in order to derive their ratio. If we define the two integrals

$$I(a) = \int_{-\infty}^{\infty} e^{-ax^2}\, dx \qquad J(a) = \int_{-\infty}^{\infty} x^2 e^{-ax^2}\, dx \tag{40}$$

as functions of the parameter a, then it is easily seen that

$$J(a) = -\frac{\partial I(a)}{\partial a} \qquad \frac{J(a)}{I(a)} = -\frac{\partial \log I(a)}{\partial a} \tag{41}$$

We can write the integral $I(a)$ as

$$I(a) = \int_{-\infty}^{\infty} e^{-ax^2}\, dx = \frac{1}{\sqrt{a}} \int_{-\infty}^{\infty} e^{-y^2}\, dy \tag{42}$$

and it follows that

$$\frac{J(a)}{I(a)} = -\frac{\partial}{\partial a} \log a^{-1/2} = \frac{1}{2a} \tag{43}$$

VIBRATION OF DIATOMIC MOLECULES, AN EXAMPLE

If we apply this to equation (39), we find

$$(\Delta q)^2 = \tfrac{1}{2}q_0^2 \qquad \Delta q = \tfrac{1}{2}q_0\sqrt{2} \tag{44}$$

The calculation of Δq boils down to the evaluation of the unit of length q_0, which is given by equation (18),

$$q_0^2 = \frac{\hbar\omega}{k} = \frac{\hbar}{\sqrt{kM}} = \frac{\hbar}{M\omega} \tag{45}$$

Let us now consider the specific case of the CO molecule. Here, it is known from the infrared spectrum that the transition from the vibrational ground state to the first excited state has a wave number σ of 2,170 cm^{-1}. This corresponds to a frequency ν_0, which is given by

$$\nu_0 = \frac{c}{\lambda} = 2.998 \times 10^{10} \times 2{,}170 \text{ sec}^{-1} = 6.506 \times 10^{13} \text{ sec}^{-1} \tag{46}$$

The energy difference ΔE between the two states is given by

$$\Delta E = \hbar\omega = \frac{h\omega}{2\pi} \tag{47}$$

and from the relation

$$\Delta E = \frac{h\omega}{2\pi} = h\nu_0 \tag{48}$$

we derive that

$$\omega = 2\pi\nu_0 = 4.088 \times 10^{14} \text{ sec}^{-1} \tag{49}$$

We obtain the reduced mass M from Eq. (35) by substituting $M_C = 12.01 \times 1.6748 \times 10^{-24}$ g and $M_O = 16.00 \times 1.6748 \times 10^{-24}$ g,

$$M = \frac{12.01 \times 16.00 \times 1.6748 \times 10^{-24}}{12.01 + 16.00} = 1.149 \times 10^{-23} \text{ g} \tag{50}$$

We determine q_0 from Eq. (45) by substituting the results for ω and M. The result is

$$q_0^2 = \frac{\hbar}{\omega M} = \frac{1.054 \times 10^{-27}}{4.088 \times 10^{14} \times 1.149 \times 10^{-23}} = 0.2244 \times 10^{-18} \tag{51}$$

It follows that

$$q_0 = 0.474 \times 10^{-9} \text{ cm} = 0.0474 \text{ Å} \tag{52}$$

The average deviation Δq of the nuclei with respect to their equilibrium position is now obtained by substituting the above result for q_0 into Eq. (44). The result is

$$\Delta q = 0.0335 \text{ Å} \tag{53}$$

This is about the magnitude that we would expect.

4. The Rigid Rotor

In the previous section we have discussed the vibrational motion of a diatomic molecule and its relation to the harmonic oscillator. In the present section we shall discuss the rotational motion of a diatomic molecule, which is related to the quantum mechanical model of the rigid rotor.

Let us first consider the rotational motion of a diatomic molecule from a classical point of view. It may be shown that the rotational motion of the two nuclei A and B may be considered separately and that it may be represented by a model in which each nucleus rotates around the center of gravity. During this motion the distance between the nuclei remains fixed, and it is as if the two nuclei were connected by an invisible rigid rod of length r. We have sketched this situation in Fig. 2.6. It should be realized that during this motion the center of gravity C of the two nuclei (with masses M_a and M_b) remains fixed and we have taken this center of gravity as the origin in both Figs. 2.6a and 2.6b. In Fig. 2.6a we represent

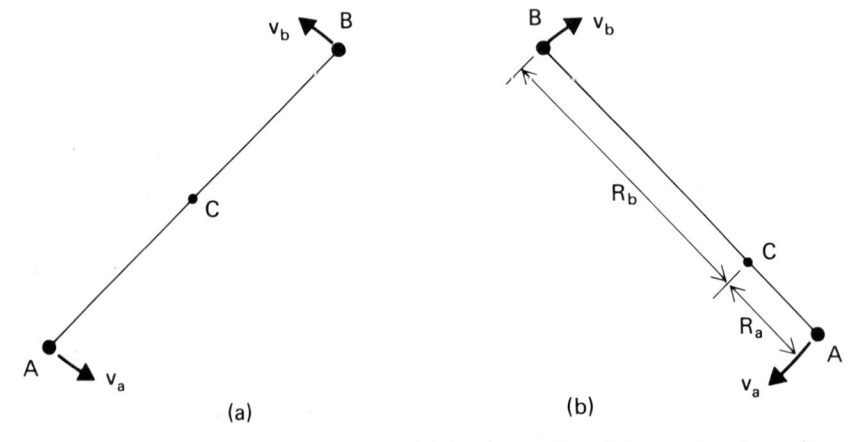

FIGURE 2.6 ● (a) The masses M_a and M_b are equal, and the center of gravity C is in the middle. (b) M_a is larger, and C is closer to A.

the situation where the two masses M_a and M_b are equal to one another, in which case the center of gravity C is exactly at the middle of the rod. In Fig. 2.6b we have assumed that one of the two nuclei, A, is much heavier than the other one, or that M_a is much larger than M_b. In that case the center of gravity C is still on the line between the two nuclei, but it is much closer to the heavier nucleus A than to the other one. The exact ratio between the two distances r_a and r_b is given by

$$\frac{r_a}{r_b} = \frac{M_b}{M_a} \tag{54}$$

We see that if M_a is much larger than M_b, the distance r_a becomes smaller. This means that the center of gravity is closer to the heavier nucleus. We may express the two distances r_a and r_b in terms of the distance r between the two nuclei. Because

$$r = r_a + r_b \tag{55}$$

it is easily derived from Eq. (54) that

$$r_a = \frac{M_b}{M_a + M_b} r \qquad r_b = \frac{M_a}{M_a + M_b} r \tag{56}$$

The position of the point A is given by the vector $\mathbf{r}_a$ and it moves with a velocity $\mathbf{v}_a$ that is perpendicular to $\mathbf{r}_a$. The magnitude of v_a may be written as the product of the length r_a and the angular velocity ω_a, which describes the rate at which the orientation of the vector $\mathbf{r}_a$ changes with time,

$$v_a = \omega_a r_a \tag{57}$$

Similarly, we write the velocity v_b of the point B as

$$v_b = \omega_b r_b \tag{58}$$

and we note again that $\mathbf{v}_b$ must be perpendicular to $\mathbf{r}_b$.

It is obvious that the two angular velocities ω_a and ω_b must be equal to each other,

$$\omega_a = \omega_b = \omega \tag{59}$$

otherwise the rod between the two nuclei would bend while they are moving.

50 EXACT SOLUTIONS OF THE SCHRÖDINGER EQUATION

Let us now consider the energy of the rigid rotor. Since there is no potential energy, the energy is simply equal to the sum of the kinetic energies of the two nuclei,

$$E = \tfrac{1}{2}M_a v_a^2 + \tfrac{1}{2}M_b v_b^2 \tag{60}$$

It is convenient to rewrite this in another form. Since the center of gravity of the system is fixed, the position of the rigid rotor is completely determined by the vector $\mathbf{r}$, which is defined as the line segment from the point A to the point B. The magnitude of r is then given by Eq. (55); it is the sum of r_a and r_b. We define the velocity v now as

$$\mathbf{v} = \frac{d\mathbf{r}}{dt} \qquad v = \omega r \tag{61}$$

The kinetic energy E of Eq. (60) may then be written as

$$E = \tfrac{1}{2}\mu v^2 \tag{62}$$

where

$$\frac{1}{\mu} = \frac{1}{M_a} + \frac{1}{M_b} \qquad \mu = \frac{M_a M_b}{M_a + M_b} \tag{63}$$

We see that if we define the momentum p as

$$\mathbf{p} = \mu \mathbf{v} \tag{64}$$

then the energy of Eq. (62) may also be written as

$$E = \frac{p^2}{2\mu} \tag{65}$$

In this way the motion of the rigid rotor is represented by the motion of a particle of mass μ over the surface of a sphere with radius r. It is fairly easy to prove that Eq. (60) is equivalent with Eq. (62) by rewriting it in terms of the angular velocity ω. If we substitute Eqs. (57) and (58) into Eq. (60), we obtain

$$E = \tfrac{1}{2}\omega^2(M_a r_a^2 + M_b r_b^2) \tag{66}$$

where we have used Eq. (59). The two distances r_a and r_b are expressed in terms of r by means of Eq. (56). By substituting this into Eq. (66), we obtain

$$E = \frac{\omega^2}{2}\left[\frac{M_a M_b^2}{(M_a+M_b)^2}r^2 + \frac{M_b M_a^2}{(M_a+M_b)^2}r^2\right] = \frac{M_a M_b}{M_a+M_b}\frac{\omega^2 r^2}{2} \quad (67)$$

The first term is equal to the reduced mass, defined by Eq. (63), and the second term is expressed in terms of the velocity $v = \omega r$. It thus follows that

$$E = \frac{\mu v^2}{2} = \frac{p^2}{2\mu} \quad (68)$$

The Hamiltonian that corresponds to the above energy expression is

$$\mathcal{H}_{op} = \frac{p^2}{2\mu} = -\frac{\hbar^2}{2\mu}\left(\frac{\partial^2}{\partial x^2} + \frac{\partial^2}{\partial y^2} + \frac{\partial^2}{\partial z^2}\right) \quad (69)$$

The Schrödinger equation for the rigid rotor is therefore

$$-\frac{\hbar^2}{2\mu}\left(\frac{\partial^2}{\partial x^2} + \frac{\partial^2}{\partial y^2} + \frac{\partial^2}{\partial z^2}\right)\psi = E\psi \quad (70)$$

We should remember, though, that during the motion the distance r between the two nuclei must remain constant and we must impose this constraint on the solution. This is mathematically equivalent to considering the Schrödinger equation (70) as describing the motion of a particle that is constrained to moving over the surface of a sphere with radius r. The position of the particle is determined by the two polar angles only, and the wave function should not contain the variable r. We must bear this in mind while solving the equation.

Let us first describe the eigenvalues and eigenfunctions of the rigid rotor equation (70); then we can show that these eigenfunctions and eigenvalues do satisfy the equation. It is customary to use a quantum number J to denote the eigenstates of the rigid rotor. The eigenvalues E_J are then given by

$$E_J = \frac{\hbar^2}{2\mu r^2}J(J+1) \qquad J = 0, 1, 2, 3, \ldots \quad (71)$$

In molecular spectroscopy this is often written as

$$E_J = BJ(J+1) \quad (72)$$

where B is known as the rotational constant of the diatomic molecule. By comparing the two equations (71) and (72), we find that B is given by

$$B = \frac{\hbar^2}{2\mu r^2} \quad (73)$$

In all these expressions r represents the distance between the two nuclei of the rigid rotor and it is assumed that r is a constant throughout the motion. The quantity

$$I = \mu r^2 \tag{74}$$

is the moment of inertia of the molecule, and this also remains constant during the motion.

Each eigenvalue E_J is $(2J + 1)$-fold degenerate. This means that there are $2J + 1$ different, linearly independent eigenfunctions belonging to the same eigenvalue E_J, or that the eigenfunction ψ_J contains $2J + 1$ arbitrary parameters. We write the eigenfunction ψ_J in the form

$$\psi_J = r^{-J} F_J(x, y, z) \tag{75}$$

where F_J is an Euler polynomial of Jth degree in the three variables x, y, and z.

Let us now discuss what an Euler polynomial is and what its properties are. An Euler polynomial is a special case of a homogeneous polynomial of the three variables x, y, and z. A homogeneous polynomial is a linear combination of terms of the type $x^k y^l z^m$ with the condition that for each term the sum of the three powers must have the same value. If that value is n, then we have a homogeneous polynomial of the nth degree. Such a polynomial may be written as

$$f_n(x, y, z) = \sum_k \sum_l \sum_m a_{k,l,m} x^k y^l z^m \tag{76}$$

where each term must satisfy the condition

$$k + l + m = n \tag{77}$$

Homogeneous polynomials have a useful property that we wish to use later. It is easily verified that

$$\Omega(x^k y^l z^m) = \left(x \frac{\partial}{\partial x} + y \frac{\partial}{\partial y} + z \frac{\partial}{\partial z}\right)(x^k y^l z^m)$$
$$= (k + l + m)(x^k y^l z^m) = n(x^k y^l z^m) \tag{78}$$

Consequently,

$$\Omega f_n(x, y, z) = \left(x \frac{\partial}{\partial x} + y \frac{\partial}{\partial y} + z \frac{\partial}{\partial z}\right) f_n(x, y, z) = n f_n(x, y, z) \tag{79}$$

We may say that f_n is an eigenfunction of the operator Ω with eigenvalue n.

An Euler polynomial $F_n(x, y, z)$ is now defined as a special case of a homogeneous polynomial $f_n(x, y, z)$, namely it is a homogeneous polynomial that satisfies the condition

$$\Delta F_n(x, y, z) = \left(\frac{\partial^2}{\partial x^2} + \frac{\partial^2}{\partial y^2} + \frac{\partial^2}{\partial z^2}\right) F_n(x, y, z) = 0 \quad (80)$$

It may be useful to give an illustration of this definition. The first three homogeneous polynomials are given by

$$\begin{aligned} f_0(x, y, z) &= a \\ f_1(x, y, z) &= ax + by + cz \\ f_2(x, y, z) &= ax^2 + bxy + cxz + dy^2 + pyz + qz^2 \end{aligned} \quad (81)$$

We have

$$\begin{aligned} \Delta f_0 &= 0 \\ \Delta f_1 &= 0 \\ \Delta f_2 &= 2(a + d + q) \end{aligned} \quad (82)$$

Since f_0 and f_1 both satisfy the condition (80), they are Euler polynomials and we have

$$\begin{aligned} F_0(x, y, z) &= f_0(x, y, z) = a \\ F_1(x, y, z) &= f_1(x, y, z) = ax + by + cz \end{aligned} \quad (83)$$

The Euler polynomial F_2 is obtained by imposing the condition

$$a + d + q = 0 \qquad q = -a - d \quad (84)$$

By substituting this into the general expression for f_2 we obtain

$$F_2(x, y, z) = a(x^2 - z^2) + d(y^2 - z^2) + bxy + cxz + pyz \quad (85)$$

We see that F_2 contains five parameters—it is a linear combination of five functions. It may be shown that F_J has $2J + 1$ parameters. An arbitrary homogeneous polynomial f_n contains

$$q_n = \tfrac{1}{2}(n + 1)(n + 2) \quad (86)$$

arbitrary parameters. If we now determine Δf_n, we obtain another homogeneous polynomial of degree $(n - 2)$, which contains q_{n-2} terms. Since Δf_n must be identically zero, this gives q_{n-2} relations between the coefficients. It follows thus that F_J has

$$q_J - q_{J-2} = \tfrac{1}{2}(J + 1)(J + 2) - \tfrac{1}{2}(J - 1)J = (2J + 1) \quad (87)$$

parameters.

54 EXACT SOLUTIONS OF THE SCHRÖDINGER EQUATION

It is relatively easy to show that the functions ψ_J of Eq. (75) satisfy the Schrödinger equation (70) for the rigid rotor. We have

$$\frac{\partial \psi_J}{\partial x} = \frac{\partial}{\partial x}(r^{-J}F_J) = -Jxr^{-J-2}F_J + r^{-J}\frac{\partial F_J}{\partial x}$$

$$\frac{\partial^2 \psi_J}{\partial x^2} = -Jr^{-J-2}F_J + J(J+2)x^2 r^{-J-4}F_J - 2Jxr^{-J-2}\frac{\partial F_J}{\partial x} + r^{-J}\frac{\partial^2 F_J}{\partial x^2} \qquad (88)$$

It may be seen that

$$\Delta \psi_J = \frac{\partial^2 \psi_J}{\partial x^2} + \frac{\partial^2 \psi_J}{\partial y^2} + \frac{\partial^2 \psi_J}{\partial z^2}$$

$$= [-3J + J(J+2)]r^{-J-2}F_J - 2Jr^{-J-2}$$

$$\left(x\frac{\partial}{\partial x} + y\frac{\partial}{\partial y} + z\frac{\partial}{\partial z}\right)F_J + r^{-J}\Delta F_J \qquad (89)$$

The last term of this expression is zero because of the definition (80) of the Euler polynomials, and the next-to-last term may be reduced by using Eq. (79). We then obtain

$$\Delta \psi_J = (-3J + J^2 + 2J - 2J^2)r^{-J-2}F_J = (-J^2 - J)r^{-2}\psi_J \qquad (90)$$

If we substitute this into the Schrödinger equation (70), it follows that

$$-\frac{\hbar^2}{2\mu}\Delta \psi_J = \frac{\hbar^2 J(J+1)}{2\mu r^2}\psi_J \qquad (91)$$

This means that ψ_J is an eigenfunction of the rigid rotor belonging to the eigenvalue E_J as defined by Eq. (71).

Finally, as an application of the rigid rotor, we calculate the rotational constant B, as given by Eq. (73) for two different diatomic molecules, namely H_2 and HBr.

In the case of H_2, the internuclear distance $r = 0.7416$ Å and the proton mass $M = 1.6724 \times 10^{-24}$ g. The reduced mass μ is half the proton mass. We should realize that we obtain the rotational constant B in terms of ergs if we use cgs units. In order to obtain B in terms of cm^{-1}, we must use the conversion factor

$$1 \text{ erg} = 5.0358 \times 10^{15} \text{ cm}^{-1} \qquad (92)$$

It follows now that for H_2 the rotational constant B is given by

$$B = \frac{\hbar^2}{2\mu r^2} \text{ ergs} = \frac{(1.05443 \times 10^{-27})^2 \times 5.0358 \times 10^{15}}{2 \times \frac{1}{2} \times 1.6724 \times 10^{-24} \times (0.7416)^2 \times 10^{-16}} \text{ cm}^{-1}$$

$$= 60.87 \text{ cm}^{-1} \qquad (93)$$

In the case of HBr, we note that the mass of the Br atom is 79.909 times the proton mass. The reduced mass is thus

$$\mu = \frac{1 \times 79.909}{1 + 79.909} \times 1.6724 \times 10^{-24} \text{ g} = 1.6517 \times 10^{-24} \text{ g} \quad (94)$$

The internuclear distance is 1.414 Å. The rotational constant is given by

$$B = \frac{(1.05443 \times 10^{-27})^2 \times 5.0358 \times 10^{15}}{2 \times 1.6517 \times 10^{-24} \times (1.414)^2 \times 10^{-16}} \text{ cm}^{-1} = 8.477 \text{ cm}^{-1} \quad (95)$$

Of course, in practice people perform the opposite of the above calculations because the rotational constants can be measured and Eq. (73) is used to derive the internuclear distance r, which cannot be measured directly.

5. Angular Momentum

We mentioned in Chapter I that in quantum mechanics we use the momentum p of a particle rather than its velocity v. The momentum is a useful quantity for describing the motion of a particle that moves more or less in a straight line. However, if the particle moves around in a closed orbit, it is more useful to have another quantity that represents the extent of the rotational motion. For this purpose the concept of angular momentum was introduced in classical mechanics. In the case of a circular motion where the velocity **v** is always perpendicular to the position vector **r** of the particle, the angular momentum M is defined as

$$M = pr = mvr \quad (96)$$

In other cases, where **v** and **r** are not perpendicular to each other, the magnitude of the angular momentum is given by

$$M = pr \sin \theta = mvr \sin \theta \quad (97)$$

where θ is the angle between **v** and **r**. This expression describes the extent (or the magnitude) of the rotational motion.

In general it is useful to know not only the extent of the rotational motion but also its direction. The easiest way to do this is to know the direction of the rotation axis; the plane of the rotational motion is then perpendicular to this rotation axis. We can also describe this direction by means of the angular momentum, but we must then define the angular momentum **M** as a vector. Its magnitude is still given by Eq. (97) and its direction is taken as the axis of rotation, which is the line perpendicular to the plane of the two vectors **r** and **p** (see Fig. 2.7). The way we define **M** then is known in vector analysis as the cross product of the two vectors **r** and **p**, and it is written as

$$\mathbf{M} = \mathbf{r} \times \mathbf{p} \tag{98}$$

Again, the magnitude of **M**, as defined by Eq. (98), is given by Eq. (97), and the direction of **M** is perpendicular to the plane of the two vectors **r** and **p**. The sign of the direction is determined by the corkscrew rule: If we rotate a corkscrew from **r** to **p** over the smallest angle, then **M** points in the direction that the corkscrew moves (see Fig. 2.7).

There is also an algebraic definition of the vector **M**. Let us first consider the simple case where the two vectors **p** and **r** are both located in the XY plane, where **r** has the components x and y and **p** has the components p_x and p_y (see Fig. 2.8). The vector **M** is then directed along the Z axis, and its magnitude M_z may be defined as

$$M_z = xp_y - yp_x \tag{99}$$

It is easily shown that this definition gives the same result as Eq. (97). If α is the polar angle for the vector **r** and β is the angle for the vector **p** (see Fig. 2.8), then we have

$$\begin{aligned} x &= r\cos\alpha & p_x &= p\cos\beta \\ y &= r\sin\alpha & p_y &= p\sin\beta \end{aligned} \tag{100}$$

Consequently,

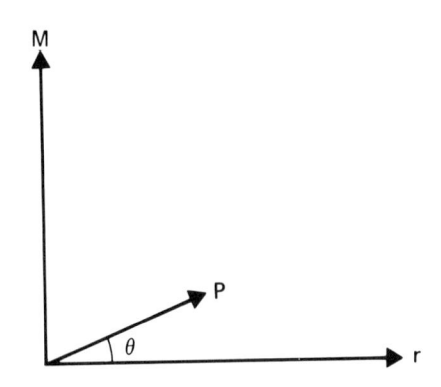

FIGURE 2.7 ● Definition of the angular momentum vector **M**.

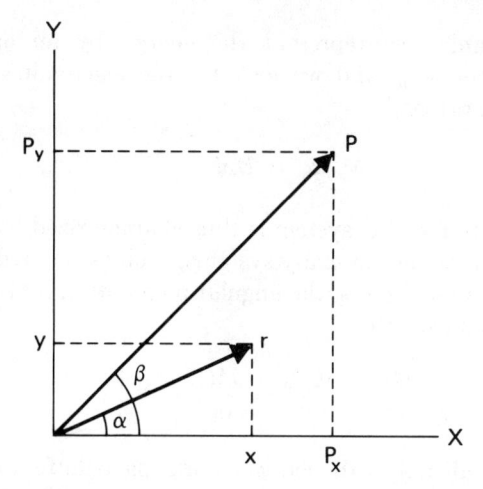

FIGURE 2.8 ● Definition of the angular momentum M_z if **r** and **ρ** are in the XY plane.

$$M_z = pr(\cos \alpha \sin \beta - \sin \alpha \cos \beta) = pr \sin(\beta - \alpha) \qquad (101)$$

This is equivalent to Eq. (97), because $(\beta - \alpha)$ is the angle between the two vectors. Equation (101) automatically gives the right sign for M_z. In Fig. 2.8 we have sketched the situation where $\beta > \alpha$. In that case it follows from the corkscrew rule that **M** must point upward, in the positive Z direction; Eq. (101) gives the same result.

In the general case, the vector **M** is not necessarily oriented along the Z axis, but the vector has three components, M_x, M_y, and M_z. These vector components are defined similarly to Eq. (99), namely

$$\begin{aligned} M_x &= yp_z - zp_y \\ M_y &= zp_x - xp_z \\ M_z &= xp_y - yp_x \end{aligned} \qquad (102)$$

This definition of the vector **M** is again equivalent to Eq. (97).

Since the rigid rotor has a purely rotational motion, there should be a close correlation between its eigenfunctions and eigenvalues and the properties of the angular momentum. It is useful to discuss this correlation, because it helps to interpret the eigenvalues and eigenfunctions of the rigid rotor and it also helps us understand the quantum mechanical description of the angular momentum.

In classical mechanics the energy E is a constant of the motion. This means that if there are no outside influences on a system, its energy does not change in time. We can write this in mathematical form as

$$\frac{dE}{dt} = 0 \qquad (103)$$

58 EXACT SOLUTIONS OF THE SCHRÖDINGER EQUATION

In quantum mechanics we represent the energy by an operator, the Hamiltonian operator $\mathcal{H}_{op}$, and we look for the eigenvalues and eigenfunctions of this operator,

$$\mathcal{H}_{op}\psi_n = E_n\psi_n \qquad (104)$$

Each stationary state n of the system is thus characterized by a constant.

It may be shown now that in many systems, such as the hydrogen atom, the rigid rotor, and many others, the angular momentum is also a constant of the motion. This means that

$$\frac{dM_x}{dt} = \frac{dM_y}{dt} = \frac{dM_z}{dt} = 0 \qquad (105)$$

The above systems all fall in the category of a particle in a central force field. This means that at any point **r** the force acting on the particle is directed toward the origin so that the force **F** has the same direction as the vector **r**. We can then write

$$F_x : F_y : F_z : F = x : y : z : r \qquad (106)$$

The three components F_x, F_y, and F_z may then be represented as

$$F_x = \frac{xF}{r} \qquad F_y = \frac{yF}{r} \qquad F_z = \frac{zF}{r} \qquad (107)$$

Since the acceleration **a** is proportional to **F**,

$$\mathbf{F} = m\mathbf{a} \qquad (108)$$

The same relationship as (107) holds for a,

$$a_x = \frac{xa}{r} \qquad a_y = \frac{ya}{r} \qquad a_z = \frac{za}{r} \qquad (109)$$

Let us now consider M_x, as defined by Eq. (102). If we differentiate M_x with respect to t, we obtain

$$\frac{dM_x}{dt} = \frac{dy}{dt}p_z + y\frac{dp_z}{dt} - \frac{dz}{dt}p_y - z\frac{dp_y}{dt}$$
$$= m(v_y v_z - v_z v_y) + m(ya_z - za_y) = 0 \qquad (110)$$

We may prove in the same way that the time derivatives of M_y and M_z are zero, so that the vector **M** remains at a constant value at all times. This means that the angular momentum is a constant of the motion, just as is the energy. This result leads us to believe that each stationary state of the two systems is also characterized by a specific value for the angular momentum. We shall investigate this question for the rigid rotor because here the problem is relatively simple.

In the case of the rigid rotor the angular momentum is the sum of the two contributions from each nucleus. According to the definition (96), it is given by

$$M = M_a v_a r_a + M_b v_b r_b \tag{111}$$

We showed in Section 4 that the rigid rotor also may be represented as a particle with mass μ moving over the surface of a sphere with radius r. Its velocity **v** is then given by Eq. (61). In this second model the angular momentum is given by

$$M = \mu v r \tag{112}$$

It is easily shown that the two equations (111) and (112) are identical. It may be recalled that the energy E of the rigid rotor is given by

$$E = \tfrac{1}{2}\mu v^2 \tag{113}$$

It then follows from Eqs. (112) and (113) that the energy may also be written as

$$E = \frac{M^2}{2\mu r^2} \tag{114}$$

where μ and r are constants. If we represent both E and M^2 by operators, namely $\mathcal{H}_{op}$ and $(M^2)_{op}$, respectively, then it follows from Eq. (114) that

$$\begin{aligned} \mathcal{H}_{op} &= (2\mu r^2)^{-1} (M^2)_{op} \\ (M^2)_{op} &= (2\mu r^2) \mathcal{H}_{op} \end{aligned} \tag{115}$$

Since the two operators are proportional, they must have the same eigenfunctions. We have seen that

$$\mathcal{H}_{op}\psi_J = E_J\psi_J = \frac{\hbar^2}{2\mu r^2}J(J+1)\psi_J \qquad J = 0, 1, 2, 3, \ldots \tag{116}$$

Obviously,

$$(M^2)_{op}\psi_J = 2\mu r^2 E_J\psi_J = J(J+1)\hbar^2\psi_J \qquad J = 0, 1, 2, 3, \ldots \tag{117}$$

We see that in the quantum state J the quantum number J describes the magnitude of the angular momentum vector.

The results of the quantum mechanical description of the angular momentum may be summarized in the form of a simple nonmathematical model. This is the old vector model, which was used by spectroscopists to explain and interpret the results of atomic spectroscopy. In this model it is assumed that the angular momentum is quantized and that it is determined by an integer quantum number, in our case the quantum number J, which has the values $J = 0, 1, 2, 3, \ldots$. Initially it was assumed that the magnitude of the angular momentum was equal to $\hbar J$, but it soon became apparent that this was not true. The magnitude $|M|$ of the angular momentum must be slightly larger than $\hbar J$, and it is given by

$$|M| = \hbar\sqrt{J(J+1)} \tag{118}$$

This is consistent with the result we derived from the Schrödinger equation in Eq. (117).

The vector model also makes predictions about the possible orientations of the angular momentum vector. We show in Fig. 2.9 how this works. We consider the possible values of the projection of the angular momentum vector along a given direction in space. Physically speaking, all directions in space are equivalent so that we can take any direction that we want, but it turns out that it is convenient to take the Z axis. According to the vector model, the projection M_z must have an integer value,

$$M_z = m\hbar \qquad m = 0, \pm 1, \pm 2, \ldots, \pm J \tag{119}$$

if the system is in a stationary state. The quantum number m has the possible values

$$m = -J, -J+1, -J+2, \ldots, J-1, J \tag{120}$$

In Fig. 2.9 we have drawn the situation for $J = 2$. Here the possible values for the projection are $-2, -1, 0, 1$, and 2. In one of the stationary stated, say $m = 1$, we know only the value of the projection of the vector **M** with respect to the Z axis. In our model we imagine that the vector rotates around the Z axis at a given angle in such a way that its projection on the Z axis is always one unit of $\hbar$. In this way we can visualize the quantization of the angular momentum.

It is worth noting that there is no stationary state where the angular momentum vector is directed exactly along the Z axis. The maximum value for the projection is J, and since the length of the vector is slightly larger (see Eq. 118), the vector **M** also rotates around the Z axis in this

ANGULAR MOMENTUM 61

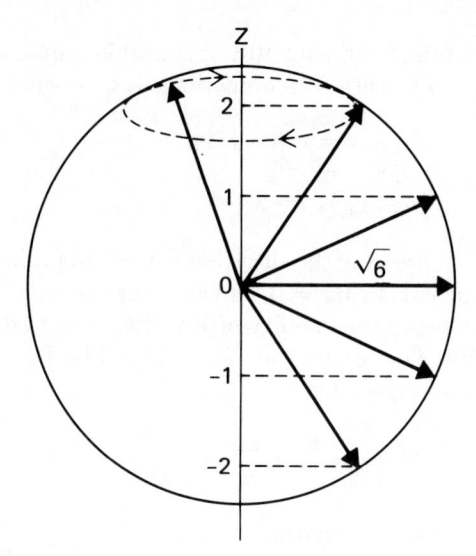

FIGURE 2.9 ● Quantization of the angular momentum along the Z axis for $J = 2$.

stationary state. This aspect of the quantization of the angular momentum vector can be explained from the uncertainty principle.

Let us consider, finally, the mathematical description of the vector model. We have said that for a given value for the total angular momentum or for a given value of J, the possible values of the projection M_z must be given by Eq. (119). In quantum mechanical language this means that the operator $(M_z)_{op}$ representing M_z must have a set of eigenvalues and eigenfunctions that are given by

$$(M_z)_{op}\psi_{J,m} = m\hbar\psi_{J,m} \qquad m = 0, \pm 1, \pm 2, \ldots, \pm J \qquad (121)$$

In the case where $J = 0$, the quantum number m can only be equal to zero. The corresponding eigenfunction is given by Eq. (83),

$$\psi_{0,0} = F_0(x, y, z) = a \qquad (122)$$

where a is a constant.

In the case where $J = 1$, the quantum number m can have the three values $m = 0, \pm 1$, and there are three eigenfunctions, $\psi_{1,0}$, $\psi_{1,1}$, and $\psi_{1,-1}$. We have seen in the previous section, Eq. (83), that

$$r\psi_1 = F_1(x, y, z) = ax + by + cz \qquad (123)$$

Each of the three eigenfunctions $\psi_{1,0}$, $\psi_{1,1}$, and $\psi_{1,-1}$ must be an eigenfunction of $(M^2)_{op}$ belonging to $J = 1$. They must therefore be of the form

62   EXACT SOLUTIONS OF THE SCHRÖDINGER EQUATION

(83), and they are obtained by substituting suitable values of the three parameters a, b, and c. The three eigenfunctions $\psi_{1,m}$ are given by

$$\begin{aligned}\psi_{1,1} &= r^{-1}(x + iy) \\ \psi_{1,0} &= r^{-1}z \\ \psi_{1,-1} &= r^{-1}(x - iy)\end{aligned} \qquad (124)$$

The first function $\psi_{1,1}$ represents the situation where the projection M_z has the value $\hbar$, $\psi_{1,0}$ corresponds to $M_z = 0$, and $\psi_{1,-1}$ corresponds to $M_z = -\hbar$.

It is not difficult to prove the validity of Eq. (124) and to derive as well the more general result. The component M_z is defined by Eq. (99), and the corresponding operator is given by

$$(M_z)_{\text{op}} = \frac{\hbar}{i}\left(x\frac{\partial}{\partial y} - y\frac{\partial}{\partial x}\right) \qquad (125)$$

If we now introduce polar coordinates,

$$\begin{aligned} x &= r\sin\theta\cos\phi \\ y &= r\sin\theta\sin\phi \\ z &= r\cos\theta \end{aligned} \qquad (126)$$

it is easily verified that

$$\frac{\partial}{\partial\phi} = \frac{\partial x}{\partial\phi}\frac{\partial}{\partial x} + \frac{\partial y}{\partial\phi}\frac{\partial}{\partial y} + \frac{\partial z}{\partial\phi}\frac{\partial}{\partial z} = x\frac{\partial}{\partial y} - y\frac{\partial}{\partial x} \qquad (127)$$

Consequently we may write the operator (125) as

$$(M_z)_{\text{op}} = \frac{\hbar}{i}\frac{\partial}{\partial\phi} \qquad (128)$$

Since it depends only on the variable ϕ, its eigenfunctions $\chi(\phi)$ depend only on the angle ϕ. The equation for the eigenvalues and eigenfunctions is thus

$$\frac{\hbar}{i}\frac{\partial\chi}{\partial\phi} = \lambda\chi \qquad (129)$$

and the solutions are

$$\chi(\phi) = \exp\frac{i\lambda\phi}{\hbar} \qquad (130)$$

We obtain the eigenvalues from the condition that the eigenfunction is single-valued, that is,

$$\chi(\phi) = \chi(\phi + 2\pi) \qquad (131)$$

The result is

$$\lambda = m\hbar \quad m = 0, \pm 1, \pm 2, \ldots \quad (132)$$

in agreement with what we mentioned earlier. The corresponding eigenfunctions are

$$\chi(\phi) = e^{im\phi} \quad m = 0, \pm 1, \pm 2, \ldots \quad (133)$$

We may summarize these results in the form

$$(M_z)_{\text{op}} e^{im\phi} = m\hbar e^{im\phi} \quad m = 0, \pm 1, \pm 2, \ldots \quad (134)$$

The eigenfunctions of $(M_z)_{\text{op}}$ may be multiplied by an arbitrary function of r and θ, and it is thus easily seen that the three functions of Eq. (124) are eigenfunctions of M_z, corresponding to the eigenvalues $m = 1, 0$, and -1, respectively.

As a final example, let us determine the five eigenfunctions of $(M_z)_{\text{op}}$ that belong to the state $J = 2$. Here the general eigenfunction ψ_2 has the form

$$\psi_2 = r^{-2}[axy + bxz + cyz + p(x^2 - z^2) + q(x^2 - y^2)] \quad (135)$$

We must find the linear combinations that contain the factors $e^{\pm i\phi}$ and $e^{\pm 2i\phi}$, respectively, if we express these functions in terms of polar coordinates. It may be verified that the five eigenfunctions of $(M_z)_{\text{op}}$ are given by

$$\begin{aligned}
\psi_{2,2} &= r^{-2}(x + iy)^2 = \sin^2\theta \, e^{2i\phi} \\
\psi_{2,1} &= r^{-2}(x + iy)z = \sin\theta \cos\theta \, e^{i\phi} \\
\psi_{2,0} &= r^{-2}(r^2 - 3z^2) = 1 - 3\cos^2\theta \\
\psi_{2,-1} &= r^{-2}(x - iy)z = \sin\theta \cos\theta \, e^{-i\phi} \\
\psi_{2,-2} &= r^{-2}(x - iy)^2 = \sin^2\theta \, e^{-2i\phi}
\end{aligned} \quad (136)$$

6. The Hydrogen Atom

It is important to know the eigenvalues and eigenfunctions of the hydrogen atom, especially because the wave functions of the ground state and the lower excited states are used as the basis for almost all valence theories. In previous sections we have considered only one-dimensional problems and the rigid rotor, which is a two-dimensional problem. The hydrogen atom is a three-dimensional problem, which means that the Schrödinger equation contains three variables; we expect that such an equation may be somewhat difficult to solve. Fortunately, we can simplify the equation

64 EXACT SOLUTIONS OF THE SCHRÖDINGER EQUATION

and it can then be solved exactly. We shall not give a purely mathematical analysis of this problem; instead we shall consider some of the more physical aspects which are related to the angular momentum. In this way it will be possible to get a better understanding of the general features of the quantum mechanical description.

The hydrogen atom consists of a proton with mass M and a positive charge e and an electron with mass m and a negative charge $-e$.

The proton mass is almost 2,000 times larger than the electron mass. The potential energy of this system is the Coulomb attraction between the two particles, which is given by

$$V = -\frac{e^2}{r} \quad (137)$$

where r is the distance between the electron and the proton. Since the proton is so much heavier than the electron, it is a good approximation to assume that the proton remains in a fixed position and the electron moves around it. The Schrödinger equation of the hydrogen atom is then

$$\left(-\frac{\hbar^2}{2m}\Delta - \frac{e^2}{r}\right)\psi(x, y, z) = E\psi(x, y, z) \quad (138)$$

with

$$\Delta = \frac{\partial^2}{\partial x^2} + \frac{\partial^2}{\partial y^2} + \frac{\partial^2}{\partial z^2} \quad (139)$$

Here, the vector **r**, which runs from the nucleus to the electron, has the components (x, y, z). If we want to be more precise, we have to realize that the proton does not remain in a fixed position. Instead, both proton and electron move around the center of gravity, which remains in a fixed position. The net effect of this correction is that in Eq. (138) we must replace the electron mass m by the reduced mass μ,

$$\frac{1}{\mu} = \frac{1}{M} + \frac{1}{m} \quad (140)$$

which is slightly different from the electron mass. The Schrödinger equation then takes the form

$$\left(-\frac{\hbar^2}{2\mu}\Delta - \frac{e^2}{r}\right)\psi(x, y, z) = E\psi(x, y, z) \quad (141)$$

In classical mechanics the motion of the electron around the nucleus is in every respect the same as the motion of a satellite around the earth (or the earth around the sun). In all these cases there is a force that tends to pull the orbiting particle toward the center. In our case the force is electrostatic and in astronomy the force is gravitational, but both forces are given by the same type of mathematical expression (137). We know that the earth describes an orbit around the sun that is elliptical and that this type of behavior may in general be derived from classical mechanics—the moving particle describes either a circular or an elliptical orbit around the center as long as its energy is below a certain limit (otherwise it is not in a closed orbit).

In order to understand the behavior of the moving orbital, it is useful to consider the two quantities that are constants of the motion, namely the energy and the angular momentum. The fact that

$$\frac{dE}{dt} = 0 \tag{142}$$

follows from the conservation of energy principle. We have shown in the previous section that the angular momentum vector **M** must also stay the same as a function of time,

$$\frac{d\mathbf{M}}{dt} = \mathbf{0} \tag{143}$$

It follows from Eqs. (142) and (143) that a stationary state of the hydrogen atom in classical mechanics is characterized by a constant, time-independent value of the energy E and the angular momentum vector **M**. Conversely, if E and **M** are given, then it is possible to calculate the classical orbit of the electron. It is then found that the electron moves in an elliptical orbit around the origin in such a way that the angular momentum remains constant.

Naturally, the quantum mechanical description of the hydrogen atom is quite different from the classical description, but the two theories have one thing in common: In both cases the energy E and the angular momentum **M** are constants of the motion. We have seen what this means in classical mechanics. In quantum mechanics the consequences are a little different. Here, a stationary state is represented by a wave function, which is an eigenfunction ψ_n of the Hamiltonian operator $\mathcal{H}_{op}$,

$$\mathcal{H}_{op}\psi_n = E_n\psi_n \tag{144}$$

It should be noted that the wave function of a stationary state must be an eigenfunction of $\mathcal{H}_{op}$ because $\mathcal{H}_{op}$ represents the energy, which is a constant of the motion. It follows, then, that the wave functions of the stationary states must also be eigenfunctions of the angular momentum operators, because the angular momentum is also a constant of the motion.

We have seen in the previous section that the angular momentum is represented in quantum mechanics by the two operators $(M^2)_{op}$ and $(M_z)_{op}$. The first operator $(M^2)_{op}$ represents the magnitude of the angular momentum vector (to be precise, the square of the magnitude), and the second operator $(M_z)_{op}$ represents the projection of **M** on the Z axis and describes the orientation of the angular momentum vector. It follows that each stationary state of the hydrogen atom is characterized by three quantum numbers, n, l, and m. The first quantum number n labels the value of the energy E_n, the second quantum number l gives us the magnitude of the angular momentum vector, and the third quantum number m describes the orientation of the angular momentum vector.

We can also express this statement in terms of mathematics. It means that the wave function $\Phi_{n,l,m}$ of the stationary state (n, l, m) must be an eigenfunction of the three operators $\mathcal{H}_{op}$, $(M^2)_{op}$, and $(M_z)_{op}$,

$$\begin{aligned}
\mathcal{H}_{op}\Phi_{n,l,m} &= E_n\Phi_{n,l,m} \\
(M^2)_{op}\Phi_{n,l,m} &= (M^2)_l\Phi_{n,l,m} \\
(M_z)_{op}\Phi_{n,l,m} &= (M_z)_m\Phi_{n,l,m}
\end{aligned} \qquad (145)$$

In this way the value E_n of the energy depends on the quantum number n, the value $(M^2)_l$ of the magnitude of the angular momentum depends on l, and the value $(M_z)_m$ is determined by m.

Once we understand the general behavior of the eigenvalues and eigenfunctions of the hydrogen atom, we can derive their specific forms without too much difficulty. We should remember that we derived the eigenvalues and eigenfunctions of the angular momentum in the previous section, and we can also use these results for the hydrogen atom by making a few minor changes.

In the description of the hydrogen atom, it is helpful to make use of the polar coordinates shown in Fig. 2.10. The variable r is the distance OP, the angle θ is the angle between the vector OP and the Z axis, and ϕ is the angle between OP' and the X axis. The relation between the Cartesian coordinates (x, y, z) and the polar coordinates (r, θ, ϕ) is given by the equations

$$\begin{aligned}
x &= r \sin \theta \cos \phi \\
y &= r \sin \theta \sin \phi \\
z &= r \cos \theta
\end{aligned} \qquad (146)$$

We have seen in the previous section, Eq. (128), that the operator $(M_z)_{op}$ is a function only of the polar angle ϕ. It is given by

THE HYDROGEN ATOM   67

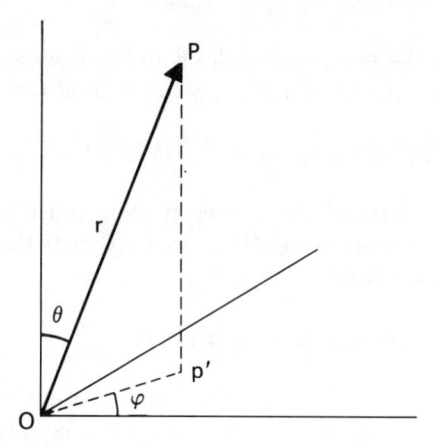

FIGURE 2.10 ● Definition of polar coordinates. P' is the projection of P in the XY plane.

$$(M_z)_{op} = \frac{\hbar}{i} \frac{\partial}{\partial \phi} \qquad (147)$$

Its eigenvalues and eigenfunctions are given by Eq. (134), namely

$$(M_z)_{op} e^{im\phi} = m\hbar e^{im\phi} \qquad (148)$$

It should be realized that these eigenfunctions may be multiplied by an arbitrary function ψ of the other polar coordinates r and θ. We have, namely,

$$(M_z)_{op}[\psi(r, \theta)e^{im\phi}] = m\hbar[\psi(r, \theta)e^{im\phi}] \qquad (149)$$

because the operator does not contain the two variables r and θ. This makes it possible to find a set of functions that are eigenfunctions of both $(M^2)_{op}$ and $(M_z)_{op}$. For example, in the previous section, Eq. (124), we mentioned that the set of functions $\psi_{1,1}$, $\psi_{1,0}$, and $\psi_{1,-1}$ are eigenfunctions of both operators $(M^2)_{op}$ and $(M_z)_{op}$.

The same argument can be applied to the eigenfunctions of the hydrogen atom as compared to the eigenfunctions of the angular momentum operator $(M^2)_{op}$. We have seen in the previous section that the eigenvalues and eigenfunctions of $(M^2)_{op}$ are given by

$$(M^2)_{op}[r^{-l}F_l(x, y, z)] = l(l+1)\hbar^2[r^{-l}F_l(x, y, z)] \qquad (150)$$

where $F_l(x, y, z)$ is the Euler polynomial of the lth degree in the three variables x, y, and z. We note that $(M^2)_{op}$ does not contain the variable r, so it depends only on the two polar angles θ and ϕ. Consequently, we may multiply its eigenfunctions by an arbitrary function of r, and we also have

$$(M^2)_{\text{op}}[g(r)F_l(x, y, z)] = l(l + 1)\hbar^2[g(r)F_l(x, y, z)] \quad (151)$$

Since each stationary state of the hydrogen atom must be characterized by an eigenfunction of the operator $(M^2)_{\text{op}}$, we may write the eigenfunctions $\Phi_{n,l,m}$ of Eq. (145) in the form

$$\Phi(x, y, z) = g(r)F_l(x, y, z) \quad (152)$$

By substituting this function into the Schrödinger equation (141), we can reduce the Schrödinger equation to a differential equation in the one variable r only.

In order to substitute the function Φ of Eq. (152) into the Schrödinger equation, we must determine its derivatives with respect to x, y, and z. We have

$$\begin{aligned}\frac{\partial \Phi}{\partial x} &= \frac{\partial g}{\partial x} F_l + g \frac{\partial F_l}{\partial x} = \frac{x}{r} \frac{dg}{dr} F_l + g \frac{\partial F_l}{\partial x} \\ \frac{\partial^2 \Phi}{\partial x^2} &= \left(\frac{1}{r} - \frac{x^2}{r^3}\right) \frac{dg}{dr} F_l + \frac{x^2}{r^2} \frac{d^2 g}{dr^2} F_l + \frac{2x}{r} \frac{dg}{dr} \frac{\partial F_l}{\partial x} + g \frac{\partial^2 F_l}{\partial x^2}\end{aligned} \quad (153)$$

It follows that

$$\begin{aligned}\Delta \Phi &= \frac{\partial^2 \Phi}{\partial x^2} + \frac{\partial^2 \Phi}{\partial y^2} + \frac{\partial^2 \Phi}{\partial z^2} \\ &= \left(\frac{d^2 g}{dr^2} + \frac{2}{r} \frac{dg}{dr}\right) F_l + \frac{2}{r} \frac{dg}{dr} \left(x \frac{\partial F_l}{\partial x} + y \frac{\partial F_l}{\partial y} + z \frac{\partial F_l}{\partial z}\right) + g \Delta F_l \quad (154)\end{aligned}$$

The last term is zero because of the definition of the Euler polynomials, and the next-to-last term may be reduced by using Eq. (79). The result is

$$\Delta \Phi = F_l \left(\frac{d^2 g}{dr^2} + \frac{2l + 2}{r} \frac{dg}{dr}\right) \quad (155)$$

If we substitute this into the Schrödinger equation, then it reduces to the following differential equation for the function $g(r)$,

$$-\frac{\hbar^2}{2\mu} \left(\frac{d^2 g}{dr^2} + \frac{2l + 2}{r} \frac{dg}{dr}\right) - \frac{e^2 g}{r} = Eg \quad (156)$$

Just as in the case of the harmonic oscillator, it is convenient here to introduce new units of length and energy because in this way we can

THE HYDROGEN ATOM

eliminate all the physical constants in the equation. These units are widely used in atomic calculations, and they are known as atomic units. The atomic unit of length, a_0, is given by

$$a_0 = \frac{\hbar^2}{\mu e^2} \tag{157}$$

and it is approximately equal to 0.529 Å = 0.529 × 10⁻⁸ cm. The corresponding atomic unit of energy is given by

$$\epsilon_0 = \frac{e^2}{a_0} \tag{158}$$

and it is approximately equal to 27.2 eV. Introducing these units is equivalent to the transformation

$$r = s a_0 \qquad E = \epsilon \cdot \epsilon_0 \tag{159}$$

of Eq. (156). The result is

$$-\frac{1}{2}\left(\frac{d^2 g(s)}{ds^2} + \frac{2l+2}{s}\frac{dg(s)}{ds}\right) - \frac{g(s)}{s} = \epsilon g(s) \tag{160}$$

where all the physical constants have been transformed away.

The eigenvalues of the above equation are given by

$$\epsilon_n = -\frac{1}{2n^2} \qquad E_n = -\frac{1}{2n^2}\frac{e^2}{a_0} \tag{161}$$

where the quantum number n can take the values

$$n = l+1, l+2, l+3, \ldots \tag{162}$$

The corresponding eigenfunctions $g_{n,l}(s)$ depend on both quantum numbers n and l. They have the form

$$g_{n,l}(s) = \exp\left(\frac{-s}{n}\right) \Phi\left(l+1-n, 2l+2; \frac{2s}{n}\right) \tag{163}$$

By introducing the new variable t, which is defined as

$$t = \frac{2s}{n} \tag{164}$$

we can also write this as

$$g_{n,l}(t) = \exp\left(\frac{-t}{2}\right)\Phi(l+1-n, 2l+2; t) \qquad (165)$$

The function Φ is the hypergeometric series. Its general definition is by means of an infinite sum or power series in the variable t, namely

$$\Phi(\alpha, \beta; t) = 1 + \frac{\alpha}{\beta}\frac{t}{1!} + \frac{\alpha(\alpha+1)}{\beta(\beta+1)}\frac{t^2}{2!} + \frac{\alpha(\alpha+1)(\beta+2)}{\alpha(\alpha+1)(\beta+2)}\frac{t^3}{3!} + \cdots \qquad (166)$$

However, in Eq. (165) the hypergeometric series always reduces to a finite polynomial, because this happens when the variable α is equal to a negative integer. It follows from the definition (166) that

$$\Phi(0, \beta; t) = 1$$

$$\Phi(-1, \beta; t) = 1 - \frac{t}{\beta}$$

$$\Phi(-2, \beta; t) = 1 - \frac{2t}{\beta} + \frac{2t^2}{\beta(\beta+1)\cdot 2} \qquad (167)$$

and so on.

We can use Eq. (167) to give the specific form of the radial eigenfunctions $g_{n,l}(s)$ for some specific cases, namely $n = l+1$, $n = l+2$, and $n = l+3$. We have listed these functions in Table 2.1, and we shall use these results in the following section where we discuss the specific forms of some of the hydrogen atom eigenfunctions.

In the remainder of this section we shall show that the eigenfunctions (165) or (166) do indeed satisfy the Schrödinger equation. A more detailed discussion of the properties of the atomic eigenvalues and eigenfunctions is postponed to the next section.

TABLE 2.1 ● Radial eigenfunctions $g_{n,l}(r)$ for various values of the quantum numbers n and l. The variable r is expressed in terms of the atomic unit of length a_0.

n	$g_{n,l}(r)$
$l+1$	$\exp\left(-\dfrac{r}{n}\right)$
$l+2$	$\left[\dfrac{r}{n} - (l+1)\right]\exp\left(-\dfrac{r}{n}\right)$
$l+3$	$\left[\left(\dfrac{r}{n}\right)^2 - (2l+3)\left(\dfrac{r}{n}\right) + \dfrac{1}{2}(l+1)(2l+3)\right]\exp\left(-\dfrac{r}{n}\right)$

THE HYDORGEN ATOM 71

We want to show that the functions (163) or (165) satisfy the Schrödinger equation (160). This is somewhat more complicated than in the case of the harmonic oscillator, and it is necessary to do it in two stages. The first stage is to show that the hypergeometric series (166) satisfies the differential equation

$$x \frac{d^2\Phi}{dx^2} + (\beta - x) \frac{d\Phi}{dx} - \alpha \Phi = 0 \tag{168}$$

We write the confluent hypergeometric series as

$$\Phi(\alpha, \beta; x) = \sum_{n=0}^{\infty} c_n \left(\frac{x^n}{n!}\right) \tag{169}$$

where the definition of the coefficients c_n follows from Eq. (166). We have now

$$\frac{d\Phi}{dx} = \sum_{n=1}^{\infty} \frac{n c_n x^{n-1}}{n!} = \sum_{n=0}^{\infty} \frac{c_{n+1} x^n}{n!}$$

$$x \frac{d\Phi}{dx} = \sum_{n=1}^{\infty} \frac{n c_n x^n}{n!} \tag{170}$$

$$x \frac{d^2\Phi}{dx^2} = \sum_{n=2}^{\infty} \frac{n(n-1) c_n x^{n-1}}{n!} = \sum_{n=2}^{\infty} \frac{(n-1) c_n x^{n-1}}{(n-1)!} = \sum_{n=1}^{\infty} \frac{n c_{n+1} x^n}{n!}$$

It follows that Φ satisfies Eq. (168) if it satisfies the equation

$$\sum_{n=0}^{\infty} \frac{n c_{n+1} x^n}{n!} + \beta \sum_{n=0}^{\infty} \frac{c_{n+1} x^n}{n!} - \sum_{n=0}^{\infty} \frac{n c_n x^n}{n!} - \alpha \sum_{n=0}^{\infty} \frac{c_n x^n}{n!} = 0 \tag{171}$$

or

$$\sum_{n=0}^{\infty} [(n + \beta) c_{n+1} - (n + \alpha) c_n] \left(\frac{x^n}{n!}\right) = 0 \tag{172}$$

It follows from Eq. (166) that

$$\frac{c_{n+1}}{c_n} = \frac{n + \alpha}{n + \beta} \tag{173}$$

and we see that every term of the series (172) is zero so that $\Phi(\alpha, \beta; x)$ satisfies the differential equation (168).

In the second stage of our proof we show that substitution of the eigenfunction $g_{n,l}(s)$ into the Schrödinger equation (160) leads to the differential equation (168) for the confluent hydergeometric function. We write $g_{n,l}(s)$ as

$$g(s) = \chi(t) \exp\left(-\frac{s}{n}\right) \qquad (174)$$

We have then

$$\frac{dg}{ds} = \frac{d\chi}{dt}\frac{dt}{ds}\exp\left(-\frac{s}{n}\right) - \frac{1}{n}\chi\exp\left(-\frac{s}{n}\right)$$

$$= \left(\frac{2}{n}\frac{d\chi}{dt} - \frac{1}{n}\chi\right)\exp\left(-\frac{s}{n}\right) \qquad (175)$$

$$\frac{d^2g}{ds^2} = \left(\frac{4}{n^2}\frac{d^2\chi}{dt^2} - \frac{4}{n^2}\frac{d\chi}{dt} + \frac{1}{n^2}\chi\right)\exp\left(-\frac{s}{n}\right)$$

Substitution into the differential equation (160) gives

$$\left(-\frac{2}{n^2}\frac{d^2\chi}{dt^2} + \frac{2}{n^2}\frac{d\chi}{dt} - \frac{1}{2n^2}\chi\right) - \frac{2(l+1)}{nt}\left(\frac{2}{n}\frac{d\chi}{dt} - \frac{1}{n}\chi\right) - \frac{2\chi}{nt} = \epsilon\chi \qquad (176)$$

or

$$t\frac{d^2\chi}{dt^2} + (2l + 2 - t)\frac{d\chi}{dt} - (l + 1 - n)\chi + (1 + 2\epsilon n^2)t\chi = 0 \qquad (177)$$

It is easily seen that if we take

$$1 + 2\epsilon n^2 = 0 \qquad \epsilon = \frac{-1}{2n^2} \qquad (178)$$

then the equation reduces to

$$t\frac{d^2\chi}{dt^2} + (2l + 2 - t)\frac{d\chi}{dt} - (l + 1 - n)\chi = 0 \qquad (179)$$

This is identical with Eq. (168) for the confluent hypergeometric series and its solution is

$$\chi(t) = \Phi(l + 1 - n, 2l + 2; t) \qquad (180)$$

We have thus shown that the functions $g_{n,l}(s)$ of Eq. (163) are indeed the eigenfunctions of the Schrödinger equation (160) for the hydrogen atom.

7. Properties of the Eigenvalues and Eigenfunctions of the Hydrogen Atom

Before we consider the detailed form of the eigenfunctions, it may be useful to look into the bookkeeping in classifying them. We have mentioned that each stationary state of the hydrogen atom is characterized by the

EIGENVALUES AND EIGENFUNCTIONS OF HYDROGEN ATOM 73

quantum numbers, n, which refers to the energy, l, which refers to the magnitude of the angular momentum, and m, which refers to the orientation of the angular momentum. The possible values of these quantum numbers are

$$\begin{aligned} l &= 0, 1, 2, 3, 4, \ldots \\ m &= 0, \pm 1, \pm 2, \pm 3, \ldots, \pm l \\ n &= l+1, l+2, l+3, \ldots \end{aligned} \tag{181}$$

We see that the possible values of n and l must satisfy the condition

$$n \geq l + 1 \tag{182}$$

In Eq. (181) we let the quantum number l take all possible values and then we limit the possible values of n to those that are allowed by the condition (182). We rather classify the quantum states in a different way where we let n take all possible values and restrict the values of l. In this way we have

$$\begin{aligned} n &= 1, 2, 3, 4, \ldots \\ l &= 0, 1, 2, \ldots, n-1 \\ m &= 0, \pm 1, \pm 2, \ldots, \pm l \end{aligned} \tag{183}$$

This is the customary way to classify the quantum states of the hydrogen atom.

The symbols used to describe these various quantum states date from before quantum mechanics; they are derived from the symbols the experimental spectroscopists used to describe the various terms in an atomic spectrum. In this terminology the value of the quantum number l, the magnitude of the angular momentum, is given by a letter. States with $l = 0$ are described by a letter s, states with $l = 1$ with a letter p, $l = 2$ is d, $l = 3$ is f, $l = 4$ is g, and so on. The value of the quantum number n is then put in front of this letter.

Since the energy eigenvalues are given by

$$\epsilon_n = -\frac{1}{2n^2} \tag{184}$$

we see that the lowest energy occurs for the value $n = 1$ of the quantum number n. It follows from Eq. (183) that the quantum number l must then have the value $l = 0$, and this is called the $1s$ state of the hydrogen atom. The quantum number m can have only one value, $m = 0$, and it follows that the $1s$ state is nondegenerate. The eigenfunction of the $1s$ state ($n = 1$, $l = 0$, $m = 0$) follows from Eq. (163) and Table 2.1,

$$\phi(1s) = F_0(x, y, z) g_{1,0}(r) = \exp(-r) \tag{185}$$

74 EXACT SOLUTIONS OF THE SCHRÖDINGER EQUATION

We obtain the first excited state of the hydrogen atom if we take $n = 2$. Now it follows from Eq. (183) that the quantum number l can have two possible values, namely $l = 0$ and $l = 1$. Both states have the same energy but different values of the angular momentum. The eigenfunctions of these states are again derived from Eq. (163) and Table 2.1,

$$\phi(2s) = F_0(x, y, z)g_{2,0}(r) = \left(\frac{1}{2}r - 1\right)\exp\left(\frac{-r}{2}\right)$$

$$\phi(2p) = F_1(x, y, z)g_{2,1}(r) = (ax + by + cz)\exp\left(\frac{-r}{2}\right) \qquad (186)$$

The energy eigenvalue $n = 2$ is fourfold degenerate because the function $\phi(2p)$ contains three parameters so that there are four functions that belong to the same energy eigenvalue. We could have predicted this degeneracy by considering the possible values of the quantum numbers l and m,

$$\begin{array}{ll} l = 0 & m = 0 \\ l = 1 & m = 0, \pm 1 \end{array} \qquad (187)$$

Since there are four possibilities, the energy eigenvalue $n = 2$ must be fourfold degenerate.

The next higher energy eigenvalue occurs for $n = 3$. The quantum numbers l and m can have the values

$$\begin{array}{ll} l = 0 & m = 0 \\ l = 1 & m = 0, \pm 1 \\ l = 2 & m = 0, \pm 1, \pm 2 \end{array} \qquad (188)$$

and the energy eigenvalue is ninefold degenerate. We see that we have three different states if we just look at the quantum number l, namely a $3s$ state for $l = 0$, a $3p$ state for $l = 1$, and a $3d$ state for $l = 2$. The corresponding eigenfunctions follow again from Eq. (163) and Table 2.1:

$$\phi(3s) = F_0(x, y, z)g_{3,0}(r) = \left(1 - \frac{2r}{3} + \frac{2r^2}{27}\right)\exp\left(\frac{-r}{3}\right)$$

$$= \frac{1}{27}(2r^2 - 18r + 27)\exp\left(\frac{-r}{3}\right)$$

$$\phi(3p) = F_1(x, y, z)g_{3,1}(r) = (ax + by + cz)\left(\frac{r}{3} - 2\right)\exp\left(\frac{-r}{3}\right) \qquad (189)$$

$$= \frac{1}{3}(r - 6)(ax + by + cz)\exp\left(\frac{-r}{3}\right)$$

$$\phi(3d) = F_2(x, y, z)g_{3,2}(r) = [c_1(x^2 - z^2) + c_2(y^2 - z^2) + c_3xy + c_4xz + c_5yz]\exp\left(\frac{-r}{3}\right)$$

EIGENVALUES AND EIGENFUNCTIONS OF HYDROGEN ATOM

The $3s$ function is nondegenerate, the $3p$ function is threefold degenerate because it contains three parameters, and the $3d$ function is fivefold degenerate because it contains five parameters. We can predict these degeneracies from the angular momentum. If $l = 0$ (in the $3s$ state) there is no degeneracy, if $l = 1$ then the projections of **M** can have the three values 0, ± 1, so there is a threefold degeneracy for the $3p$ state, and if $l = 2$ (in a d state) the projection has the five values 0, ± 1, ± 2 so there must be a fivefold degeneracy (see Fig. 2.9).

The final question we want to consider is how to select the basis functions that constitute a p or a d state. There is no clear-cut answer to this question, because we can make a number of different choices and the answer depends on what we want to do. We discuss the two most common procedures.

The first possible choice is the one we discussed at the beginning of this section. Here we choose the wave functions so that they are eigenfunctions of the three operators $\mathcal{H}_{op}$, $(M^2)_{op}$, and $(M_z)_{op}$. We note that the homogeneous polynomials are all eigenfunctions of the operator $(M^2)_{op}$. For p and d states we have

$$(M^2)_{op} F_1(x, y, z) = 2\hbar^2 F_1(x, y, z) \quad (l = 1)$$
$$(M^2)_{op} F_2(x, y, z) = 6\hbar^2 F_1(x, y, z) \quad (l = 2) \tag{190}$$

and the same relationship holds for the eigenfunctions $\phi(np)$ and $\phi(nd)$,

$$(M^2)_{op} \phi(np) = 2\hbar^2 \phi(np)$$
$$(M^2)_{op} \phi(nd) = 6\hbar^2 \phi(nd) \tag{191}$$

We have seen in the previous section that each function $F_l(x, y, z)$ may be decomposed into a set of functions $\psi_{l,m}(x, y, z)$ so that

$$(M^2)_{op} \psi_{l,m}(x, y, z) = l(l+1)\hbar^2 \psi_{l,m}(x, y, z)$$
$$(M_z)_{op} \psi_{l,m}(x, y, z) = m\hbar \psi_{l,m}(x, y, z) \quad m = 0, \pm 1, \ldots, \pm l \tag{192}$$

We listed the specific forms of these functions for $l = 1$ and for $l = 2$. In the former case they are given by

$$\psi_{1,1}(x, y, z) = x + iy$$
$$\psi_{1,0}(x, y, z) = z \tag{193}$$
$$\psi_{1,-1}(x, y, z) = x - iy$$

according to Eq. (124). The $(2p)$ wave function is given by Eq. (186). It follows from Eq. (193) that it can be decomposed into the three functions

$$\phi(2p_1) = (x + iy) \exp\left(\frac{-r}{2}\right)$$

$$\phi(2p_0) = z \exp\left(\frac{-r}{2}\right) \quad (194)$$

$$\phi(2p_{-1}) = (x - iy) \exp\left(\frac{-r}{2}\right)$$

which are eigenfunctions of the three operators $\mathcal{H}_{op}$, $(M^2)_{op}$, and $(M_z)_{op}$. In this way we have implemented our plan that we discussed in the previous section, that is, to classify the eigenstates by means of the quantum numbers n, l, and m which refer to the eigenvalues of $\mathcal{H}_{op}$, $(M^2)_{op}$, and $(M_z)_{op}$, respectively. This procedure is useful if there is a physical reason to single out the Z axis as the quantization axis for the angular momentum axis—for example, if we have an electric or a magnetic field along that axis.

Ordinarily, there is no physical reason to prefer one specific direction in space over any other direction, and we do not necessarily have to use the above classification of the eigenfunctions. It may be seen that the functions of Eq. (194) are complex functions, and in many cases this gives a mathematical complexity that can easily be avoided. Instead of Eq. (194), we can break down the $2p$ functions into

$$\phi(2p_x) = x \exp\left(\frac{-r}{2}\right)$$

$$\phi(2p_y) = y \exp\left(\frac{-r}{2}\right) \quad (195)$$

$$\phi(2p_z) = z \exp\left(\frac{-r}{2}\right)$$

which are easier to work with than the functions (194) because they are all real. It should be noted that the functions (195) are not eigenfunctions of the operator $(M_z)_{op}$ (except for the $2p_z$ function), but that does not really matter. The various quantum mechanical theories of the chemical bond customarily use the functions (195) as a basis.

It is easily seen that the $3p$ functions can be classified by analogy with Eq. (195) as

$$\phi(3p_x) = x(r - 6) \exp\left(\frac{-r}{3}\right)$$

$$\phi(3p_y) = y(r - 6) \exp\left(\frac{-r}{3}\right) \quad (196)$$

$$\phi(3p_z) = z(r - 6) \exp\left(\frac{-r}{3}\right)$$

Here we have used the expression (189) for the $3p$ wave function.

EIGENVALUES AND EIGENFUNCTIONS OF HYDROGEN ATOM

The classification of the various $3d$ functions (defined by Eq. 189) is a somewhat more complicated one. In Eq. (189) we have the three functions

$$\phi(3d_{xy}) = xy \exp\left(\frac{-r}{3}\right)$$
$$\phi(3d_{xz}) = xz \exp\left(\frac{-r}{3}\right) \tag{197}$$
$$\phi(3d_{yz}) = yz \exp\left(\frac{-r}{3}\right)$$

There are three more functions, namely

$$\phi(3d_1) = (x^2 - y^2) \exp\left(\frac{-r}{3}\right)$$
$$\phi(3d_2) = (y^2 - z^2) \exp\left(\frac{-r}{3}\right) \tag{198}$$
$$\phi(3d_3) = (z^2 - x^2) \exp\left(\frac{-r}{3}\right)$$

Two of these functions occur in Eq. (189), and we have added the third one by cyclic substitution. The two equations (197) and (198) contain six $3d$ functions and we know that there are only five, so we have one function too many. This discrepancy is easily explained if we note that the three functions (198) are linearly dependent. It may be seen that

$$\phi(3d_1) + \phi(3d_2) + \phi(3d_3) = 0 \tag{199}$$

Actually, Eq. (198) contains only two linearly independent functions, and it is customary to select the functions

$$\phi(3d_0) = (x^2 - y^2) \exp\left(\frac{-r}{3}\right)$$
$$\phi(3d_{zz}) = (r^2 - 3z^2) \exp\left(\frac{-r}{3}\right) \tag{200}$$

as the basis functions. We have listed all the hydrogen atom eigenfunctions with quantum numbers $n = 1, 2,$ and 3 in Table 2.2.

Let us finally try to visualize the various functions that we have listed in Table 2.2. Here it is useful to consider the s, p, and d functions separately. The s functions depend only on the variable r, and in an s state the probability density function is radially symmetric. The probability that the electron is found between the distances r and $r + dr$ from the origin is given by

$$P(r) = 4\pi r^2 [\phi(ns)]^2 \, dr \tag{201}$$

78 EXACT SOLUTIONS OF THE SCHRÖDINGER EQUATION

TABLE 2.2 • Eigenfunctions of the hydrogen atom. All lengths are expressed in terms of atomic units, and the eigenfunctions are normalized to unity.

State	Eigenfunction
$1s$	$\phi(1s) = (1/\sqrt{\pi})^{-r}$
$2s$	$\phi(2s) = (1/\sqrt{\pi})(r - 2)e^{-(r/2)}$
$2p$	$\phi(2p_x) = (1/4\sqrt{2\pi})xe^{-(r/2)}$
	$\phi(2p_y) = (1/4\sqrt{2\pi})ye^{-(r/2)}$
	$\phi(2p_z) = (1/4\sqrt{2\pi})ze^{-(r/2)}$
$3s$	$\phi(3s) = (1/81\sqrt{3\pi})(2r^2 - 18r + 27)e^{-(r/3)}$
$3p$	$\phi(3p_x) = (\sqrt{2}/81\sqrt{\pi})(r - 6)xe^{-(r/3)}$
	$\phi(3p_y) = (\sqrt{2}/81\sqrt{\pi})(r - 6)ye^{-(r/3)}$
	$\phi(3p_z) = (\sqrt{2}/81\sqrt{\pi})(r - 6)ze^{-(r/3)}$
$3d$	$\phi(3d_{xy}) = (\sqrt{2}/81\sqrt{\pi})xye^{-(r/3)}$
	$\phi(3d_{yz}) = (\sqrt{2}/81\sqrt{\pi})yze^{-(r/3)}$
	$\phi(3d_{xz}) = (\sqrt{2}/81\sqrt{\pi})xze^{-(r/3)}$
	$\phi(3d_{zz}) = (1/81\sqrt{6\pi})(r^2 - 3z^2)e^{-(r/3)}$
	$\phi(3d_0) = (1/81\sqrt{2\pi})(x^2 - y^2)e^{-(r/3)}$

and we write this as

$$P(r) = R(ns; r)\, dr \qquad (202)$$

where $R(ns)$ is known as the radial distribution function. It follows easily from Eqs. (201) and (202) that the radial distribution function is given by

$$R(ns) = 4\pi r^2 [\phi(ns)]^2 \qquad (203)$$

We show the radial distribution functions for the $1s$ and $2s$ states in Fig. 2.11. It may be seen that these functions have fairly sharp maxima, and we can visualize ns states as a probability density that looks roughly like an orange peel. The radius of this orange peel is approximately given by the expectation value of the variable r,

$$\bar{r}(ns) = \int_0^\infty r R(ns; r)\, dr \qquad (204)$$

It is then found that

$$\begin{aligned}\bar{r}(1s) &= 1.5 \\ \bar{r}(2s) &= 6 \\ \bar{r}(3s) &= 13.5\end{aligned} \qquad (205)$$

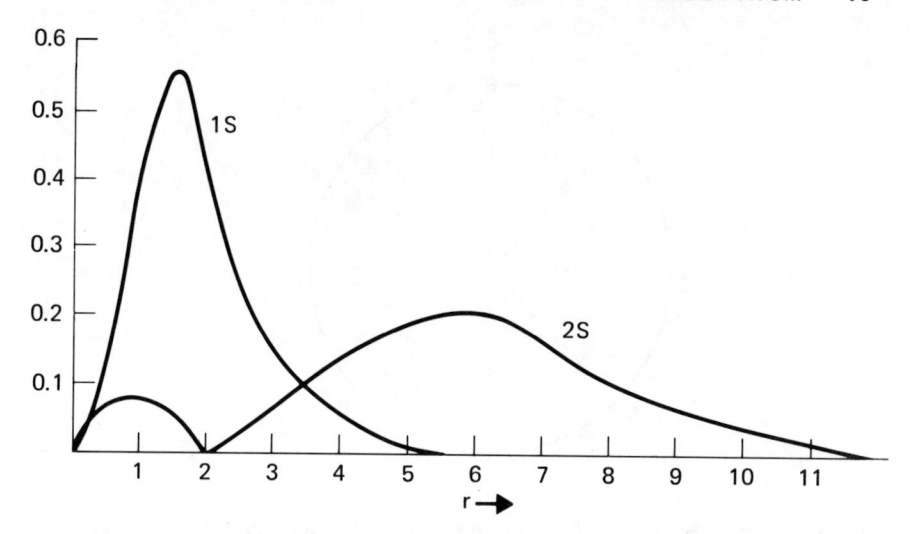

EIGENVALUES AND EIGENFUNCTIONS OF HYDROGEN ATOM 79

FIGURE 2.11 ● Radial distribution functions $R(ns)$ for the 1s and 2s states of the hydrogen atom.

In general,

$$\bar{r}(ns) = \frac{3n^2}{2} \tag{206}$$

and the radius of the charge shell increases quadratically with the quantum number n.

It is a bit more difficult to visualize the behavior of the functions $\phi(np)$ because these functions depend both on the variable r and on the polar angles. Let us consider, for example, the function $\phi(2p_z)$, which is given by

$$\phi(2p_z) = \left(\frac{1}{4\sqrt{2\pi}}\right) z e^{-(r/2)} \tag{207}$$

If we write this in terms of polar coordinates, then the probability density function is given by

$$[\phi(2p_z)]^2 = \frac{1}{32\pi} r^2 \cos^2\theta\, e^{-r} \tag{208}$$

We can represent this function by means of Figs. 2.12a and 2.12b. For a given value of r the probability density function depends on the polar angle θ, which is 0° or 180° along the Z axis and 90° in the XY plane. We have listed the values of the angular function $\cos^2\theta$ at some points along the circle. The radial dependence of the function may again be given by means of the radial distribution function $R(np)$; we show these functions

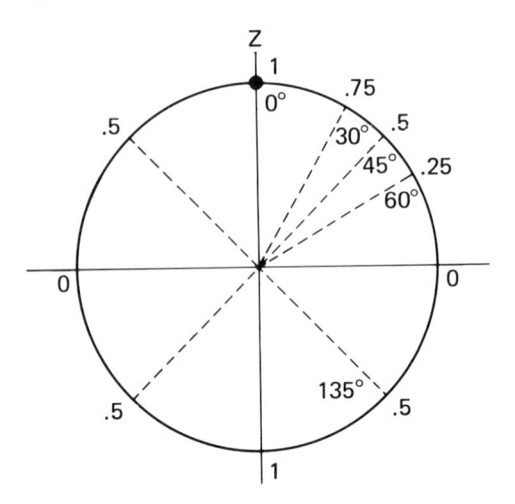

80 EXACT SOLUTIONS OF THE SCHRÖDINGER EQUATION

FIGURE 2.12a ● Angular dependence of probability density in 2p function.

in Fig. 2.12b for the $2p$ and $3p$ states. These functions have again fairly well-defined maxima. The values of the expectation values $\bar{r}(np)$ indicate where these maxima are located. They are

$$\bar{r}(2p) = 5$$
$$\bar{r}(3p) = 12.5$$
(209)

The functions $\phi(np_z)$ are zero when $z = 0$. This means that the functions are zero in the XY plane. Such a plane, where the wave function is zero, is called a nodal plane.

If we could take a picture of the charge density, we would see something like Fig. 2.13. It is easily seen that the probability density has two maxima

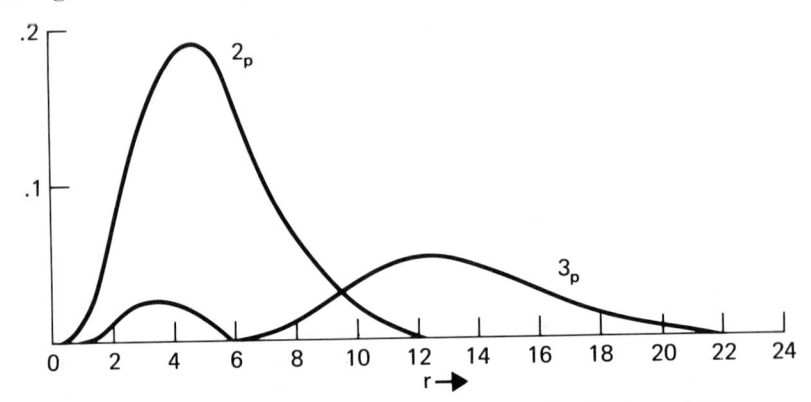

FIGURE 2.12b ● Radial distribution functions $R(np)$ for the $2p$ and $3p$ states of the hydrogen atom.

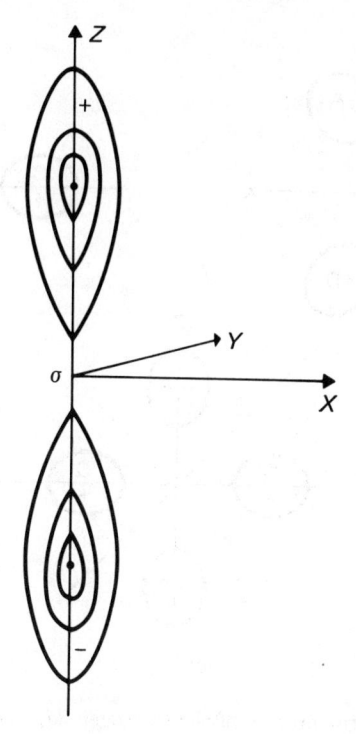

EIGENVALUES AND EIGENFUNCTIONS OF HYDROGEN ATOM 81

FIGURE 2.13 ● Representation of the hydrogen $2p_z$ wave function.

which are situated on the Z axis [at the points (0, 0, 5) and (0, 0, −5), roughly speaking]. We observe two charge blobs around these two points.

The other p functions, np_x and np_y, have the same behavior as the function $2p_z$, the only difference being that the charge density is now along the X axis or the Y axis, respectively.

The behavior of the $3d$ functions is best understood if we concentrate on the angular dependence of these functions. The expectation value $\bar{r}(3d)$ is given by

$$\bar{r}(3d) = 10.5 \tag{210}$$

and the probability density is mostly located on or near the surface of a sphere with radius 10.5.

Let us consider first the function $\phi(3d_{xy})$, which is given by

$$\phi(3d_{xy}) = \frac{\sqrt{2}}{81\sqrt{\pi}} xy \exp\left(\frac{-r}{3}\right) \tag{211}$$

This function is zero in the two nodal planes $x = 0$ and $y = 0$, and the Z axis is the intersection between these two planes. The wave function has

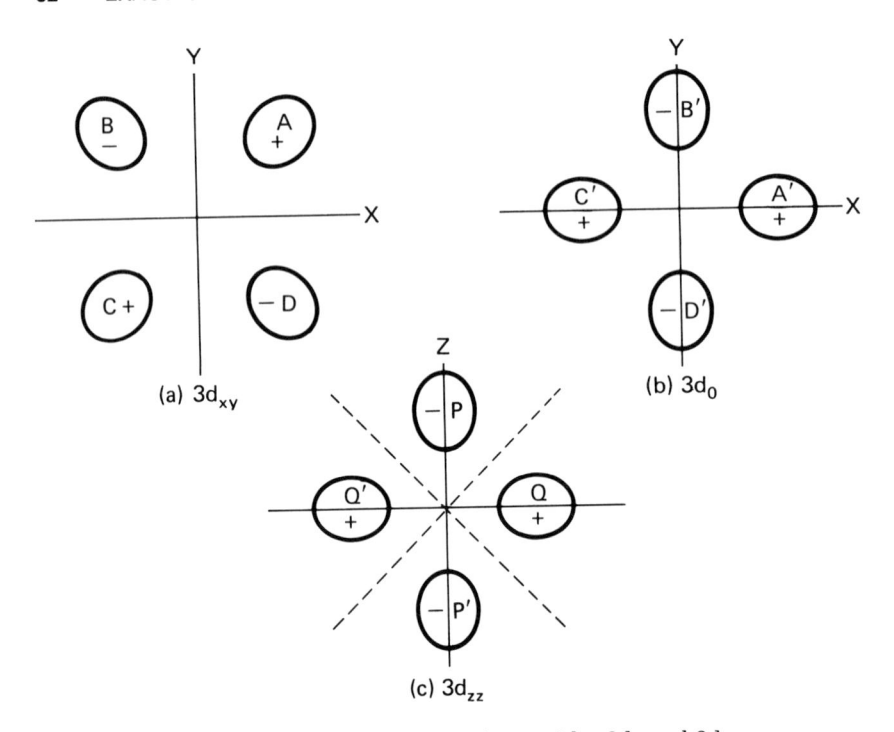

82   EXACT SOLUTIONS OF THE SCHRÖDINGER EQUATION

FIGURE 2.14 ● Representation of the hydrogen $3d_{xy}$, $3d_0$, and $3d_{zz}$ wave functions.

its maximum values in the XY plane, and in Fig. 2.14a we have sketched the behavior of the wave function in the XY plane. It may be seen that the maximum values of the wave function are at the points A and B and the points C and D. All these points are on the two lines that bisect the X and the Y axes, at a distance of about 10.5 from the origin. At A and B the function is positive, and at C and D the function is negative. The probability density of the function $\phi(3d_{xy})$ consists of four charge blobs around these four points. They are separated by the nodal planes $x = 0$ and $y = 0$.

It is interesting to compare this behavior with the behavior of the function

$$\phi(3d_0) = \frac{1}{81\sqrt{\pi}} (x^2 - y^2) \exp\left(\frac{-r}{3}\right) \tag{212}$$

which we have sketched in Fig. 2.14b. Here the nodal planes are $x = \pm y$ and the maximum values of the function occur at the points A', B', C', and D', which are located on the X axis and the Y axis. It follows that the function $\phi(3d_0)$ may be obtained from $\phi(3d_{xy})$ by rotating around the Z axis through an angle of $45°$. After this rotation the two functions are identical.

It is obvious that the functions $\phi(3d_{xz})$ and $\phi(3d_{yz})$ have the same behavior as $\phi(3d_{xy})$; they may be derived from each other by renaming the coordinate axes.

The most complicated of the $3d$ functions is $\phi(3d_{zz})$, which is given by

$$\phi(3d_{zz}) = \frac{1}{81\sqrt{6\pi}} (r^2 - 3z^2) \exp\left(\frac{-r}{3}\right) \quad (213)$$

This function may be expressed in terms of r and the polar angle θ,

$$\phi(3d_{zz}) = \frac{1}{81\sqrt{6\pi}} r^2(1 - 3\cos^2\theta) \exp\left(\frac{-r}{3}\right) \quad (214)$$

It has cylindrical symmetry around the Z axis and it has the same behavior in every plane that contains the Z axis. We have sketched one of these planes in Fig. 2.14c.

It follows from Eq. (214) that the function $\phi(3d_{zz})$ is zero when

$$1 - 3\cos^2\theta = 0 \quad \cos^2\theta = \tfrac{1}{3} \quad (215)$$

As a result, the nodal surface for this function is a cone around the Z axis with an angle $\theta_0 = 54°45'$; this is the angle for which $\cos^2\theta_0 = \tfrac{1}{3}$. In the cross section of Fig. 2.14c we have drawn the two lines where this cone intersects with the plane. We have also indicated the points where the wave function has its maximum values, the two points P and P' on the Z axis and the two points Q and Q' in the XY plane. All these points are at a distance of about 10.5 from the origin. From Fig. 2.14c we can visualize the three-dimensional charge density, which consists of two charge blobs around the points P and P' on the Z axis and of a ring of charge in the XY plane at a distance of about 10.5 from the Z axis.

We have limited this discussion to the states $n = 1, 2,$ and 3 of the hydrogen atom. The higher states can be discussed in a similar way, but their behavior is more complicated and we feel that there is no need to discuss them here.

Problems

1. The vibrational transitions for the molecules H_2 and HBr are 4395 cm^{-1} and 2650 cm^{-1}, respectively. Determine the force constants for the harmonic oscillators that represent these vibrations, they are expressed in terms of dynes per centimeter.
2. Show that the two functions $\psi_0 = \exp(-\alpha x^2)$ and $\psi_1 = x \exp(-\alpha x^2)$ are eigenfunctions for the harmonic oscillator if we take the proper value for the param-

eter α. What is the value of α for which these two functions are eigenfunctions, and what are the corresponding eigenvalues?

3. The vibrational transition in H_2 occurs at 4395 cm^{-1}. If we assume that the vibrational motion in D_2 has the same force constant as in H_2, where does the vibrational transition in D_2 occur?

4. Assuming that the two molecules HF and DF have the same force constant for their vibrational motion, determine the wave number of the vibrational transition in DF from the value 4139 cm^{-1} for the wave number of the vibrational transition in HF.

5. Calculate the expectation values of x^2 and p^2 for the ground state of a harmonic oscillator (x is the coordinate and p is the momentum). Determine the product of the square roots of x^2 and p^2 and relate the result to Heisenberg's uncertainty principle.

6. Calculate the expectation value of $(\Delta q)^2$, the square of the displacement coordinate for the ground state of a harmonic oscillator. What are the numerical values for $(\Delta q)^2$ and Δq for the HF molecule, where the vibrational transition is 4139 cm^{-1}?

7. In the ground state of the $^1H^{35}Cl$ molecule, the equilibrium internuclear distance R_0 is 1.2746 Å. What is the position of the center of gravity?

8. The internuclear distance for H_2 is 0.7416 Å, and its rotational constant is 60.809 cm^{-1}. The rotational constants for HD and D_2 are 45.655 cm^{-1} and 30.429 cm^{-1}, respectively. What are the internuclear distances for these two molecules? Do they differ to an appreciable extent from H_2?

9. Calculate the rotational constant B for the HB molecule, where the internuclear distance is 1.2325 Å.

10. The rotational constant for the ground state of the BF molecule ($^{11}B^{19}F$) is 1.518 cm^{-1}. Determine the internuclear distance R_0 for the ground state. For the first excited electronic state the rotational constant is 1.423 cm^{-1}. Determine the internuclear distance R_1 for this excited state.

11. Write the general expression for the homogeneous polynomial $f_3(x, y, z)$ of the third degree in x, y, and z. Derive the general form of the corresponding Euler polynomial $F_3(x, y, z)$.

12. Consider an electron that moves with a constant angular velocity around a circle with radius 0.5292 Å (the atomic unit of length for the hydrogen atom). If the angular momentum of this electron is equal to Dirac's constant $\hbar$, what is the velocity of the electron and what is its momentum? What is its kinetic energy, and how does this compare with the ground-state energy (for the state $n = 1$) of the hydrogen atom?

13. Show explicitly that the three functions $x \pm iy$ and z are eigenfunctions of $(M_z)_{op}$ and derive the corresponding eigenvalues.

14. Calculate the rotational constant for the motion of an electron on a sphere of radius 0.5292 Å. Is the rotational excitation energy for this system, groing from $J = 0$ to $J = 1$, comparable to the energy of the hydrogen atom $1s$ state?

15. The eigenvalues and eigenfunctions of the He$^+$ atom may be derived in the same way as for the H atom. What are the values for the reduced masses μ for these two systems? What are the values for the units of length and energy a_0 and ϵ_0 that we use in the two systems?

16. What is the value of the Coulomb attraction between an electron and a proton that are a distance 0.529 Å apart?
17. Before the discovery of quantum mechanics, the energy levels of the hydrogen atom were found to be $E_n = R_H/n^2$, where the Rydberg constant R_H is expressed in terms of cm^{-1}. Determine the value of R_H from the results of the quantum mechanical calculation.
18. Show explicitly that the eigenfunctions for the 1s and the 2s states of the hydrogen atom are orthogonal.
19. List the various eigenfunctions that belong to the eigenvalue E_4 of the hydrogen atom. What is the total degeneracy of E_4?
20. What is the value of the angular momentum for the 1s state of the hydrogen atom according to quantum mechanics? Can this result be explained by means of classical mechanics?
21. If we use atomic units, then the Hamiltonian of the hydrogen atom is $-1/2\Delta - (1/r)$ and its lowest energy eigenvalue is -0.5. Calculate the expectation value of the energy from the variational function $\psi = \exp[-(1+\delta)r]$. Show that this energy has a minimum of -0.5 for $\delta = 0$.
22. Calculate the expectation value of the energy of the hydrogen atom from the Hamiltonian of Problem 21 by using the variational function $\psi = \exp(-\alpha r^2)$. Determine the energy minimum and the corresponding value of α and compare it with the exact value -0.5.
23. Show explicitly that the five functions $(x \pm iy)^2$, $z(x \pm iy)$, and $(r^2 - 3z^2)$ are eigenfunctions of $(M_z)_{op}$ and derive the corresponding eigenvalues.

Bibliography

In Chapters I and II we have given a brief outline of the general theory of quantum mechanics and of some of its applications. For a more detailed treatment of the subject we refer to any of the numerous textbooks on quantum mechanics that are now available. Our bibliography gives a selected representation of these books. We have divided them into five categories: A are historical descriptions of the developments of the basic ideas underlying quantum mechanics; B are introductory treatments for beginning students; C are "classics," that is, the fundamental texts that were written in the 1930s; D are texts that were written specifically for chemists; and E are the texts that are most commonly used in physics courses. Very likely we have omitted some good books on the subject, but we do not claim that our list is complete.

A1. M. Jammer. *The Conceptual Development of Quantum Mechanics.* McGraw-Hill, New York (1966).
A2. E. Whittaker. *A History of the Theories of Aether and Electricity,* Vols. I and II. Harper & Row, New York (1960).
B1. W. Heitler. *Elementary Wave Mechanics.* Oxford University Press, London (1945).
B2. V. Rojanski. *Introductory Quantum Mechanics.* Prentice-Hall, Englewood Cliffs, N.J. (1938).
B3. A. Sommerfeld. *Wave Mechanics.* Methuen, London (1930).
C1. P. A. M. Dirac. *The Principles of Quantum Mechanics,* 4th ed. Oxford University Press, London (1958).

86 EXACT SOLUTIONS OF THE SCHRÖDINGER EQUATION

C2. E. C. Kemble. *The Fundamental Principles of Quantum Mechanics*. McGraw-Hill, New York (1937).
C3. H. A. Kramers. *Quantum Mechanics*. North-Holland, Amsterdam (1958).
D1. H. F. Hameka. *Introduction to Quantum Theory*. Harper & Row, New York (1967).
D2. L. Pauling and E. B. Wilson, Jr. *Introduction to Quantum Mechanics*. McGraw-Hill, New York (1935).
E1. A. S. Davydov. *Quantum Mechanics*. Addison-Wesley, Reading Mass. (1965).
E2. R. H. Dicke and J. P. Wittke. *Introduction to Quantum Mechanics*. Addison-Wesley, Reading, Mass. (1960).
E3. P. Fong. *Elementary Quantum Mechanics*. Addison-Wesley, Reading, Mass. (1962).
E4. L. D. Landau and E. M. Lifshitz. *Quantum Mechanics*. Pergamon Press, Oxford (1958).
E5. J. L. Powell and B. Crasemann. *Quantum Mechanics*. Addison-Wesley, Reading, Mass. (1961).
E6. L. Schiff. *Quantum Mechanics*, 3rd ed. McGraw-Hill, New York (1968).

CHAPTER 3   ATOMIC STRUCTURE

1. Introduction

In the previous chapters there was no particular need to consider experimental information. We discussed only simple one-electron systems for which the Schrödinger equation could be solved exactly, and our mathematical derivations provided a complete description for every system that we considered. Now that we wish to proceed to more complicated systems, it becomes desirable to include experimental results in our discussion. We should realize that the Schrödinger equation cannot be solved for any atom other than the hydrogen atom. This means that a quantum mechanical discussion of atomic structure must necessarily be based on approximations. Especially for the more complicated atoms, we need some guidelines for devising suitable approximate methods, and the obvious place to look for such guidelines is the experimental information.

The bulk of the experimental information on atomic structure is derived from atomic spectra. Some aspects of atomic spectra had already been interpreted by means of classical, semiclassical, or semiempirical arguments before the formal introduction of the Schrödinger equation. All these semiempirical descriptions are based on two broad assumptions, (1) in each atomic eigenstate it is possible to recognize the individual electrons, and (2) each individual electron may be identified by a set of hydrogen-type quantum numbers. The validity of these assumptions is justified by the agreement between the semiempirical theories and the experiments. By using them it is possible to explain the general features of most atomic spectra. A more detailed interpretation of the atomic spectra is based on what is known as the vector model of the atom.

88 ATOMIC STRUCTURE

We must also discuss some additional postulates of quantum theory. The first one deals with the electron spin. Even though it also applies to one-electron systems, it is particularly relevant to many-electron systems. The second one is known as the Pauli exclusion principle, and it applies only to many-particle systems. Again, the concept of angular momentum plays an important role in the theory of electron spin.

To illustrate these assumptions we consider the He atom. This is the simplest atom with more than one electron, and it is the easiest system to understand. The Hamiltonian of the He atom is given by

$$\mathcal{H} = -\frac{\hbar^2}{2m}\Delta_1 - \frac{\hbar^2}{2m}\Delta_2 - \frac{Ze^2}{r_1} - \frac{Ze^2}{r_2} + \frac{e^2}{r_{12}} \tag{1}$$

The first two terms are the kinetic energies for electrons 1 and 2, respectively, the third term is the Coulomb attraction between electron 1 and the nucleus (with charge $Z = 2$), the next term is the Coulomb attraction between the second electron and the nucleus, and the last term is the Coulomb repulsion between the two electrons. If we did not have the last term, we could solve the corresponding Schrödinger equation; but since we do have the last term, we have a problem that cannot be solved exactly. The only thing we can say is that the exact eigenfunctions of the He atom must be rather involved functions of the six variables that denote the positions of the two electrons.

To find an approximate description for the He atom we consider only the behavior of one of the two electrons, say electron 1, rather than consider the behavior of both electrons at the same time. Electron 1 moves in a potential field that is the sum of the Coulomb attraction of the nucleus and of the Coulomb repulsion of the other electron. We assume now that the position of the second electron is described by a probability density function $\rho(\mathbf{r}_2)$, and we assume further (and this is the important assumption) that this probability density function $\rho(\mathbf{r}_2)$ is independent of the position of the first electron. We then have the situation that we have sketched in Fig. 3.1, where electron 1 moves through the charge cloud that

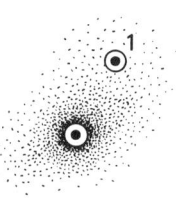

FIGURE 3.1 ● Sketch of the potential field acting on one of the two electrons, electron 1, in the helium atom. Electron 1 experiences the field due to the nucleus and the field due to the charge cloud of the second electron.

represents the probability density of the second electron. At this point we do not know the exact form of $\rho(\mathbf{r}_2)$, but since the total probability of finding the second electron must be unity we know that

$$\int \rho(\mathbf{r}_2) \, d\mathbf{r}_2 = 1 \qquad (2)$$

Our next assumption is that the charge cloud of electron 2 is spherically symmetric, which means that $\rho(\mathbf{r}_2)$ is a function of the radius r_2 only and is independent of the angles. This simplifies the situation considerably, because we now can use a property of electrostatics that we have sketched in Fig. 3.2. If we have a sphere of radius R which is covered with a homogeneous layer of charge, total charge q, then the electric field strength is zero inside the sphere and it is equal to

$$\mathbf{F} = \frac{q\mathbf{r}}{r^3} \qquad (3)$$

outside the sphere. So outside the sphere the field strength is the same as the field strength due to a charge q, located at the origin (the center of the sphere), and inside the sphere the field strength is zero.

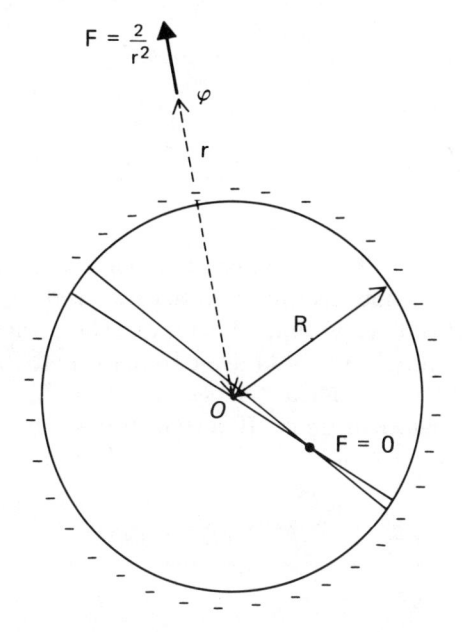

FIGURE 3.2 ● Field strength due to a sphere of radius R covered with a homogeneous layer of charge. The total charge is q. Inside the sphere the field strength is zero, because the effects of opposite surface elements cancel. Outside the sphere the field strength is equal to the field strength due to a charge q located at the center of the sphere.

If we apply this to the helium atom, we see that the part δ of the charge cloud that is between electron 1 and the nucleus acts on electron 1 as if the whole charge δ were concentrated at the origin. Mathematically, δ is given by

$$\delta(r_1) = 4\pi \int_0^{r_1} \rho(r_2) r_2^2 \, dr_2 \tag{4}$$

The rest of the charge cloud, which is $1 - \delta(r_1)$, has no effect on electron 1. It is easily seen that $\delta(r_1)$ is zero when r_1 is zero and increases with increasing values of r_1 until it approaches the value 1 asymptotically for large values of r_1.

It thus follows that electron 1 moves in an effective potential field $V_{\text{eff}}(r_1)$, given by

$$V_{\text{eff}}(r_1) = -\frac{Z - \delta(r_1)}{r_1} \tag{5}$$

with

$$Z = 2 \quad 0 < \delta(r_1) < 1 \tag{6}$$

The corresponding Schrödinger equation for electron 1 is then

$$-\frac{\hbar^2}{2m} \Delta_1 + V_{\text{eff}}(r_1) \Psi(1) = \epsilon \Psi(1) \tag{7}$$

We say that the quantity $\delta(r_1)$ describes the amount of shielding of the nuclear charge by the second electron. It varies from zero near the nucleus, where there is no shielding, to unity far away, where there is complete shielding by the second electron. As a final approximation, we can now replace the function $\delta(r_1)$ by an effective average value δ_{eff}, which must be somewhere between zero and unity. If we substitute this into the Schrödinger equation (7), we obtain

$$-\frac{\hbar^2}{2m} \Delta_1 - \frac{Z_{\text{eff}} e^2}{r_1} \Psi(1) = \epsilon \Psi(1) \tag{8}$$

with

$$Z_{\text{eff}} = Z - \delta_{\text{eff}} \tag{9}$$

Here Z_{eff} is called the effective nuclear charge and is equal to 2 minus the effective shielding constant due to the other electron.

The Schrödinger equation (8) is very similar to the Schrödinger equation for the hydrogen atom; in fact, we shall show in the next section that it can be reduced to the hydrogen atom equation by introducing new units of length and energy. The eigenvalues of Eq. (8) have the form

$$E_n = -\frac{Z^2}{2n^2} \epsilon_0 = -\frac{Z^2}{2n^2} \cdot \frac{e^2}{a_0} \tag{10}$$

so that they are all larger by a factor Z^2 than the corresponding eigenvalues of the hydrogen atom. The eigenfunctions of Eq. (8) have the same form as the hydrogen atom eigenfunctions if we take the unit of length now as (a_0/Z) instead of just a_0 as in the case of the hydrogen atom. This means that the $(1s)$ and $(2p)$ functions are now given as

$$\phi(1s) = \exp\left(\frac{-Zr}{a_0}\right)$$
$$\phi(2p_x) = x \exp\left(\frac{-Zr}{2a_0}\right) \tag{11}$$

and so on. The nuclear charge Z has the effect of contracting the eigenfunctions by a factor Z. We shall prove all this in the following section.

To find the total wave function of the He atom, we note that electron 2 moves in the effective field of the nucleus, shielded by electron 1. We shall see later that the total ground-state eigenfunction should be symmetric with respect to the two electrons 1 and 2. Our approximate discussion leads thus to the following expression for the ground-state wave function of the He atom:

$$\Psi(1, 2) = \exp\left(\frac{-Zr_1}{a_0}\right) \exp\left(\frac{-Zr_2}{a_0}\right) \tag{12}$$

which is the product of the $(1s)$ function (11) for electron 1 and of the same function for electron 2.

The best possible value of the effective nuclear charge Z may be determined by means of the variational principle, as we shall show in the next section. We find that $Z = 1.6875$, which seems reasonable if we consider Eq. (9) and our previous arguments.

Let us now proceed to discuss the Pauli exclusion principle. The original formulation of this principle was based on some approximations and was applicable only to atoms. Subsequently it was formulated in a more rigorous mathematical form that is generally valid; we discuss this in Section 4.

In writing the wave function of the He atom in the form of Eq. (12), we have made the approximation of taking it as a product of two one-electron functions. Of course, the exact wave function has a more complicated form, but it is common practice to take atomic or molecular wave functions as products of one-electron functions, because it becomes quite complicated if we want to be more precise than that. A one-electron function is called an orbital, and we speak of an atomic orbital if the one-electron function belongs to one specific atom. In Eq. (12) we have approximated the total wave function as a product of atomic orbitals. It has been found that even in complicated atoms, where the precise form of the atomic orbitals may not be known, each atomic orbital can at least be identified by means of a set of hydrogen-like quantum numbers. In this way we can describe an atomic state by means of its configuration, where we indicate how the electrons are distributed over the various orbitals.

For example, in the ground state of the He atom, both electrons occupy a $(1s)$ orbital, and we denote this by the symbol $(1s)^2$.

The excited states are obtained by exciting one of the two electrons to a higher orbital, which are atomic configurations such as $(1s)(2s)$, $(1s)(2p)$, $(1s)(3s)$, and so on. The excited states where both electrons are in higher orbitals are ordinarily not observed in atomic spectra. In Fig. 3.3 we show the energy levels of the He atom, which have been obtained from the atomic spectrum. In this energy level diagram we can recognize the configurations we mentioned above. We note that the eigenstate $(1s)(2s)$ has a slightly lower energy than $(1s)(2p)$. We can understand this if we realize that the effective nuclear charge Z_{eff} should be a little larger for the $2s$ than for the $2p$ state, because in the $2s$ orbital the electron is on the average closer to the nucleus. In helium a (ns) state has a lower energy than a (np) state, which has a lower energy than a (nd) state, and so on, even though in hydrogen all these states have the same energy.

If we proceed to higher atoms, then, according to the above argument their ground states should be obtained by placing all electrons in $(1s)$ orbitals, since that gives the lowest energies. However, this is inconsistent with the experimental information. In order to resolve this controversy, Pauli introduced in 1924 the exclusion principle, by which he imposes the condition that in an atomic eigenstate there cannot be more than two electrons with the same set of quantum numbers n, l, and m.

According to the exclusion principle, the ground state of Li must have a configuration $(1s)^2(2s)$, and the ground state of Be must have a configuration $(1s)^2(2s)^2$. In the $2p$ state the quantum number m can have three possible values, namely $m = -1, 0, 1$.

According to the exclusion principle, we can place six electrons in the $2p$ orbitals before they are filled up. Hence, the ground-state configuration of B is $(1s)^2(2s)^2(2p)$, for C is $(1s)^2(2s)^2(2p)^2$, and so on, until we reach Ne, where we have $(1s)^2(2s)^2(2p)^6$ and the $2p$ orbitals are all filled. We have

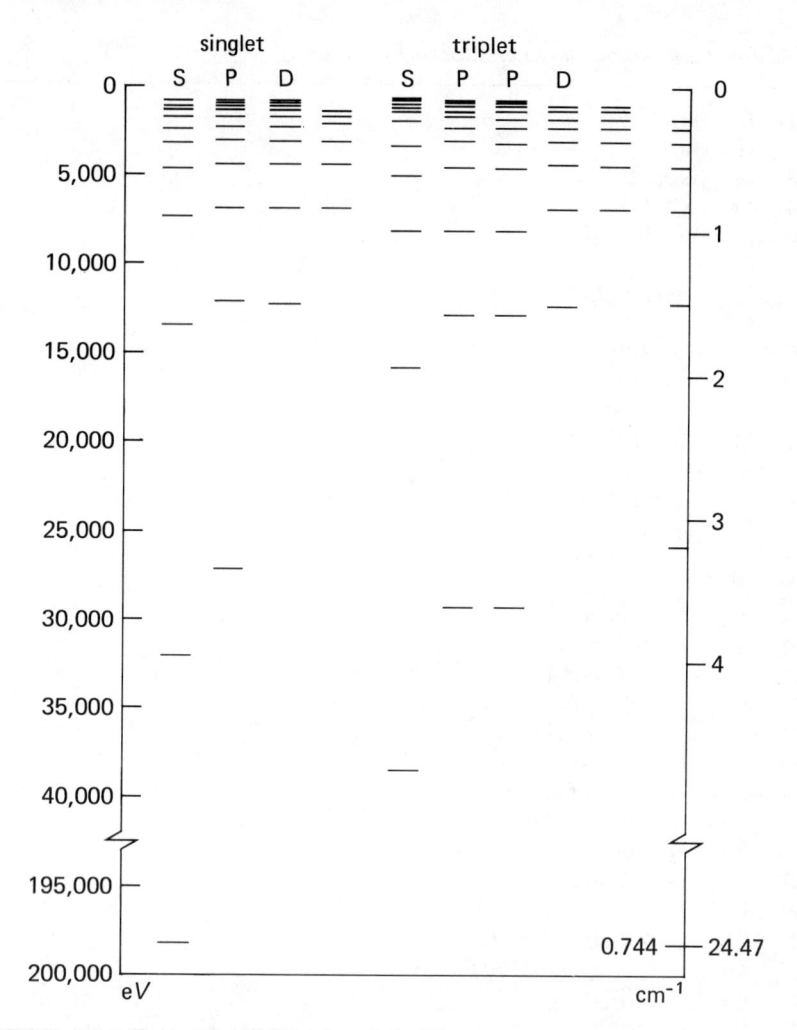

FIGURE 3.3 ● Energy level diagram of the He atom.

listed all atomic configurations in Table 3.1, and it may be seen that with increasing atomic number we add an electron every time we proceed to the next element. The configuration of argon is $(1s)^2(2s)^2(2p)^6(3p)^6$, and it may be seen from the table that the next element, K, has the ground-state configuration $(1s)^2(2s)^2(2p)^6(3s)^2(3p)^6(4s)$. It follows that in K the $4s$ orbital has a lower energy than the $3d$ orbital. Apparently the differences in energy between orbitals with different l values become larger with increasing quantum numbers, and in potassium these energy differences become larger than the differences in energy between states with different values of the quantum number n. In Ca the $(4s)$ orbital is filled and then, moving

TABLE 3.1 • Atomic ground-state configurations.

1.	H	$(1s)$
2.	He	$(1s)^2$
3.	Li	$(1s)^2(2s)$
4.	Be	$(1s)^2(2s)^2$
5.	B	$(1s)^2(2s)^2(2p)$
6.	C	$\cdots (2s)^2(2p)^2$
7.	N	$\cdots (2s)^2(2p)^3$
8.	O	$\cdots (2s)^2(2p)^4$
9.	F	$\cdots (2s)^2(2p)^5$
10.	Ne	$\cdots (2s)^2(2p)^6$
11.	Na	$\cdots (2s)^2(2p)^6(3s)$
12.	Mg	$\cdots (2s)^2(2p)^6(3s)^2$
13.	Al	$\cdots (3s)^2(3p)$
14.	Si	$\cdots (3s)^2(3p)^2$
15.	P	$\cdots (3s)^2(3p)^3$
16.	S	$\cdots (3s)^2(3p)^4$
17.	Cl	$\cdots (3s)^2(3p)^5$
18.	Ar	$\cdots (3s)^2(sp)^6$
19.	K	$\cdots (3s)^2(3p)^6(4s)$
20.	Ca	$\cdots (3s)^2(3p)^6(4s)^2$
21.	Sc	$\cdots (3s)^2(3p)^6(3d)(4s)^2$
22.	Ti	$\cdots (3s)^2(3p)^6(3d)^2(4s)^2$
23.	V	$\cdots (3s)^2(3p)^6(3d)^3(4s)^2$
24.	Cr	$\cdots (3s)^2(3p)^6(3d)^5(4s)$
25.	Mn	$\cdots (3s)^2(3p)^6(3d)^5(4s)^2$
26.	Fe	$\cdots (3s)^2(3p)^6(3d)^6(4s)^2$
27.	Co	$\cdots (3s)^2(3p)^6(3d)^7(4s)^2$
28.	Ni	$\cdots (3s)^2(3p)^6(3d)^8(4s)^2$
29.	Cu	$\cdots (3s)^2(3p)^6(3d)^{10}(4s)$
30.	Zn	$\cdots (3s)^2(3p)^6(3d)^{10}(4s)^2$
31.	Ca	$\cdots (4s)^2(4p)$
32.	Ge	$\cdots (4s)^2(4p)^2$
33.	As	$\cdots (4s)^2(4p)^3$
34.	Se	$\cdots (4s)^2(4p)^4$
35.	Br	$\cdots (4s)^2(4p)^5$
36.	Kr	$\cdots (4s)^2(4p)^6$
37.	Rb	$\cdots (4s)^2(4p)^6(5s)$
38.	Sr	$\cdots (4s)^2(4p)^6(5s)^2$
39.	Y	$\cdots (4s)^2(4p)^6(4d)(5s)^2$
40.	Zr	$\cdots (4s)^2(4p)^6(4d)^2(5s)^2$
41.	Nb	$\cdots (4s)^2(4p)^6(4p)^4(5s)$
42.	Mo	$\cdots (4s)^2(4p)^6(4d)^5(5s)$
43.	Tc	$\cdots (4s)^2(4p)^6(4d)^5(5s)^2$
44.	Ru	$\cdots (4s)^2(4p)^6(4d)^7(5s)$
45.	Rh	$\cdots (4s)^2(4p)^6(4d)^8(5s)$

TABLE 3.1 ● Atomic ground-state configurations (continued).

46.	Pd	$\cdots$ $(4s)^2(4p)^6(4d)^{10}$
47.	Ag	$\cdots$ $(4s)^2(4d)^6(4d)^{10}(5s)$
48.	Cd	$\cdots$ $(4s)^2(4p)^6(4d)^{10}(5s)^2$
49.	In	$\cdots$ $(4d)^{10}(5s)^2(5p)$
50.	Sn	$\cdots$ $(4d)^{10}(5s)^2(5p)^2$
51.	Sb	$\cdots$ $(4d)^{10}(5s)^2(5p)^3$
52.	Te	$\cdots$ $(4d)^{10}(5s)^2(5p)^4$
53.	I	$\cdots$ $(4d)^{10}(5s)^2(5p)^5$
54.	Xe	$\cdots$ $(4d)^{10}(5s)^2(5p)^6$
55.	Cs	$\cdots$ $(4d)^{10}(5s)^2(5p)^6(6s)$
56.	Ba	$\cdots$ $(4d)^{10}(5s)^2(5p)^6(6s)^2$
57.	La	$\cdots$ $(4d)^{10}(5s)^2(5p)^6(5d)(6s)^2$
58.	Ce	$\cdots$ $(4d)^{10}(4f)(5s)^2(5p)^6(5d)(6s)^2$
.		
.		
.		
70.	Yb	$\cdots$ $(4d)^{10}(4f)^{13}(5s)^2(5p)^6(5d)(6s)^2$
71.	Lu	$\cdots$ $(4d)^{10}(4f)^{14}(5s)^2(5p)^6(5d)(6s)^2$
72.	Hf	$\cdots$ $(5s)^2(5p)^6(5d)^2(6s)^2$
73.	Ta	$\cdots$ $(5s)^2(5p)^6(5d)^3(6s)^2$
74.	W	$\cdots$ $(5s)^2(5p)^6(5d)^4(6s)^2$
75.	Re	$\cdots$ $(5s)^2(5d)^6(5d)^5(6s)^2$
76.	Os	$\cdots$ $(5s)^2(5p)^6(5d)^6(6s)^2$
77.	Ir	$\cdots$ $(5s)^2(5p)^6(5d)^7(6s)^2$
78.	Pt	$\cdots$ $(5s)^2(5p)^6(5d)^9(6s)$
79.	Au	$\cdots$ $(5s)^2(5p)^6(5d)^{10}(6s)$
80.	Hg	$\cdots$ $(5s)^2(5p)^6(5d)^{10}(6s)^2$
81.	Tl	$\cdots$ $(5d)^{10}(6s)^2(6p)$
82.	Pb	$\cdots$ $(5d)^{10}(6s)^2(6p)^2$
83.	Bi	$\cdots$ $(5d)^{10}(6s)^2(6p)^3$
84.	Po	$\cdots$ $(5d)^{10}(6s)^2(6p)^4$
85.	At	$\cdots$ $(5d)^{10}(6s)^2(6p)^5$
86.	Rn	$\cdots$ $(5d)^{10}(6s)^2(6p)^6$
87.	Fr	$\cdots$ $(5d)^{10}(6s)^2(6p)^6(7s)$
88.	Ra	$\cdots$ $(5d)^{10}(6s)^2(6s)^6(7s)^2$
89.	Ac	$\cdots$ $(5d)^{10}(6s)^2(6p)^6(6d)(7s)^2$
90.	Th	$\cdots$ $(5d)^{10}(6s)^2(6p)^6(6d)^2(7s)^2$
91.	Pa	$\cdots$ $(5d)^{10}(6s)^2(6p)^6(6d)^3(7s)^2$
92.	U	$\cdots$ $(5d)^{10}(6s)^2(6p)^6(6d)^4(7s)^2$

from Sc to Ni, the $(3d)$ orbital is filled before the $4p$ orbital is filled. It follows from Table 3.1 that the order in which the orbitals are filled up is

$$(1s) \to (2s) \to (2p) \to (3s) \to (3p) \to (4s) \to (3d) \to (4p) \to (5s)$$
$$\to (4d) \to (5p) \to (6s) \to (4f) \to (5d) \to (6p) \to (7s) \to (6d) \quad (13)$$

In general, a completely filled atomic orbital has a relatively lower energy and for that reason there are some minor exceptions to the above rules. For example, in Cu it is possible to fill up the (5d) orbital with ten electrons if we take one of the electrons out of the (4s) orbital and place it in the (5d) orbital, and it appears that this is the ground-state configuration of Cu.

The above formulation of the exclusion principle is not too accurate, because it depends on the assumption that atomic configurations can be constructed from atomic orbitals. In addition it is limited to atoms only. A more rigorous formulation of the exclusion principle will be given in the next section, after we have discussed electron spin.

2. Calculations on the He Atom

We wish to discuss a simple calculation for deriving the approximate ground-state eigenfunction of the He atom. In this way we can illustrate how to apply quantum mechanics to atomic structure. Also, we used some of the results that we obtain here in the previous section.

We start with the Schrödinger equation (8) that represents the motion of one of the electrons in the effective field of the other, and we show that its eigenfunctions have indeed the form of Eq. (11). The equation is

$$-\frac{\hbar^2}{2m}\left(\frac{\partial^2}{\partial x^2}+\frac{\partial^2}{\partial y^2}+\frac{\partial^2}{\partial y^2}\right)\Psi(x,y,z) - \frac{Ze^2}{r}\Psi(x,y,z) = E\Psi(x,y,z) \quad (14)$$

We introduce new variables

$$x = \frac{x_0}{Z} \qquad y = \frac{y_0}{Z} \qquad z = \frac{z_0}{Z} \quad (15)$$

and we write the energy as

$$E = Z^2 E_0 \quad (16)$$

The equation then takes the form

$$-\frac{\hbar^2 Z^2}{2m}\left(\frac{\partial^2}{\partial x_0^2}+\frac{\partial^2}{\partial y_0^2}+\frac{\partial^2}{\partial z_0^2}\right)\Psi(x_0,y_0,z_0) - \frac{Z^2 e^2}{r_0}\Psi(x_0,y_0,z_0) = Z^2 E_0 \Psi(x_0,y_0,z_0) \quad (17)$$

If we divide this equation by Z^2, it becomes identical with the Schrödinger equation for the H atom of Chapter II, Section 6. The eigenvalues of Eq. (17) are thus identical with the hydrogen atom eigenvalues, which means that the eigenvalues of the original equation (14) are the hydrogen atom

CALCULATIONS ON THE HE ATOM

eigenvalues multiplied by a factor Z^2. The eigenfunctions of Eq. (17) are identical with the hydrogen atom eigenfunctions, which means that the eigenfunctions of Eq. (14) are the hydrogen atom eigenfunctions with all the lengths contracted by a factor Z. The $(1s)$ and $(2p)$ eigenfunctions of Eq. (14) are thus given by Eq. (11).

Let us now proceed to the second part of the calculation. We approximate the ground-state wave function of the He atom by Eq. (12), and we use the variational principle to find the most suitable value for the parameter Z. For convenience's sake, we use atomic units of length and energy. Then the Hamiltonian of the He atom is given by

$$\mathcal{H} = -\frac{1}{2}\Delta_1 - \frac{2}{r_1} - \frac{1}{2}\Delta_2 - \frac{2}{r_2} + \frac{1}{r_{12}} \quad (18)$$

and the wave function $\Psi(1, 2)$ is

$$\Psi(1, 2) = e^{-Zr_1} e^{-Zr_2} \quad (19)$$

First we normalize the wave function (19) to unity. We introduce polar coordinates r, θ, ϕ for each electron:

$$\begin{aligned} x_i &= r_i \sin\theta_i \cos\phi_i \\ y_i &= r_i \sin\theta_i \sin\phi_i \\ z_i &= r_i \cos\theta_i \end{aligned} \quad (20)$$

The integration over all space transforms then as

$$\int_{-\infty}^{\infty} dx \int_{-\infty}^{\infty} dy \int_{-\infty}^{\infty} dz \to \int_0^{\infty} r^2\, dr \int_0^{\pi} \sin\theta\, d\theta \int_0^{2\pi} d\phi \quad (21)$$

We obtain then

$$\begin{aligned} \int [\Psi(1,2)]^2\, dv_1\, dv_2 &= \int e^{-2Zr_1}\, dv_1 \int e^{-2Zr_2}\, dv_2 \\ &= \left(\int_0^{\infty} r_1^2 e^{-2Zr_1}\, dr_1 \int_0^{\pi} \sin\theta_1\, d\theta_1 \int_0^{2\pi} d\phi_1 \right) \\ &\quad \times \left(\int_0^{\infty} r_2^2 e^{-2Zr_2}\, dr_2 \int_0^{\pi} \sin\theta_2\, d\theta_2 \int_0^{2\pi} d\phi_2 \right) \\ &= \left(4\pi \int_0^{\infty} r_1^2 e^{-2Zr_1}\, dr_1 \right) \left(4\pi \int_0^{\infty} r_2^2 e^{-2Zr_2}\, dr_2 \right) \end{aligned} \quad (22)$$

98   ATOMIC STRUCTURE

We note that

$$\int_0^\infty x^n e^{-\alpha x}\, dx = \frac{n!}{\alpha^{n+1}} \tag{23}$$

Consequently,

$$\int [\Psi(1,2)]^2\, dv_1\, dv_2 = \left[\frac{8\pi}{(2Z)^3}\right]^2 = \left(\frac{\pi}{Z^3}\right)^2 \tag{24}$$

The normalized wave function is thus

$$\Psi(1,2) = s(r_1)s(r_2) = \left(\frac{Z^3}{\pi}\right)^{1/2} e^{-Zr_1} \left(\frac{Z^3}{\pi}\right)^{1/2} e^{-Zr_2} \tag{25}$$

To make use of the variational principle, we must determine the expectation value of the Hamiltonian (18) with respect to the function (25),

$$E(Z) = \langle \Psi(1,2) \mid \mathcal{H} \mid \Psi(1,2) \rangle \tag{26}$$

We substitute (18) and we split the total expression (26) into three parts:

$$\begin{aligned}
E(Z) &= E_1(Z) + E_2(Z) + E_3(Z) \\
E_1(Z) &= \langle \Psi(1,2) \mid -\tfrac{1}{2}\Delta_1 - 2r_1^{-1} \mid \Psi(1,2) \rangle \\
E_2(Z) &= \langle \Psi(1,2) \mid -\tfrac{1}{2}\Delta_2 - 2r_2^{-1} \mid \Psi(1,2) \rangle \\
E_3(Z) &= \langle \Psi(1,2) \mid r_{12}^{-1} \mid \Psi(1,2) \rangle
\end{aligned} \tag{27}$$

The calculation of the two parts $E_1(Z)$ and $E_2(Z)$ is relatively easy. We note that

$$\left(-\tfrac{1}{2}\Delta_1 - \frac{Z}{r_1}\right)s(r_1) = \left(\frac{-Z^2}{2}\right)s(r_1) \tag{28}$$

because $s(r_1)$ is an eigenfunction of the operator. Hence

$$\begin{aligned}
(-\tfrac{1}{2}\Delta_1 - 2r_1^{-1})s(r_1) &= (-\tfrac{1}{2}\Delta - Zr_1^{-1})s(r_1) + (Z-2)r_1^{-1}s(r_1) \\
&= -\tfrac{1}{2}Z^2 s(r_1) + (Z-2)r_1^{-1}s(r_1)
\end{aligned} \tag{29}$$

We find now that

CALCULATIONS ON THE HE ATOM

$$E_1(Z) = \iint s(r_1)s(r_2)\left[-\frac{1}{2}Z^2 + (Z-2)r_1^{-1}\right]s(r_1)s(r_2)\, dv_1\, dv_2$$

$$= \int s(r_1)\left[-\frac{1}{2}Z^2 + (Z-2)r_1^{-1}\right]s(r_1)\, dv_1$$

$$= -\frac{1}{2}Z^2 + \frac{Z^3}{\pi}(Z-2)\int_0^\infty \int_0^\infty \int_0^{2\pi} r_1^{-1}e^{-2Zr_1}r_1^2 \sin\theta\, dr_1\, d\theta\, d\phi$$

$$= -\frac{1}{2}Z^2 + \frac{Z^3}{\pi}(Z-2)4\pi \int_0^\infty r_1 e^{-2Zr_1}\, dr_1$$

$$= -\frac{1}{2}Z^2 + \frac{Z^3}{\pi}(Z-2)4\pi\left(\frac{1}{4Z^2}\right)$$

$$= -\frac{1}{2}Z^2 + Z(Z-2) = \frac{1}{2}Z^2 - 2Z \tag{30}$$

It is easily seen that the second contribution $E_2(Z)$ is equal to $E_1(Z)$, because the two expressions differ only in the labeling of the integration variables. Hence

$$E_2(Z) = E_1(Z) = \tfrac{1}{2}Z^2 - 2Z \tag{31}$$

The calculation of the remaining term $E_3(Z)$ is much more difficult than the others. We write it as

$$E_3(Z) = \frac{Z^3}{\pi^2} \iint e^{-2Zr_1}\left(\frac{1}{r_{12}}\right)e^{-2Zr_2}\, dv_1\, dv_2 \tag{32}$$

Here r_{12} is the distance between the two electrons as we have sketched in Fig. 3.4. It may be seen in Fig. 3.4 that r_{12} can be expressed in terms of the

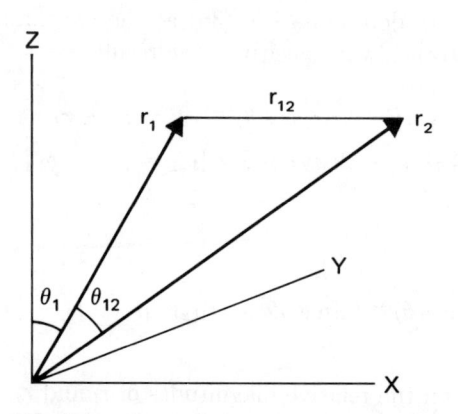

FIGURE 3.4 ● Definition of coordinates for two electrons.

distances r_1 and r_2 and of the angle θ_{12} between the two vectors $\mathbf{r}_1$ and $\mathbf{r}_2$. According to the cosine rule, the result is

$$r_{12} = (r_1^2 + r_2^2 - 2r_1r_2 \cos \theta_{12})^{1/2} \tag{33}$$

In Eq. (32) we must integrate over all space for each electron; that is, we must integrate over six variables. We do this in two stages: We imagine that the position of the first electron is fixed and is determined by the vector $\mathbf{r}_1$. We then integrate over the coordinates of the second electron and introduce polar coordinates, taking the vector $\mathbf{r}_1$ as the Z axis to define the polar coordinates of the second electron. The angle θ_{12} now becomes the polar angle for the second electron, and we have

$$\int r_{12}^{-1} e^{-2Zr_2} dv_2 = \iiint (r_1^2 + r_2^2 - \cos\theta)^{-1/2} e^{-2Zr_2} r_2^2 \sin\theta \, dr_2 \, d\theta \, d\phi$$

$$= 2\pi \int_0^\infty r_2^2 e^{-2Zr_2} \int_0^\pi (r_1^2 + r_2^2 - 2r_1r_2 \cos\theta)^{-1/2} \sin\theta \, d\theta \tag{34}$$

The last integration may be performed analytically, because

$$\frac{\partial}{\partial \theta}(r_1^2 + r_2^2 - 2r_1r_2 \cos\theta)^{1/2} = r_1r_2 \sin\theta (r_1^2 + r_2^2 - 2r_1r_2 \cos\theta)^{-1/2} \tag{35}$$

and

$$\int_0^\pi (r_1^2 + r_2^2 - 2r_1r_2 \cos\theta)^{-1/2} \sin\theta \, d\theta = (r_1r_2)^{-1}[(r_1^2 + r_2^2 - 2r_1r_2 \cos\phi)^{1/2}]_0^\pi \tag{36}$$

We must be careful to determine Eq. (36) at the two limits, because we should realize that r_{12} is always positive. The result is

$$[(r_1^2 + r_2^2 - 2r_1r_2 \cos\theta)^{1/2}]_{\theta=\pi} = r_1 + r_2$$
$$[(r_1^2 + r_2^2 - 2r_1r_2 \cos\theta)^{1/2}]_{\theta=0} = |r_1 - r_2| \tag{37}$$

so that

$$\int_0^\pi (r_1^2 + r_2^2 - 2r_1r_2 \cos\theta)^{-1/2} \sin\theta \, d\theta = (r_1r_2)^{-1}[r_1 + r_2 - |r_1 - r_2|] \tag{38}$$

The result depends on the relative magnitudes of r_1 and r_2.
Let us now substitute the above result into Eq. (34). We obtain

CALCULATIONS ON THE HE ATOM

$$\int r_{12}^{-1} e^{-2Zr_2} \, dv_2 = 2\pi \int_0^\infty r_2^2 e^{-2Zr_2} (r_1 r_2)^{-1} [r_1 + r_2 - |r_1 - r_2|] \, dr_2$$

$$= 4\pi \int_0^{r_1} r_2^2 e^{-2Zr_2} r_1^{-1} \, dr_2 + 4\pi \int_{r_1}^\infty r_2^2 e^{-2Zr_2} r_2^{-1} \, dr_2 \quad (39)$$

The remaining integrals are easily calculated. We have

$$\int_\alpha^\infty x^n e^{-\beta x} \, dx = \frac{1}{\beta} \left(\alpha^n e^{-\beta \alpha} + n \int_\alpha^\infty x^{n-1} e^{-\beta x} \, dx \right)$$

$$\int_\alpha^\infty e^{-\beta x} \, dx = \frac{e^{-\beta \alpha}}{\beta} \quad (40)$$

Consequently,

$$\int \frac{e^{-2Zr_2}}{r_{12}} \, dv_2 = \frac{4\pi}{r_1} \left[\frac{1}{4Z^3} - e^{-2Zr_1} \left(\frac{r_1^2}{2Z} + \frac{r_1}{2Z^2} + \frac{1}{4Z^3} \right) \right]$$

$$+ 4\pi \left[\left(\frac{r_1}{2Z} + \frac{1}{4Z^2} \right) e^{-2Zr_1} \right] \quad (41)$$

To find $E_3(Z)$, we must substitute the above result into Eq. (32) and integrate over the coordinates of the first electron. Since the right side of Eq. (41) depends only on r_1, the result is simply

$$E_3(Z) = \left(\frac{Z^3}{\pi} \right)^2 4\pi \int_0^\infty r_1^2 e^{-2Zr_1} \left(\int e^{-2Zr_2} r_{12}^{-1} \, dv_2 \right) dr_1 \quad (42)$$

or

$$E_3(Z) = \frac{5Z}{8} \quad (43)$$

after substituting Eq. (41) and making use of Eq. (23).

We can now calculate the total expectation value of the energy $E(Z)$ by adding the three contributions of Eqs. (30), (31), and (43). The result is

$$E(Z) = Z^2 - 4Z + \frac{5Z}{8} = Z^2 - \frac{27Z}{8} = \left(Z - \frac{27}{16} \right)^2 - \left(\frac{27}{16} \right)^2 \quad (44)$$

This expression has a minimum for $Z = \frac{27}{16} = 1.6875$. This is the most suitable value for the effective nuclear charge according to the variational principle, and it seems quite reasonable. The energy minimum is $E_{min} = -\frac{729}{256} = -2.8477(e^2/a_0)$. The corresponding experimental value is $E_{exp} = -2.9037(e^2/a_0)$.

The difference between the experimental and theoretical values is $0.056(e^2/a_0)$ or about 2%. Considering that we used a fairly crude, one-parameter variational function to get this result, a 2% error is fairly small and the agreement between theory and experiment is satisfactory. We may add that the ground-state energy of the He atom has been calculated with great accuracy (10^{-6} or better) from elaborate variational functions containing large numbers of variational parameters.

The above calculation is the simplest possible problem that we can consider in atomic structure, because the He atom is the smallest many-electron atom and because we have chosen a very simple variational function. Yet the mathematical analysis is fairly involved. This illustrates that in the case of larger atoms or more complex variational functions, it becomes impractical to perform the whole calculation by hand using analytical methods. Such calculations are now performed with the aid of electronic computers. In subsequent sections we shall mention the computational methods that are used in such calculations, but it would lead us too far to discuss them in detail.

3. The Electron Spin

It is customary to explain the structure of an atom by comparing it with the solar system. The motion of the various electrons around the nucleus is then compared with the motion of the planets around the sun. We know that the earth describes an elliptic orbit around the sun with a time period of a year. In addition, the earth rotates around its axis and completes a full rotation in one day. The orbit of the electron around the nucleus may be compared with the yearly motion of the electron around the sun. Once we begin to think about all these analogies, we may wonder whether there is a motion of the electron that is comparable with the daily rotation of the earth around its axis. In fact, such a motion exists, and it is called the electron spin.

The concept of the electron spin was introduced in 1925 by Goudsmit and Uhlenbeck in order to explain the fine structure that occurs in some atomic spectra. This early theoretical description of the electron spin relied heavily on the vector model of the atom, and it contained some general assumptions about the properties of the spin. Very likely these assumptions were introduced mainly for the purpose of explaining the experimental splitting in atomic spectra. At the time some of these assumptions seemed somewhat arbitrary, but they were all explained satisfactorily in later years by making use of relativity theory.

THE ELECTRON SPIN

Let us summarize the vector model for the orbital angular momentum **M** of an atom (see Fig. 3.5). The magnitude of **M** is determined from a quantum number l that must have a nonnegative, integer value. The possible projections of **M** in a given direction (say, the Z axis) are $m\hbar$, where $m = 0, \pm 1, \pm 2, \ldots, \pm l$. In each stationary state **M** can be represented as a vector that rotates around the Z axis so that its projection is constant; the length of this vector is equal to $\sqrt{l(l+1)}\hbar$.

An electron that moves in a closed orbit can be considered as a ring current, and as such it creates a magnetic field. The magnetic field is identical with the magnetic field due to a magnetic dipole **μ**. The ratio between the magnetic moment **μ** and the angular momentum **M** is easily derived from classical electromagnetic theory. It is given by

$$\mathbf{\mu} = -\frac{e}{2mc}\mathbf{M} \tag{45}$$

According to Goudsmit and Uhlenbeck, the spin angular momentum **S** may be represented in a similar fashion by a vector. For the electron spin the quantum number s, which plays the same role as the quantum number l for orbital angular momentum, must then be taken equal to $\frac{1}{2}$. Consequently, the length S of the spin vector is given by

$$S = \hbar\sqrt{\tfrac{1}{2}(\tfrac{1}{2}+1)} = \tfrac{1}{2}\hbar\sqrt{3} \tag{46}$$

The projection of the spin vector along a given direction can have two possible values, namely $\pm\tfrac{1}{2}\hbar$. If we take the Z axis as the axis of quantiza-

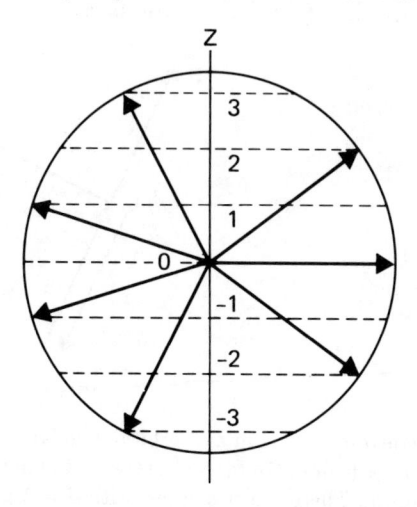

FIGURE 3.5 ● Quantization of the angular momentum vector **M** along the Z axis for $l = 3$.

ATOMIC STRUCTURE

tion, then the possible eigenvalues of S_z are $\pm\frac{1}{2}\hbar$, and the possible values of the spin quantum number m_s are $\pm\frac{1}{2}$. The spinning motion of the electron gives rise to a magnetic moment $\mathbf{\mu}_s$, which is also related to the spin angular momentum $\mathbf{S}$. The relation between $\mathbf{\mu}_s$ and $\mathbf{S}$ is given by

$$\mathbf{\mu}_s = -\frac{e}{mc}\mathbf{S} \qquad (47)$$

It should be noted that this ratio is twice as large as in the case of orbital angular momentum (compare Eq. 45). The relation (47) was introduced because it led to a correct quantitative interpretation of the fine structure of atomic spectra. It is difficult to justify Eq. (47) from classical electromagnetic theory, but eventually it was shown to be correct by making use of relativity theory.

It may be useful to illustrate how the simple vector model can be applied to the theory of atomic structure. The helium atom has two electrons, and each electron has a spin with quantum number $S = \frac{1}{2}$. We take the spin direction of the first electron as the axis of quantization for the second electron (see Fig. 3.6). The second spin than has two possible orientations: It either points in the opposite direction of the first spin or it points in the same direction. In the first case the total angular momentum of the two spins is zero, and in the second case the total spin angular momentum is represented by a quantum number $S = 1$. The stationary states of the He atom where the total spin is zero are called singlet states, because they are nondegenerate as far as the spin is concerned. The states with the total spin S equal to unity are threefold degenerate, because the spin vector can

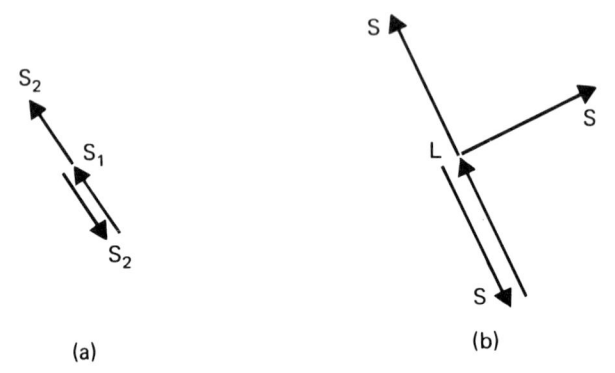

FIGURE 3.6 ● The quantization of spin and orbital angular momentum for the $(1s)(2p)$ configuration of helium. In (a) we represent the quantization of the spin angular momentum. There are two cases, with $S = 1$ and $S = 0$, respectively. The $S = 1$ state combines with $L = 1$ in (b) to give three states with total angular momentum $J = 2$, 1, or 0, respectively.

have three possible projections in a given direction. These are known as triplet states.

Let us now consider the various states of the He atom that have the configuration $(1s)(2p)$. Altogether there are 12 eigenstates with this configuration, because the $(2p)$ orbital is threefold degenerate and because there are four possible spin states. We have seen in the previous paragraph that the two electron spins can be combined to form either a nondegenerate singlet state with $S = 0$ or a threefold degenerate triplet state with $S = 1$. We may now use the vector model to derive the possible values of the total angular momentum J of the helium atom by combining the spin angular momentum S with the orbital angular momentum L. For a triplet $(2p)$ state, L is equal to unity and S is equal to unity. If we take the vector $\mathbf{L}$ as the axis of quantization for $\mathbf{S}$, then it is easily seen that J can have the values 2, 1, or 0. These three stationary states belonging to the different J values have slightly different energies, and they are fivefold, threefold, and nondegenerate, respectively. The singlet $(1s)(2p)$ state is threefold degenerate and should have an energy that differs considerably from the energies of the triplet states. It may be seen in Fig. 3.7 that these predictions agree with the experimental information.

The mathematical description of the electron spin can be derived by drawing an analogy with the angular momentum. If we take the Z axis as the axis of quantization, then there are two possible stationary states of the spin; the first has a projection $\tfrac{1}{2}\hbar$ and the second has a projection $-\tfrac{1}{2}\hbar$ along the Z axis. Both these stationary states are represented by an eigenfunction. We denote the eigenfunction of the state with projection $\tfrac{1}{2}\hbar$ by α and the eigenfunction of the state with projection $-\tfrac{1}{2}\hbar$ by β. Obviously, we must then have

$$S_z \alpha = \tfrac{1}{2}\hbar\alpha \qquad S_z \beta = -\tfrac{1}{2}\hbar\beta \tag{48}$$

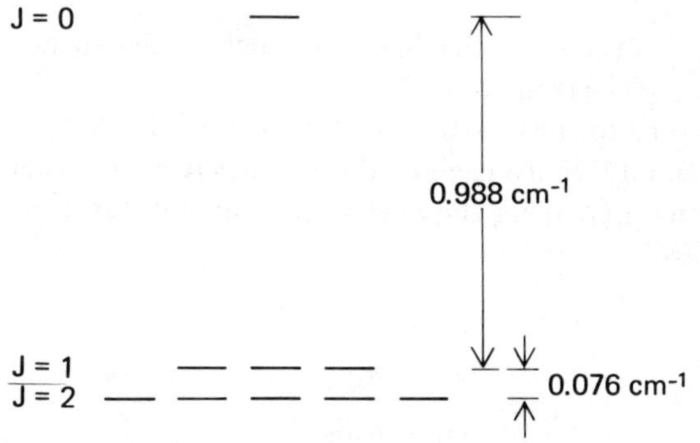

FIGURE 3.7 ● Fine structure of the triplet $(1s)(2p)$ state of the He atom.

where S_z is the operator that represents the Z component of the electron spin **S**. The spin functions α and β are taken to be orthonormal, which means that they are normalized to unity and orthogonal to one another,

$$\langle \alpha | \alpha \rangle = \langle \beta | \beta \rangle = 1 \qquad \langle \alpha | \beta \rangle = \langle \beta | \alpha \rangle = 0 \tag{49}$$

The two functions α and β must also be eigenfunctions of the operator S^2, and it follows from Eq. (46) and by drawing an analogy with the angular momentum that

$$\begin{aligned} S^2 \alpha &= \tfrac{1}{2}(\tfrac{1}{2}+1)\hbar^2 \alpha = \tfrac{3}{4}\hbar^2 \alpha \\ S^2 \beta &= \tfrac{1}{2}(\tfrac{1}{2}+1)\hbar^2 \beta = \tfrac{3}{4}\hbar^2 \beta \end{aligned} \tag{50}$$

The functions α and β are not eigenfunctions of the operators S_x and S_y. The effect of these two operators is described by the equations

$$\begin{aligned} S_x \alpha &= \tfrac{1}{2}\hbar\beta & S_y \alpha &= \tfrac{1}{2}i\hbar\beta \\ S_x \beta &= \tfrac{1}{2}\hbar\alpha & S_y \beta &= -\tfrac{1}{2}i\hbar\alpha \end{aligned} \tag{51}$$

If we consider the spin of an electron in addition to its orbital motion, then we must write the electron eigenfunctions as

$$\Psi_n(x, y, z; s) = \phi_n(x, y, z)(C_1 \alpha + C_2 \beta) \tag{52}$$

The two possible spin states of the electron are described by the functions α and β and in the general case, where we have no information with regard to the spin, we must write the spin part of the eigenfunction as an arbitrary linear combination of α and β. The probability of finding the spin parallel to the Z axis is given by $C_1 C_1^*$, and the probability of finding an antiparallel spin is given by $C_2 C_2^*$. Obviously, the parameters must satisfy the condition

$$C_1 C_1^* + C_2 C_2^* = 1 \tag{53}$$

Symbolically, we use the spin variable s to indicate the dependence of the eigenfunction on the electron spin.

Let us now proceed to the case of two electrons. We begin by considering the spin functions only. Since each of the two electron spins can have the possible projections $\pm\tfrac{1}{2}\hbar$ along the Z axis, the four spin functions of a two-electron system are

$$\begin{aligned} \phi_1 &= \alpha_1 \alpha_2 \\ \phi_2 &= \beta_1 \beta_2 \\ \phi_3 &= \alpha_1 \beta_2 \\ \phi_4 &= \beta_1 \alpha_2 \end{aligned} \tag{54}$$

The total spin of the two electrons is represented by an operator **S**, which is defined by

$$\mathbf{S} = \mathbf{S}_1 + \mathbf{S}_2 \tag{55}$$

or

$$\begin{aligned} S_x &= S_{1x} + S_{2x} \\ S_y &= S_{1y} + S_{2y} \\ S_z &= S_{1z} + S_{2z} \end{aligned} \tag{56}$$

The operator S^2 is defined as

$$S^2 = S_x^2 + S_y^2 + S_z^2 \tag{57}$$

It is now possible to construct a set of functions ζ from the four functions ϕ_k of Eq. (54) that are eigenfunctions of both operators S^2 and S_z. We write this as

$$\begin{aligned} S^2\, {}^1\zeta(1, 2) &= 0 \\ S^2\, {}^3\zeta_j(1, 2) &= 2\hbar^2\, {}^3\zeta_j(1, 2) \end{aligned} \tag{58}$$

The first function ${}^1\zeta(1, 2)$ represents the situation where the total spin is zero; obviously this state is nondegenerate. This state is called a singlet state and is denoted by a superscript 1 on the left. The second function ${}^3\zeta_j$ represents the situation where the total spin S is unity. This state is threefold degenerate; it is called a triplet state and is denoted by a superscript 3. It may be verified that the singlet spin function is given by

$$ {}^1\zeta(1, 2) = \tfrac{1}{2}\sqrt{2}(\phi_3 - \phi_4) = \tfrac{1}{2}\sqrt{2}(\alpha_1\beta_2 - \beta_1\alpha_2) \tag{59}$$

This result may be understood on the basis of the vector model that was illustrated in Fig. 3.6a. If we quantize the second spin with respect to the first spin, then we find that the total spin can be equal to zero when the two spins are antiparallel or equal to unity if they are parallel. The first case is the singlet state that is described by the function ${}^1\zeta(1, 2)$.

In the second case, where the total spin is unity, its possible projections along the Z axis are $\hbar$, 0, and $-\hbar$. These three situations are described by the eigenfunctions ${}^3\zeta_1$, ${}^3\zeta_0$, and ${}^3\zeta_{-1}$, respectively. These functions must satisfy the equations

$$\begin{aligned} S_z\, {}^3\zeta_1(1, 2) &= \hbar\, {}^3\zeta_1(1, 2) \\ S_z\, {}^3\zeta_0(1, 2) &= 0 \\ S_z\, {}^3\zeta_{-1}(1, 2) &= -\hbar\, {}^3\zeta_{-1}(1, 2) \end{aligned} \tag{60}$$

Every one of these functions corresponds to the case where the total spin is unity, so they must all satisfy the equation

$$S^2\,{}^3\zeta_k(1, 2) = 1(1 + 1)\hbar^2\,{}^3\zeta_k(1, 2) = 2\hbar^2\,{}^3\zeta_k(1, 2) \tag{61}$$

It may be verified that the various spin functions are given by

$$\begin{aligned}{}^3\zeta_1(1, 2) &= \alpha_1\alpha_2 \\ {}^3\zeta_0(1, 2) &= \tfrac{1}{2}\sqrt{2}(\alpha_1\beta_2 + \beta_1\alpha_2) \\ {}^3\zeta_{-1}(1, 2) &= \beta_1\beta_2\end{aligned} \tag{62}$$

The easiest way to prove Eqs. (59) and (62) is to derive first the effect of the operator S_z on the four spin functions of Eq. (54). We have

$$\begin{aligned}S_z\,\phi_1 &= (S_{1z} + S_{2z})\alpha_1\alpha_2 = \hbar\alpha_1\alpha_2 = \phi_1 \\ S_z\,\phi_2 &= (S_{1z} + S_{2z})\beta_1\beta_2 = -\hbar\beta_1\beta_2 = -\phi_2 \\ S_z\,\phi_3 &= (S_{1z} + S_{2z})\alpha_1\beta_2 = 0 \\ S_z\,\phi_4 &= (S_{1z} + S_{2z})\beta_1\alpha_2 = 0\end{aligned} \tag{63}$$

It follows immediately that the function ϕ_1 must be the function ${}^3\zeta_1$; because it describes the situation where the projection S_z is $\hbar$, the spin can only have a projection $\hbar$ if the total spin is at least unity, so ϕ_1 must be a triplet function. By means of the same argument, we conclude that ϕ_2 must be the function ${}^3\zeta_{-1}$.

We cannot apply this argument to the functions ϕ_3 and ϕ_4. The situation where the projection S_z is zero corresponds either to a total spin unity perpendicular to the Z axis or to a total spin zero. We must therefore determine the effect of the operator S^2 on these two functions. It follows from Eqs. (56) and (57) that S^2 may also be written as

$$S^2 = S_1^2 + S_2^2 + 2(S_{1x}S_{2x} + S_{1y}S_{2y} + S_{1z}S_{2z}) \tag{64}$$

By making use of Eqs. (48), (50), and (51), we find

$$\begin{aligned}S^2\,\phi_3 &= S^2(\alpha_1\beta_2) = \hbar^2(\alpha_1\beta_2 + \beta_1\alpha_2) = \hbar^2(\phi_3 + \phi_4) \\ S^2\,\phi_4 &= S^2(\beta_1\alpha_2) = \hbar^2(\alpha_1\beta_2 + \beta_1\alpha_2) = \hbar^2(\phi_3 + \phi_4)\end{aligned} \tag{65}$$

Consequently,

$$\begin{aligned}S^2(\phi_3 + \phi_4) &= 2\hbar^2(\phi_3 + \phi_4) \\ S^2(\phi_3 - \phi_4) &= 0\end{aligned} \tag{66}$$

or

$$\begin{aligned}{}^1\zeta(1, 2) &= \phi_3 - \phi_4 = \tfrac{1}{2}\sqrt{2}(\alpha_1\beta_2 - \beta_1\alpha_2) \\ {}^3\zeta_0(1, 2) &= \phi_3 + \phi_4 = \tfrac{1}{2}\sqrt{2}(\alpha_1\beta_2 + \beta_1\alpha_2)\end{aligned} \quad (67)$$

4. Exclusion Principle and Spin

Our formulation of the exclusion principle in Section 2 was not very precise because it was limited to atoms and was based on the assumption that each electron could be identified by means of a set of hydrogen-like quantum numbers. The introduction of the electron spin allows a much more precise formulation of the exclusion principle; namely, every eigenfunction of a many-electron system must satisfy the requirement that it is antisymmetric with respect to permutations of the electron spin and space coordinates. In order to discuss what this means, we first consider a two-electron system.

A two-electron function is called symmetric with respect to permutations if it remains the same when we permute the two electrons; that is

$$\Psi_s(1, 2) = \Psi_s(2, 1) \quad (68)$$

If the function changes sign when we permute the two electrons or

$$\Psi_a(1, 2) = -\Psi_a(2, 1) \quad (69)$$

then it is called antisymmetric.

It should be realized that an arbitrary two-electron function may be neither symmetric nor antisymmetric. For example, the function

$$\Psi(1, 2) = \exp(-r_1) \cdot z_2 \exp(-\tfrac{1}{2}r_2) \quad (70)$$

is nonsymmetric. However, it is possible to construct both a symmetric and an antisymmetric function, starting from Eq. (70). The symmetric function is

$$\Psi_s = \Psi(1, 2) + \Psi(2, 1) \quad (71)$$

and the antisymmetric function is

$$\Psi_a = \Psi(1, 2) - \Psi(2, 1) \quad (72)$$

The exclusion principle requires that the total wave function $\Psi(1, 2)$ of

110 ATOMIC STRUCTURE

a two-electron system be antisymmetric; it must therefore satisfy the condition

$$\Psi(1, 2) = -\Psi(2, 1) \tag{73}$$

It should be noted here that the function $\Psi(1, 2)$ must include both the space and spin coordinates of the two electrons.

We concluded in the previous section that the most general wave function of a two-electron system can be written in the form

$$\Psi(1, 2) = {}^3\Psi_1(\mathbf{r}_1, \mathbf{r}_2)\,{}^3\zeta_1(1, 2) + {}^3\Psi_0(\mathbf{r}_1, \mathbf{r}_2)\,{}^3\zeta_0(1, 2) \\ + {}^3\Psi_{-1}(\mathbf{r}_1, \mathbf{r}_2)\,{}^3\zeta_{-1}(1, 2) + {}^1\Psi(\mathbf{r}_1, \mathbf{r}_2)\,{}^1\zeta(1, 2) \tag{74}$$

namely as a linear combination of all possible spin states. If we permute the two electrons, we obtain

$$\Psi(2, 1) = {}^3\Psi_1(\mathbf{r}_2, \mathbf{r}_1)\,{}^3\zeta_1(2, 1) + {}^3\Psi_0(\mathbf{r}_2, \mathbf{r}_1)\,{}^3\zeta_0(2, 1) \\ + {}^3\Psi_{-1}(\mathbf{r}_2, \mathbf{r}_1)\,{}^3\zeta_{-1}(2, 1) + {}^1\Psi(\mathbf{r}_2, \mathbf{r}_1)\,{}^1\zeta(1, 2) \tag{75}$$

It follows from Eqs. (59) and (62) that the singlet spin function is antisymmetric and that the triplet spin functions are all symmetric,

$$\begin{aligned} {}^1\zeta(1, 2) &= -\,{}^1\zeta(2, 1) \\ {}^3\zeta_k(1, 2) &= {}^3\zeta_k(2, 1) \end{aligned} \tag{76}$$

The wave function (74) satisfies the antisymmetry condition (73) only if

$$\begin{aligned} {}^3\Psi_k(\mathbf{r}_1, \mathbf{r}_2) &= -\,{}^3\Psi_k(\mathbf{r}_2, \mathbf{r}_1) \\ {}^1\Psi(\mathbf{r}_1, \mathbf{r}_2) &= {}^1\Psi(\mathbf{r}_2, \mathbf{r}_1) \end{aligned} \tag{77}$$

In other words, the space part of a triplet function is antisymmetric and the space part of a singlet function is symmetric.

The orbital functions ${}^3\Psi_k$ or ${}^1\Psi$ must be eigenfunctions of the Schrödinger equation

$$\mathcal{H}(1, 2)\Psi(1, 2) = \epsilon\Psi(1, 2) \tag{78}$$

According to the exclusion principle the eigenfunction $\Psi(1, 2)$ must be either symmetric or antisymmetric with respect to permutation of the two electrons. If it is symmetric, then it is part of a singlet wave function and if it is antisymmetric, then it is part of a triplet wave function.

The definition of antisymmetry with respect to permutations is a little more complicated for many-electron systems. Here we call a wave function antisymmetric with respect to permutations if it changes sign when we permute a pair of electron coordinates. Let us illustrate this definition for a three-electron system, where the wave function is written as $\Psi(1, 2, 3)$. There are six possible ways in which we can rank-order three objects, and there must be five permuted functions in addition to $\Psi(1, 2, 3)$, namely $\Psi(1, 3, 2)$, $\Psi(2, 1, 3)$, $\Psi(2, 3, 1)$, $\Psi(3, 1, 2)$, and $\Psi(3, 2, 1)$. Only three of these functions, $\Psi(1, 3, 2)$, $\Psi(2, 1, 3)$, and $\Psi(3, 2, 1)$ may be obtained from the original function by a pairwise exchange of electron coordinates. Hence, if the function $\Psi(1, 2, 3)$ is anti-symmetric with respect to permutations, it must satisfy the conditions

$$\Psi(1, 3, 2) = -\Psi(1, 2, 3)$$
$$\Psi(2, 1, 3) = -\Psi(1, 2, 3) \quad (79)$$
$$\Psi(3, 2, 1) = -\Psi(1, 2, 3)$$

The remaining permuted functions, $\Psi(2, 3, 1)$ and $\Psi(3, 1, 2)$, may be obtained by a pairwise exchange of electron coordinates from one of the functions on the left-hand side of Eq. (79). They must satisfy the conditions

$$\Psi(2, 3, 1) = -\Psi(2, 1, 3) = \Psi(1, 2, 3)$$
$$\Psi(3, 1, 2) = -\Psi(3, 2, 1) = \Psi(1, 2, 3) \quad (80)$$

If the function $\Psi(1, 2, 3)$ is not antisymmetric to start with, then it may be used as a basis for the construction of an antisymmetric function $\Psi_a(1, 2, 3)$, namely

$$\Psi_a(1, 2, 3) = \Psi(1, 2, 3) + \Psi(2, 3, 1) + \Psi(3, 1, 2)$$
$$- \Psi(1, 3, 2) - \Psi(2, 1, 3) - \Psi(3, 2, 1) \quad (81)$$

It may be understood from the above considerations that it is useful to divide permutations into even and odd permutations. A permutation is defined as odd if it can be derived from the original sequence by means of an odd number of pairwise exchanges; similarly, a permutation is even if it is derived by means of an even number of pairwise exchanges. A permutation may be represented by a permutation operator; it is usually denoted by the symbol P. The symbol δ_P represents unity if P is even and minus unity if P is odd. If we use this notation, we can abbreviate Eq. (81) as

$$\Psi_a(1, 2, 3) = \sum_P P \, \delta_P \, \Psi(1, 2, 3) \quad (82)$$

The above equation is easily generalized to an N-electron system. Let $\Psi(1, 2, 3, \ldots, N)$ be an arbitrary function of N electron coordinates. Then the function

$$\Psi_a(1, 2, 3, \ldots, N) = \sum_P P \, \delta_P \, \Psi(1, 2, 3, \ldots, N) \tag{83}$$

is antisymmetric with respect to permutations of the electrons.

Let us now consider the question of how many possible permutations of N electrons exist, or, which is the same question, in how many different ways we can rank-order N objects. We call this number S_N, and we have already seen that S_1 is equal to unity and S_2 is equal to 2. If we add an object, then the number S_3 is three times larger than S_2, because each permutation of two objects gives rise to three permutations of three objects,

$$S_3 = 3 \cdot S_2 = 1 \cdot 2 \cdot 3 = 3! \tag{84}$$

This is easily generalized to the relation

$$S_N = N \, S_{N-1} = 1 \cdot 2 \cdot 3 \cdots N = N! \tag{85}$$

We have already mentioned that it is not possible to derive the exact solutions of the Schrödinger equation for many-electron systems. It is necessary, therefore, to introduce approximations to represent atomic and molecular wave functions. In the most customary approximation, the atomic wave functions are derived from products of one-electron wave functions. We have already mentioned that these one-electron wave functions are known as orbitals. For example, in the helium atom we consider the situation where the two electrons are either in the same orbital, where the total wave function is derived from the product

$$\Psi = \phi(\mathbf{r}_1)\phi(\mathbf{r}_2) \tag{86}$$

or in different orbitals, where the total wave function is derived from the function

$$\Psi = \phi(\mathbf{r}_1)\chi(\mathbf{r}_2) \tag{87}$$

It is important to note that in the case of Eq. (86) the atomic wave function must be a singlet function,

$$\Phi(1, 2) = \phi(\mathbf{r}_1)\phi(\mathbf{r}_2) \, {}^1\zeta(1, 2) \tag{88}$$

Because the orbital part of the wave function is symmetric with respect to permutations, the spin part must be antisymmetric. We find, therefore, that if we have two electrons in the same orbital, they must have antiparallel spins. We can derive either a singlet or a triplet eigenfunction from the product of Eq. (84),

$$^1\Phi(1, 2) = \frac{1}{\sqrt{2}} [\phi(\mathbf{r}_1)\chi(\mathbf{r}_2) + \chi(\mathbf{r}_1)\phi(\mathbf{r}_2)] \,^1\zeta(1, 2)$$

$$^3\Phi(1, 2) = \frac{1}{\sqrt{2}} [\phi(\mathbf{r}_1)\chi(\mathbf{r}_2) - \chi(\mathbf{r}_1)\phi(\mathbf{r}_2)] \,^3\zeta(1, 2)$$

(89)

Let us now consider a four-electron system. Here again, we find that if we have two electrons in the same orbital, they must have antiparallel spins in order to satisfy the exclusion principle. Clearly, it is not allowed to have more than two electrons in the same orbital. In that case, there would be two electrons with the same orbital and spin functions, and then the wave function could satisfy the exclusion principle only if it were zero. If we assign an energy ϵ_i to every orbital χ_i, then we can have only two electrons in the orbital χ_1 with the lowest energy ϵ_1. The atomic ground state is derived from the configuration where we have two electrons each in the orbitals χ_1 and χ_2 with the lowest energy ϵ_1 and the next-lowest energy ϵ_2. The atomic eigenfunction is then obtained as

$$\Psi = \sum_P P \,\delta_P \,[\chi_1(\mathbf{r}_1)\alpha(1)\chi_1(\mathbf{r}_2)\beta(2)\chi_2(\mathbf{r}_3)\alpha(3)\chi_2(\mathbf{r}_4)\beta(4)] \qquad (90)$$

This is called a closed-shell state. It should be noted that the total spin is zero because it consists of pairs of electrons that have antiparallel spins, so that each of such an electron pair has spin zero and the sum of these spins is also zero.

It may be seen that the early formulation of the exclusion principle that we discussed in Section 1 is a special case of the general definition. If we derive the total wave function from a product of one-electron orbitals, then we are allowed to have no more than two electrons in any one orbital. If there are two electrons in one orbital, they must have antiparallel spins. In atomic systems it is usually possible to identify the orbitals by means of hydrogen-like quantum numbers. If we do that, then we are back at the original formulation of the exclusion principle, which we discussed in Section 1.

The lower excited states of the four-electron system may be derived from Eq. (90) by taking one of the electrons out of the orbital χ_2 and by placing it in an excited orbital χ_3. The configuration of this state is $(\chi_1)^2(\chi_2)(\chi_3)$. The two electrons in the orbitals χ_2 and χ_3 have either parallel or antiparallel spins. In the first case the total antisymmetric eigenfunction is

$$^3\Psi = \sum_P P \,\delta_P \,[\chi_1(\mathbf{r}_1)\alpha(1)\chi_1(\mathbf{r}_2)\beta(2)\chi_2(\mathbf{r}_3)\chi_3(\mathbf{r}_4) \,^3\zeta(3, 4)] \qquad (91)$$

The total spin of the system is unity, and we have a triplet state. If the two spins are antiparallel, the total wave function is

$$^1\Psi = \frac{1}{\sqrt{2}} \sum_P P\,\delta_P \{\chi_1(\mathbf{r}_1)\alpha(1)\chi_1(\mathbf{r}_2)\beta(2)$$
$$\times [\chi_2(\mathbf{r}_3)\chi_3(\mathbf{r}_4) + \chi_3(\mathbf{r}_3)\chi_2(\mathbf{r}_4)]\alpha(3)\beta(4)\} \quad (92)$$

and we have a singlet state. Both wave functions of Eqs. (91) and (92) satisfy the exclusion principle, because they are antisymmetric with respect to all possible electron permutations.

All systems with more than two electrons may be described in a similar manner. In writing the antisymmetrized total wave functions, it is advantageous to take the one-electron orbitals orthonormal,

$$\langle \chi_i \mid \chi_j \rangle = \delta_{ij} \quad (93)$$

It is not strictly necessary that the above orthogonality condition be satisfied, but it may be shown that the nonorthogonal parts of the orbitals cancel out anyway in the antisymmetrized wave function, and that these parts do not contribute to the expectation values of Hermitian operators. For that reason we may just as well take the orbitals orthonormal, because this simplifies the mathematics considerably. If the orbitals are orthonormal, then it is easily shown that the eigenfunctions in Eqs. (90), (91), and (92) should all be multiplied by a factor $(1/\sqrt{24})$ in order to be normalized to unity. For an N-electron system the normalization factor is $(1/N!)^{1/2}$ because here there are $N!$ terms in the antisymmetrized wave function.

5. Atomic Orbitals

It is an approximation to write an atomic wave function as a product (or an antisymmetrized product) of one-electron orbitals, but it is the only practical method in the theory of atomic structure. The approximation is fairly satisfactory, and it gives us a good overall understanding of the structure of an atom. Besides, if we want to make use of more precise theoretical descriptions, the theory becomes so complicated that it is difficult to appreciate what the results mean.

Once we have made the approximation of writing an atomic wave function as a product of one-electron orbitals, we are still left with the problem of determining the specific form of these orbitals. We have already discussed some aspects of this problem in Sections 1 and 2, where we considered the He atom. It may be recalled that we derived an approximate expression for the atomic wave function. Let us summarize this discussion here, as an introduction to the theory for more complex atoms.

The Hamiltonian of the He atom was given in Eq. (1). It is convenient to write it in terms of atomic units of length and energy. It then takes the form

$$\mathcal{H} = -\tfrac{1}{2}\Delta_1 - \tfrac{1}{2}\Delta_2 - \frac{2}{r_1} - \frac{2}{r_2} + \frac{1}{r_{12}} \tag{94}$$

In Section 2 we derived an approximate eigenfunction of this Hamiltonian for the ground state where the atomic configuration is $(1s)^2$. In this treatment we made use of two approximations. In the first approximation we wrote the ground-state eigenfunction Ψ as a product of two identical one-electron orbitals. If we include the spin functions, then Ψ_0 takes the form

$$\Psi_0 = s(r_1)s(r_2)2^{-\tfrac{1}{2}}[\alpha(1)\beta(2) - \beta(1)\alpha(2)] \tag{95}$$

In the second approximation we approximated the atomic orbitals as

$$s(r_1) = \left(\frac{Z^3}{\pi}\right)^{1/2} e^{-Zr} \tag{96}$$

We obtained the value of the effective nuclear charge Z by means of the variational principle. The value was $Z = 1.6875$.

We ought to point out now that it is possible to derive a more precise form of the atomic orbitals than the expression (96). In fact, once we have made the approximation (95) of expressing the atomic wave function in terms of atomic orbitals, it is possible to do the rest of the calculation exactly without making use of any additional approximations. The orbitals derived in this way are the most precise atomic orbitals that can be obtained. They are known as Hartree-Fock orbitals or SCF orbitals (self-consistent field), and they have been tabulated in numerical form for the majority of atomic ground states.

The general derivation of the Hartree-Fock method is fairly complicated, but it is possible to get a general understanding of the method by considering again the ground state of the He atom.

We follow the same approach as in Section 1, where we considered one of the two electrons, 1, and where we try to determine the effective potential field $V_{\text{eff}}(r)$ in which the electron moves. If we know this effective potential field, then $s(r_1)$ should be the solution of the Schrödinger equation

$$[-\tfrac{1}{2}\Delta_1 - V_{\text{eff}}(r_1)]s(r_1) = \epsilon s(r_1) \tag{97}$$

To be more precise, $s(r_1)$ should be the eigenfunction belonging to the lowest eigenvalue of this equation.

In Section 1 we used an approximate expression for the effective potential V_{eff}, but there is no need to resort to approximations because there is an exact expression for the potential. We again consider the situation that we sketched in Fig. 3.1, where electron 1 moves in the charge cloud of the second electron. The behavior of electron 2 is characterized by the orbital $s(r_2)$, so that its probability density is given by the expression

$$\rho(\mathbf{r}_2) = [s(r_2)]^2 \tag{98}$$

The total interaction between electron 1 and the charge cloud of electron 2 is then given by

$$V_{\text{int}}(r_1) = \int \frac{\rho(\mathbf{r}_2)}{r_{12}} d\mathbf{r}_2 = \int \frac{[s(r_2)]^2}{r_{12}} d\mathbf{r}_2 \tag{99}$$

This is a function of the coordinate r_1 only, because the charge cloud of the second electron is spherically symmetric. It follows now easily that the effective potential $V_{\text{eff}}(r_1)$ is the sum of the Coulomb attraction of the nucleus and the interaction term of Eq. (99),

$$V_{\text{eff}}(r_1) = -\frac{2}{r_1} + V_{\text{int}}(r_1) = -\frac{2}{r_1} + \int \frac{[s(r_2)]^2}{r_{12}} d\mathbf{r}_2 \tag{100}$$

This is an exact expression for the effective potential, and if we substitute it into the Schrödinger equation (97) for the atomic orbital $s(r_1)$, we can attempt to solve it. Unfortunately, there is one difficulty: The function $s(r)$ that we are trying to find occurs in the potential function V_{eff}, which we have to know in order to solve the equation. Due to this difficulty, we cannot solve the equation in a straightforward manner. However, the equation can be solved by means of the self-consistent field method. Here, we start with an approximate form $s_0(r)$ of $s(r)$ and we substitute this into Eq. (100) for V_{eff}. Now we can solve the Schrödinger equation for this approximate expression for V_{eff}. The solution that we obtain in this way, $s_1(r)$, should be a better approximation than $s_0(r)$. If we substitute this new expression into Eq. (100), then we should obtain a more accurate expression for V_{eff} and we can solve the Schrödinger equation again. The new solution, $s_2(r)$, should be better than the previous one, and if we substitute it, then we should obtain a better approximation, $s_3(r)$. We keep repeating this procedure, which is known as an iterative method, until the function $s_n(r_1)$ that we obtain is identical with the function $s_{n-1}(r_1)$ that we substitute into V_{eff}. At that point we have obtained self-consistency of the equation, and the function $s_n(r_1)$ is the solution. The method is known either as the Hartree-Fock method, named after the two scientists who proposed it, or as the self-consistent field or SCF method.

It is clear that the above method is fairly involved and laborious, and that it does not lead to simple analytical expressions for the orbitals. On the other hand, it is feasible to do the calculation if we make use of an electronic computer. Hartree calculated the SCF orbitals for a large number of molecules long before electronic computers were even invented. He made use of analog computers. These are electronic devices that were specially designed to deal with one specific problem. These analog computers may be considered predecessors of the present-day electronic computers.

It should be realized that a one-dimensional differential equation, no matter how complicated, can usually be solved numerically if we use an analog computer or an electronic computer. The solutions that were derived by Hartree are all in numerical form; they come in the form of a table where the values of the orbital for various values of the variable are listed.

The Hartree-Fock method can also be applied to the excited states of the He atom and to the ground and excited states of more complicated atoms. However, we must know the configuration of the state with which we are dealing, so that we know enough about the orbitals that they can be written as functions of one variable only. For example, the $(1s)^2(2s)^2$ state of Be can be studied in this way. Here we have to find two different orbitals for the $1s$ and for the $2s$ states, and each of the orbitals is a function of the variable r only. In the case of neon, the ground-state configuration is $(1s)^2(2s)^2(2p)^6$. The various p orbitals are then written as

$$p_x = xp(r) \qquad p_y = yp(r) \qquad p_z = zp(r) \tag{101}$$

and we have to determine the three functions $s_1(r)$, $s_2(r)$, and $p(r)$, corresponding to the $(1s)$, $(2s)$, and $(2p)$ orbitals. In all these cases the Hartree-Fock orbitals are known, and they can be found in the literature in numerical form.

Even though the exact Hartree-Fock orbitals are available for most atoms, the form in which they are given is not too convenient. Also, it is not easy to relate a numerical table to the chemical properties of an atom. Consequently, there have been various attempts to construct simple analytical expressions for the atomic orbitals, which would form reasonable approximations to the exact Hartree-Fock orbitals. The best known of these are the Slater orbitals, which were derived by making use of the concept of the effective nuclear charge, Z_{eff}. In fact, the He orbitals that we discussed in Section 1 are the Slater $(1s)$ orbitals for the He atom. It may be recalled that we wrote the $(1s)$ orbital of the He atom as

$$s(r) = e^{-Zr} \tag{102}$$

where Z is the effective nuclear charge. We derived from the variation principle that Z should be taken as $Z = 1.6875$.

We argued that each of the two electrons in the He atom experiences the

118   ATOMIC STRUCTURE

Coulomb attraction of the nucleus with charge 2, shielded by the charge cloud of the other electron with effective charge δ. In the case of He we have

$$Z = 1.6875 \qquad \delta = 0.3125 \qquad (103)$$

and this value of δ represents the amount of shielding.

It is possible to perform similar calculations for all other atoms. For example, in the case of neon we may introduce a variational function of the form

$$\Psi = (10!)^{-1/2} \sum_p P\, \delta_P[(1s)_1\alpha_1(1s)_2\beta_2(2s)_3\alpha_3(2s)_4\beta_4(2p_x)_5 \\ \times \alpha_5(2p_x)_6\beta_6(2p_y)_7\alpha_7(2p_y)_8\beta_8(2p_z)_9\alpha_9(2p_z)_{10}\beta_{10}] \qquad (104)$$

where we introduce an effective nuclear charge Z_1 for the $(1s)$ electrons, a charge Z_2 for the $(2s)$ electrons, and a charge Z_3 for the $(2p)$ electrons. The values for the quantities Z_i are obtained again by minimizing the energy that is derived from the function Ψ. It turns out that the various effective nuclear charges for the electrons in different atoms may be estimated fairly accurately by making use of a set of simple semiempirical rules. These rules were first proposed by Slater, and the resulting orbitals are called Slater orbitals.

In order to find the effective nuclear charge Z' for an electron in an atom, we write it as

$$Z' = Z - \sigma \qquad (105)$$

where Z is the charge of the nucleus and σ is the amount of shielding of the nuclear potential due to the other electrons in the atom. It is assumed now that the contributions to the shielding from the other electrons are additive and that it is possible to assign a constant amount of shielding to each type of electrons. These amounts are determined from the following rules.

1. Nothing from any electron that has a principal quantum number n that is higher than the one we consider.
2. An amount 0.35 from each electron that has the same principal quantum number n as the electron we consider, except that when we consider a $(1s)$ electron, then the contribution from the other $(1s)$ is 0.30.
3. An amount 0.85 from each electron that has a principal quantum number n that is 1 less than the quantum number of the electron we consider if the latter is an s or a p electron and an amount 1.00 from each electron whose principal quantum number is less by 1 than the electron we consider if the latter is in a d, f, or g state.
4. An amount 1.00 from each electron with a principal quantum number that is less by 2 or more than the quantum number of the electron considered.

According to the above rules, the effective nuclear charge for the $(1s)^2$ state in He is 1.70. This is fairly close to the value $Z = 1.6875$ that follows from a variational calculation.

Let us illustrate the Slater rules by considering the $(1s)^2(2s)^2(2p)^2$ configuration of the carbon atom. The nuclear charge is 6, and the amount of shielding is 0.30 for each $1s$ electron and $2 \times 0.85 + 3 \times 0.35 = 2.75$ for the $2s$ and $2p$ electrons. It follows that the effective nuclear charge is 5.70 for the $1s$ electrons and 3.25 for the other electrons.

There are some differences between the Slater orbitals and the hydrogen atom eigenfunctions, and these are related to the orthogonality properties of these functions. It follows from the properties of Hermitian operators (Chapter I, Section 6) that all the hydrogen atom eigenfunctions are orthogonal to each other, but we should realize that in the hydrogen atom the nuclear charge is the same for each eigenstate. If, for example, we substitute different nuclear charges Z_1 and Z_2 in the $(1s)$ and $(2s)$ eigenfunctions, then these functions are no longer orthogonal. Since the orthogonality is destroyed anyway, the Slater $(2s)$ orbital is defined as

$$\Phi(2s) = r \exp(-\tfrac{1}{2}Z_2 r) \tag{106}$$

which is different from the hydrogen-type $2s$ eigenfunction

$$\Psi(2s) = (Z_2 r - 2) \exp(-\tfrac{1}{2}Z_2 r) \tag{107}$$

In performing calculations we should impose the condition that the Slater $(2s)$ orbital is orthogonal to the Slater $(1s)$ orbital

$$\Phi(1s) = \exp(-Z_1 r) \tag{108}$$

This is achieved by replacing $\Phi(2s)$ of Eq. (106) by the function

$$\Phi'(2s) = (2s) - a\Phi(1s) \tag{109}$$

where a is chosen in such a way that

$$\langle \Phi'(2s) \mid \Phi(1s) \rangle = 0 \tag{110}$$

Accordingly, the Slater (ns) orbitals are defined as

$$\Phi(ns) = r^n \exp\left(\frac{-Z_n r}{n}\right) \tag{111}$$

the Slater (np) orbitals are defined as

$$\Phi(np_x) = xr^{n-1} \exp\left(\frac{-Z_n r}{n}\right) \tag{112}$$

and so on.

The Slater orbitals are very useful for quick, order-of-magnitude calculations. They can also be used as starting orbitals for the Hartree-Fock procedures. The Slater orbitals give a fairly good approximation to the overall charge density of an atom, and they are also used in molecular calculations.

Problems

1. Use the vector model of the atom to discuss the various states and their degeneracies that may be derived from the $(1s)(3d)$ configuration of the He atom.
2. Consider the configuration $(1s)^2(2s)^2(2p)^2$ of the carbon atom. There are a number of different atomic states that have this configuration. By using the vector model, determine the possible values of the orbital angular momentum L that these states can have and derive the possible values of the spin angular momentum S for each of these states. Finally, give the possible values of the total angular momentum.
3. Determine the Slater orbitals for the $1s$ electrons in the He, O, and I atoms.
4. We have seen that the energy of a $1s$ electron is approximately equal to $-\frac{1}{2}Z_{\text{eff}}^2 \epsilon_0$. Relativistic effects may be neglected if the energy of an electron is small with respect to mc^2, where m is the mass of the electron and c is the velocity of light. Investigate whether or not it is allowed to neglect relativistic effects for the $1s$ electron in He, O, and I. Use the values of Z_{eff} that were derived in Problem 3.
5. We know that for one electron the eigenfunctions of the spin operator S_z are the functions α and β. Derive the eigenfunctions of the operator S_x.
6. Show explicitly that the triplet spin function $\alpha_1 \alpha_2$ is an eigenfunction of the operator S^2 and determine the eigenvalue.
7. Show explicitly that $S^2 \alpha_1 \beta_2 = \hbar^2(\alpha_1 \beta_2 + \beta_1 \alpha_2)$.
8. Write down all 24 possible permutations of four numbers, 1, 2, 3, 4. Determine which of these permutations are even and which are odd.
9. Construct an antisymmetric function from the two orbitals $\phi(1s) = e^{-pr}$ and $\phi(2s) = re^{-qr}$. (Do not include the spins.) Now construct an antisymmetric function from the two orbitals $\psi(1s) = e^{-pr}$ and $\psi(2s) = re^{-qr} - \lambda e^{-pr}$. Show that the two antisymmetric functions are the same.
10. Derive the Slater $(1s)$ and $(2s)$ orbitals for the $(1s)(2s)$ configuration of the He atom. Normalize these two functions and calculate their overlap integral.

$$\left[\text{Hint:} \int_0^\infty r^n e^{-\alpha r}\, dr = (n!/\alpha^{n+1}).\right]$$

11. Determine the Slater orbitals for the configuration $(1s)^2(2s)^2$ of Be.
12. Derive the Slater orbitals for the ground-state configuration $(1s)^2(2s)^2(2p)^6(3s)$ of Na and for the ground-state configuration $(1s)^2(2s)^2(2p)^6(3s)^2(3p)^6(4s)$ of K.
13. Write the complete antisymmetrized eigenfunctions, including spin, for the various eigenstates of the Li atom that have the configuration $(1s)^2(2s)$.
14. What are the possible values of the orbital angular momentum in the $(1s)^2(2s)(2p)^3$ configuration of the carbon atom?
15. Determine the Slater coefficients of the various orbitals in the $(1s)^2(2s)^2(2p)^6$ configuration of the Na$^+$ ion. Determine the expectation value of the coordinate r with respect to the $2p$ orbital.

$$\left[\text{Hint:} \int_0^\infty r^n e^{-\alpha r}\, dr = (n!/\alpha^{n+1}). \right]$$

16. Prove explicitly that the field strength is zero inside the charged sphere of Fig. 3.2.

Bibliography

The best-known books on atomic spectra and atomic structure are the book by Herzberg (1) and the book by Condon and Shortley (2). Herzberg gives an excellent account of atomic spectra, based on physical arguments, and it is easy to read because it avoids complicated mathematics. Condon and Shortly's book is written on a more sophisticated level, but is the best-known advanced treatment of atomic structure. We also list some other books on the subject.

1. G. Herzberg. *Atomic Spectra and Atomic Structure*. Prentice-Hall, Englewood Cliffs, N.J. (1937).
2. E. U. Condon and G. H. Shortly. *The Theory of Atomic Spectra*. Cambridge University Press, London (1935).
3. H. G. Kuhn. *Atomic Spectra*. Academic Press, New York (1962).
4. H. E. White. *Introduction to Atomic Spectra*. McGraw-Hill, New York (1934).

CHAPTER 4

LIGHT AND SPECTROSCOPY

1. The Nature of Light

Scientists and philosophers have expressed their ideas on the nature of light since the time of Aristotle, who lived during the fourth century B.C. The various theories on the nature of light make fascinating reading. We should realize that some of the early descriptions of light were consistent with what was known at that time, even though later developments showed that they were either partially or totally wrong. Also, a comprehensive theory had to explain more and more different features, such as the various colors, the diffraction of light through a prism and, at a later time, interference phenomena, the existence of polarized light, optical activity, and so on.

For example, during the seventeenth century Newton described a beam of light as a stream of fast-moving particles of various kinds in which each color is represented by a particular kind of particle. At the same time the Dutch scientist Huygens proposed an ondulatory description of light, and he showed how his description was also consistent with straight-line propagation. At the time it was hard to say which of the two theories was the right one, but the various interference experiments on light during the beginning of the nineteenth century showed that Huygens' ideas were preferable to Newton's.

Nowadays the established viewpoint is that light is a form of electromagnetic radiation. All types of electromagnetic radiation are mathematically described by the four Maxwell equations; to be exact, the radiation field is the solution of these equations in the vacuum. We shall just outline the results here without going into the mathematical derivations.

124 LIGHT AND SPECTROSCOPY

The properties of any electromagnetic field are described by two vector quantities, the electric field strength **E** and the magnetic field strength **H**. These two field strengths are usually related to one another, and the relation is described by the Maxwell equations.

A radiation field is a superposition of plane waves. We have sketched the behavior of such a plane wave in Fig. 4.1. For the sake of simplicity, we have taken the Z axis as the direction of propagation of the wave. Then at each point on the Z axis there is an oscillating electric field vector in the Y direction, the magnitude of which is given by

$$E_y = A \cos\left[\frac{2\pi}{\lambda}(z - ct)\right] \tag{1}$$

At the same time there is an oscillating magnetic field vector in the X direction with a magnitude

$$H_x = A \cos\left[\frac{2\pi}{\lambda}(z - ct)\right] \tag{2}$$

Here c is the velocity with which the waves propagate in the Z direction and λ is the wavelength of the wave. It follows from Maxwell's equations that all electromagnetic waves have the same velocity c, which is equal to the velocity of light.

We know now that a variety of seemingly different kinds of radiation are all electromagnetic radiation. The wavelength λ determines the kind of radiation. For example, the waves that we receive in an ordinary AM radio have wavelengths that vary between 100 and 500 m. In short-wave radio receivers capable of receiving transatlantic signals, the wavelengths vary from 5 m to 100 m. Radar waves have wavelengths of the order of 1 cm. The wavelengths of visible light vary between 4,000 Å and 7,000 Å, ultraviolet light may be taken as having a wavelength between 1,000 Å and

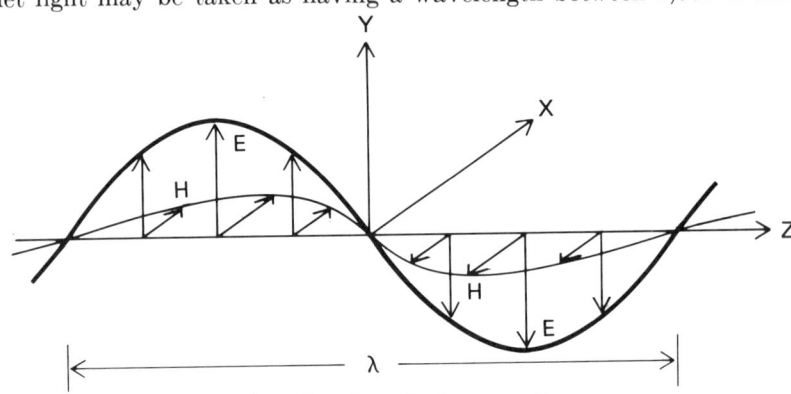

FIGURE 4.1 ● Representation of a plane electromagnetic wave.

4,000 Å, and infrared light between 7,000 Å and 10,000 Å. (These figures are not precise, but all we want to do is to give a rough idea of them.) If we proceed to shorter wavelengths, we find X rays with a wavelength of the order of 1 Å and, beyond that, γ rays, which have a wavelength of 0.01 Å and lower. We see that the color of light is determined solely by the wavelength.

Let us now return to the two equations (1) and (2). It is well known that the velocity of a wave, its wavelength λ, and its frequency ν are related to one another as follows

$$c = \lambda\nu \tag{3}$$

The frequency ν is defined as the inverse of the period T of the wave, where T is the amount of time it takes to complete one oscillation,

$$\nu = \frac{1}{T} \tag{4}$$

Similarly, the wave number σ is defined as the inverse of the wavelength λ,

$$\sigma = \frac{1}{\lambda} \tag{5}$$

By making use of the above relations, we can rewrite the two equations (1) and (2) as

$$\begin{aligned} E_y &= A\,\cos[2\pi(\sigma z - \nu t)] \\ H_x &= A\,\cos[2\pi(\sigma z - \nu t)] \end{aligned} \tag{6}$$

We should realize that an electromagnetic wave is a plane wave in three dimensions. In Fig. 4.1 we have simplified the situation to make it a little easier to understand, but in reality the electric and magnetic field strengths given by Eq. (6) are oscillating in every point in space. We see that the functions in Eq. (6) depend on the variable z only, which means that for a constant z the oscillations are in phase. The direction of propagation is perpendicular to the planes of constant phase and is therefore along the Z axis.

By rotating the coordinate system, we can transform the plane waves of Eq. (6) into waves with different directions of propagation. The general expressions for such waves are

$$\begin{aligned} E_\alpha &= A\,\cos[2\pi(\sigma_x x + \sigma_y y + \sigma_z z - \nu t)] \\ H_\beta &= A\,\cos[2\pi(\sigma_x x + \sigma_y y + \sigma_z z - \nu t)] \end{aligned} \tag{7}$$

The planes of constant phase are found from the condition that E_α and H_β must be constant in such a phase, hence these planes are determined by the condition

$$\sigma_x x + \sigma_y y + \sigma_z z = \text{constant} \qquad (8)$$

The vector

$$\boldsymbol{\sigma} = (\sigma_x, \sigma_y, \sigma_z) \qquad (9)$$

is perpendicular to all planes of equal phase, consequently $\boldsymbol{\sigma}$ represents the direction of propagation of the wave. The magnitude of $\boldsymbol{\sigma}$ is still given by Eq. (5), because it is not affected by a rotation of the coordinate system. Hence,

$$\sigma = (\sigma_x^2 + \sigma_y^2 + \sigma_z^2)^{1/2} = \frac{1}{\lambda} = \frac{\nu}{c} \qquad (10)$$

The vectors **E** and **H** must be perpendicular to each other, and both must be perpendicular to $\boldsymbol{\sigma}$. Hence E_α and H_β are the components of the electric and the magnetic field strengths in the directions perpendicular to $\boldsymbol{\sigma}$.

The direction of polarization of the light coincides with the direction of the vector **E**. Equation (6) represents a light wave that is polarized in the Y direction. It is easily seen that there must be a second light wave of the type of Eq. (6) that is polarized in the X direction. This light wave is described by the equations

$$\begin{aligned}\mathbf{E} &= (0, E_x, 0) & E_x &= A \cos[2\pi(\sigma z - \nu t)] \\ \mathbf{H} &= (H_y, 0, 0) & H_y &= A \cos[2\pi(\sigma z - \nu t)]\end{aligned} \qquad (11)$$

This wave also propagates in the Z direction, because the vector $\boldsymbol{\sigma} = (0, 0, \sigma)$ points in the Z direction. The vector **E** may have any arbitrary direction in the XY plane (it must be perpendicular to $\boldsymbol{\sigma}$), but any vector **E** in the XY plane may be written as a linear combination of a vector in the X direction and a vector in the Y direction. Therefore, there are two possible directions of polarization for a light wave once $\boldsymbol{\sigma}$ is specified.

In the general case, where $\boldsymbol{\sigma}$ has an arbitrary direction, it is more difficult to give a unique definition of the two basic directions of polarization. In principle we may define two directions perpendicular to $\boldsymbol{\sigma}$ and take these as the basis for defining the possible values of **E**. We see, then, that a light wave is determined by the value of $\boldsymbol{\sigma}$, which determines the direction of propagation, and by an additional symbol that tells us which of the two possible directions of polarization the light wave has. The general plane waves have the form

$$\begin{aligned}
\mathbf{E}_1(\mathbf{\sigma}) &= \mathbf{A}_1 \cos[2\pi(\sigma_x x + \sigma_y y + \sigma_z z - \nu t)] \\
\mathbf{H}_1(\mathbf{\sigma}) &= \mathbf{A}_1' \cos[2\pi(\sigma_x x + \sigma_y y + \sigma_z z - \nu t)] \\
\mathbf{E}_2(\mathbf{\sigma}) &= \mathbf{A}_2 \cos[2\pi(\sigma_x x + \sigma_y y + \sigma_z z - \nu t)] \\
\mathbf{H}_2(\mathbf{\sigma}) &= \mathbf{A}_2' \cos[2\pi(\sigma_x x + \sigma_y y + \sigma_z z - \nu t)]
\end{aligned} \qquad (12)$$

with

$$\mathbf{A}_1 \perp \mathbf{\sigma} \qquad \mathbf{A}_2 \perp \mathbf{\sigma} \qquad \mathbf{A}_1 \perp \mathbf{A}_2 \qquad (13)$$

The radiation field is always a superposition of plane waves of the type (12), because it is not possible to produce a radiation field that consists of only one wave. The simplest case that we can have is a beam of polarized light; here the direction of propagation $\mathbf{\sigma}$ must be the same for all components. We can write the corresponding radiation field as

$$\begin{aligned}
E_x &= \sum_j A_j \cos[2\pi(\sigma_j z - \nu_j t)] \\
H_y &= \sum_j A_j \cos[2\pi(\sigma_j z - \nu_j t)]
\end{aligned} \qquad (14)$$

if we take the vector $\mathbf{\sigma}$ along the Z axis. Each component plane wave is determined by the value of σ. Because the possible values of σ form a continuum, the most general superposition takes the form of an integral,

$$\begin{aligned}
E_x &= \int A(\sigma) \cos[2\pi(\sigma z - \nu t)] \, d\sigma \\
H_y &= \int A(\sigma) \cos[2\pi(\sigma z - \nu t)] \, d\sigma
\end{aligned} \qquad (15)$$

We should realize that it is not possible to have a purely monochromatic beam of light. At best we can have the situation where the function $A(\sigma)$ has a sharp maximum around a given value σ_0, but this maximum must have a finite line width. Hence, the term monochromatic is somewhat misleading, because purely monochromatic light cannot be produced.

The most general expression for the radiation field is a superposition of all possible plane waves of the type (12). It is given by

$$\begin{aligned}
\mathbf{E} &= \int [\mathbf{A}_1(\mathbf{\sigma}) + \mathbf{A}_2(\mathbf{\sigma})] \cos[2\pi(\mathbf{\sigma} \cdot \mathbf{r} - \nu t)] \, d\mathbf{\sigma} \\
\mathbf{H} &= \int [\mathbf{A}_1'(\mathbf{\sigma}) + \mathbf{A}_2'(\mathbf{\sigma})] \cos[2\pi(\mathbf{\sigma} \cdot \mathbf{r} - \nu t)] \, d\mathbf{\sigma}
\end{aligned} \qquad (16)$$

This represents a radiation field that contains light waves with all possible frequencies, directions of propagation, and directions of polarization.

We stated at the beginning of this section that light is a form of electromagnetic radiation. Its mathematical description may be derived from the Maxwell equations and leads to Eq. (16). We should mention some of the more recent developments in the theory of the radiation field, such as Planck's description of black-body radiation and Einstein's theory of the photoelectric effect because these two theories played a very important role in the discovery of quantum mechanics. Ultimately, their approach led to the quantum mechanical description of the radiation field, which is known as quantum electrodynamics. We are somewhat hesitant to discuss this description of the radiation field, because at first sight some of the basic concepts of this model seem to be inconsistent with what we have discussed previously. However, a more careful analysis will show that the two theoretical models are in accord with one another.

Let us first consider Planck's work on black-body radiation. This type of radiation can be observed by anyone who has an electric heater or an electric kitchen range. If the heating coils are turned on, they begin to glow redly when they warm up, and if they get hotter, the color becomes more intense and changes over to yellow and white. We can measure the spectral distribution of this glow, that is, the intensity of the light as a function of the frequency. If we were to measure the radiation from the kitchen stove, we would observe the sum of the light that is emitted by the stove and of the light that is reflected. By definition, a black body is an object that does not reflect any light at all, and black-body radiation is observed experimentally by eliminating all the reflected light. In practice, a black body is a hollow sphere with a little hole in it. The geometrical arrangement is such that the reflected light is dissipated within the sphere, and the black-body radiation is measured as the light that comes out of the hole when the sphere is heated up. It was found that black-body radiation depends on the temperature but is independent of the material of which the sphere is made.

The properties of black-body radiation were studied by measuring the spectral distribution of the light at various temperatures. Here the light is passed through a spectrograph and its intensity is plotted versus the wavelength. If we repeat this procedure at various temperatures, we obtain a set of curves. We have sketched two of these curves in Fig. 4.2; here the temperature T_2 is supposed to be larger than T_1. We see that both curves in Fig. 4.2 have a maximum, and if we look at the light we shall see the color that corresponds to the wavelength of this maximum. When the temperature T increases, the total intensity increases (the light becomes brighter), and at the same time the maximum shifts to the left. This is why we see the color change from red to yellow to white when the temperature of the radiating body goes up.

We see that the black-body radiation is a function of both the wavelength λ of the light and of the temperature T. After the experimental curves became available, several attempts were made to find analytical expressions that would represent the experimental results, but Planck showed finally

THE NATURE OF LIGHT   129

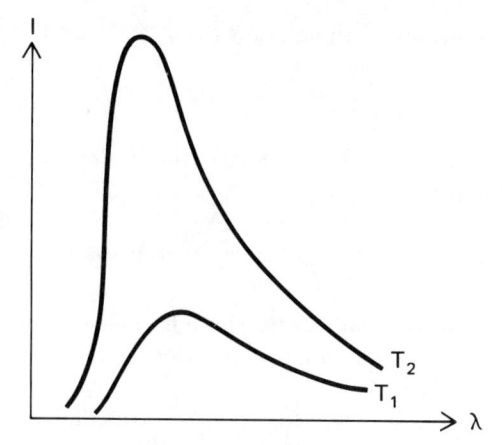

FIGURE 4.2 ● Two typical spectral distribution curves for black-body radiation at different temperatures. T_2 is larger than T_1.

that the correct expression, which is consistent with all experimental data, is given by

$$\rho(\nu, T) = \frac{8\pi h\nu}{c^3} \frac{1}{e^{h\nu/kT} - 1} \quad (17)$$

Here $\rho(\nu, T)$ is the energy density of the radiation per unit volume and per unit frequency. This means that the energy of the radiation with frequencies between ν and $\nu + d\nu$ is given by

$$dE = \rho(\nu, T)\, d\nu \quad (18)$$

Planck showed that the radiation formula (17) could be derived only if it is assumed that the radiation is quantized.

In order to understand this, we should realize that in Planck's model, just as in the classical model, it is first assumed that the radiation field is a superposition of plane waves of the type that we described by Eq. (12). The total energy of the radiation field E may then be written as the sum of the partial energies E_λ of the various components,

$$E = \sum_\lambda E_\lambda \quad (19)$$

Here we use the subscript λ to label the various plane waves, where λ is a shorthand notation for the value of the vector $\boldsymbol{\sigma}$ and for the direction of polarization. Planck's quantization rule requires that each energy E_λ can have only certain discrete energy values $E_{n\lambda}$, which are given by

$$E_{n\lambda} = n_\lambda h\nu_\lambda \quad (20)$$

where ν_λ is the frequency of the plane wave λ and n_λ must be an integer,

$$n_\lambda = 0, 1, 2, 3, \ldots \tag{21}$$

According to Planck's quantization rule, the total energy of the radiation field is then given by

$$E = \sum_\lambda E_{n\lambda} = \sum_\lambda n_\lambda h \nu_\lambda \tag{22}$$

The energy E is thus determined by the infinite set of quantum numbers $(n_1, n_2, \ldots, n_\lambda, \ldots)$ corresponding to the various plane waves λ that constitute the radiation field.

It is important to note that Planck's quantization rule refers only to the energy of the radiation field. It assumes that each component λ of the radiation field has an energy E_λ composed of energy parcels $h\nu_\lambda$, which are indivisible in the same way as a chemical substance is composed of indivisible atoms. Much later the term "photon" was introduced by G. N. Lewis, and today it is commonly used in describing these energy packages.

We mentioned before that the quantum theory of the radiation field has certain features that might be confusing. What we had in mind is that in this approach photons are often treated as particles. It is then said that the radiation field consists of a large number of particles, the photons, and that its total energy is the sum of the energies of all photons that are present. What we should realize is that this terminology is permissible as long as we consider the energy only. The energy of the radiation field is given by Eq. (22) as the sum of the photon energies, and in this equation it does not make any difference whether we think of the photons as energy parcels or as particles. It is often convenient to visualize the photons as particles, but whatever we like to think does not alter the fact that the radiation field is a superposition of electromagnetic plane waves.

Einstein made use of Planck's quantization rules and the photon concept to explain certain features of the photoelectric effect that could not be understood on the basis of the classical electromagnetic theory. The photoelectric effect is measured by placing a piece of metal in a beam of light. The light ejects electrons out of the metal, and these electrons are attracted by an electrode that is placed a little distance away and has a different potential from the metal. The number of ejected electrons is derived from measuring the electron current between the metal and the electrode. A surprising feature of the photoelectric effect is that no current is observed if the frequency of the light is below a certain frequency ν_0, no matter how intense the light is.

Einstein offered a simple explanation of the photoelectric effect. Let W be the potential barrier that the electron must pass in order to be able to leave the metal. An electron can be ejected only if it collides with a photon and absorbs the photon energy. It follows, then, that the photon must have an energy of at least eW in order to eject an electron. Therefore, if the frequency of the light is below a frequency ν_0, which is given by

$$h\nu_0 = eW \tag{23}$$

the photoelectric effect is not observed. If the light has a frequency ν that is higher than ν_0, then the electrons are ejected and their kinetic energies are given by

$$\tfrac{1}{2}mv^2 = h(\nu - \nu_0) \tag{24}$$

From this relation Einstein calculated photoelectric currents, and he obtained satisfactory agreement with the experimental results.

Planck's work on radiation is important from a historical point of view, because it marked the first time that the concept of quantization was introduced in theoretical physics. The validity of Planck's assumptions was strongly supported by Einstein's work on the photoelectric effect and by subsequent studies by Debye on the specific heat in solids. No doubt it helped Bohr in formulating his quantum theory of atomic structure and so led ultimately to the formulation of quantum mechanics as we know it today.

2. Transition Probabilities

In atomic and molecular spectroscopy we measure the intensity distribution of the light that comes out of the sample we wish to study. If we wish to measure an emission spectrum, then we have to treat the sample in such a way that we produce a large fraction of atoms or molecules in an excited state. A common procedure is to place a gaseous, low-pressure sample in a high-voltage electric discharge. When atoms return to their ground states, they emit light, and we measure this light's spectral distribution with a photographic plate or by other means. This is called an emission spectrum. In order to measure an absorption spectrum, we must have a light source that has a continuous intensity distribution over a sufficiently large range. We let the light pass through our sample, and we measure the intensity distribution of the light that has passed through the sample. In phosphorescence or fluorescence spectra the experimental arrangement is the same as in absorption spectra, but here the light is first absorbed by the sample and then reemitted. The molecule can change from one state to another before the light is reemitted, in which case the frequency of the emitted light may be different from the frequency of the light that has been absorbed.

The simplest way to measure the intensity distribution of a beam of light is to let it pass through a prism and then focus it on a photographic plate. However, this is fairly time consuming, especially if we wish our results to be reasonably accurate. In more sophisticated and more rapid experiments we make use of a monochromator. This is an arrangement of prisms, lenses, and mirrors, through which only light within a narrow frequency range can pass. At the exit of the monochromator there is a photomultiplier cell with an amplifier, by which means we can measure immediately the intensity of the light that comes out of the monochromator. By turning the prisms we can vary the frequency of the light the monochromator lets through, and in this way we can cover the whole frequency range of the spectrum we wish to measure. Nowadays the apparatus is often automated, and the absorption or emission spectrum we wish to measure is recorded on tracing paper.

Einstein gave a formal analysis of the various absorption and emission processes that can occur. In order to understand this analysis, we first consider an absorption process where a molecule is excited from a lower state k to a higher state l by absorbing light of frequency ν_{kl}. Because of the conservation of energy, we must have

$$h\nu_{kl} = E_l - E_k \tag{25}$$

In Einstein's description the absorption process is represented by a quantity $B_{k \to l}$, which is known as the Einstein transition probability of absorption. In order to define it, we consider a thin sample of molecules of thickness Δx and we assume that the sample contains N_k molecules/cm^3 in the lower state k. We call I_{in} the intensity of the light that enters the sample and I_{out} the intensity of the light that comes out of the sample. By definition, then, we have

$$\frac{I_{\text{out}}}{I_{\text{in}}} = N_k B_{k \to l} \Delta x \tag{26}$$

The intensity of a beam of light is defined here as the amount of energy that passes per second through a surface of 1 cm^2 perpendicular to the beam of light.

Let us now consider the emission of light. Einstein was the first to point out that in this case we must consider two processes rather than just one process as in the case of absorption. The first emission process is described by a quantity $B_{l \to k}$, and it is in every respect analogous to the absorption process we have described above. It may be shown in general that if the light causes a transition from a stationary state k to a stationary state l

with a probability $w_{k \to l}$, then the light also induces a transition from state l to state k with a probability $w_{l \to k}$ that is equal to $w_{k \to l}$. Hence, if we define the quantities $B_{k \to l}$ and $B_{l \to k}$ in the same way, we must have

$$B_{l \to k} = B_{k \to l} \tag{27}$$

The quantity $B_{l \to k}$ is called the Einstein coefficient of stimulated emission.

It is easy to understand the above mechanism from a theoretical point of view, but it is not so easy to appreciate it if we think about the experimental arrangements for measuring emission spectra. We know that if we have a sufficiently large number N_l of molecules in the excited state l, these molecules will return to the lower state k while emitting light. This process occurs in the absence of radiation. In fact, in measuring an emission spectrum we usually measure this spontaneous emission and not the stimulated emission that we discussed above.

It follows that there must be two possible emission mechanisms, namely the spontaneous emission, which we usually measure, and the stimulated emission, which we predict from theoretical considerations. In the most general case, where we have a sample in the presence of radiation and where we measure the emission spectrum, the intensity of the outgoing light is given by

$$I = N_l c [B_{l \to k} \rho(\nu_{kl}) + A_{l \to k}] \tag{28}$$

Here $B_{l \to k}$ is the Einstein coefficient of stimulated emission and $A_{l \to k}$ is the Einstein coefficient of spontaneous emission. We have replaced the intensity I of the light by the energy density $\rho(\nu_{kl})$ of the radiation field. Since the light moves with a velocity c, the two quantities are related by means of

$$I = c\rho \tag{29}$$

It is possible to derive two relationships among the three Einstein coefficients. We consider a transition between a lower state o and a higher state n of a set of identical molecules, and we take N_o as the number of molecules per cubic centimeter in the lower state and N_n as the same number for the higher state. The emission of light is described by Eq. (28), and we may derive that the amount of energy E_{pr} produced per unit time due to the emission of light is given by

$$E_{pr} = cN_n[B_{n \to o}\rho(\nu_{on}) + A_{n \to o}] \tag{30}$$

By analogy, the amount of energy E_{abs} that is absorbed per unit time due to the absorption of light is given by

$$E_{\text{abs}} = cN_o B_{o \to n}\rho(\nu_{on}) \tag{31}$$

134 LIGHT AND SPECTROSCOPY

If we divide Eq. (30) by $h\nu_{on}$, we obtain the number of photons emitted per unit time. We know that every time a photon is emitted there must be a transition in a molecule from an excited state n to a lower state o. This means that the emission process (30) causes changes in the numbers N_o and N_n, and the change is given by

$$\frac{dN_o}{dt} = \frac{E_{pr}}{h\nu_{on}} = N_n \frac{c}{h\nu_{on}} [B_{n \to o}\rho(\nu_{on}) + A_{n \to o}] \qquad (32)$$

In the same way, Eq. (31) represents upward transitions, which cause a decrease in the number N_o. The rate of this process is given by

$$-\frac{dN_o}{dt} = \frac{E_{abs}}{h\nu_{on}} = N_o \frac{c}{h\nu_{on}} B_{o \to n}\rho(\nu_{on}) \qquad (33)$$

If our system is in equilibrium, the rates of change due to the two processes (32) and (33) must be equal to one another, and we have

$$N_n[B_{n \to o}\rho(\nu_{on}) + A_{n \to o}] = N_o B_{o \to n}\rho(\nu_{on}) \qquad (34)$$

We know that in equilibrium the ratio between the two numbers N_n and N_o must be given by the Maxwell distribution,

$$\frac{N_n}{N_o} = \exp\left(-\frac{h\nu_{on}}{kT}\right) \qquad (35)$$

It follows from Eqs. (34) and (35) that

$$\frac{N_o}{N_n} = \exp\left(\frac{h\nu_{on}}{kT}\right) = \frac{B_{n \to o}\rho(\nu_{on}) + A_{n \to o}}{B_{o \to n}\rho(\nu_{on})} \qquad (36)$$

or

$$\rho(\nu_{on}) = \frac{A_{n \to o}}{B_{o \to n} \exp(h\nu_{on}/kT) - B_{n \to o}} \qquad (37)$$

Equation (37) expresses the energy density of the radiation field (in thermal equilibrium) in terms of the three Einstein coefficients of absorption and omission. In Eq. (17) we mentioned that the energy density ρ is also given by the expression that was derived by Planck, namely

$$\rho(\nu_{on}) = \frac{8\pi h\nu_{on}^3}{c^3} \frac{1}{\exp(h\nu_{on}/kT) - 1} \qquad (38)$$

By comparing Eqs. (37) and (38), we find that

$$A_{n \to o} = \frac{8\pi h \nu_{on}^3}{c^3} B_{n \to o} \qquad B_{n \to o} = B_{o \to n} \qquad (39)$$

As we already expected, the two coefficients $B_{n \to o}$ and $B_{o \to n}$ are equal to one another. It is important to note that the ratio between the two coefficients $A_{n \to o}$ and $B_{n \to o}$ depends very strongly on the frequency. The coefficient $A_{n \to o}$ is known as the coefficient for spontaneous emission, and $B_{n \to o}$ is the coefficient for induced emission. For optical transitions the spontaneous emission is usually much larger than the stimulated emission due to light of customary intensity (this excludes laser light). In magnetic resonance the frequencies are much smaller than in optical transitions, namely by a factor of the order of 10^4, and here the situation is just the reverse, the stimulated emission being predominant.

3. Transition Probabilities and Quantum Theory

In order to understand how the properties of spectroscopic transitions can be explained in terms of quantum theory, it is helpful to recall Bohr's theory of atomic structure, which we discussed in Chapter I. The immediate purpose of this theory was to explain, first, how an atom could remain in a stable, stationary state when it is not subject to outside perturbations and, second, why an atomic spectra consists of discrete lines. In order to explain the first difficulty, Bohr assumed that an atom could only be in certain stationary states, each having a fixed energy and angular momentum, and that the atom does not emit or absorb radiation as long as it remains in one of those stationary states. A spectroscopic transition occurs when the atom jumps from one stationary state k to another stationary state l. If the energy E_k of state k is higher than the energy E_l of state l, then this jump is accompanied by the emission of a photon with an energy

$$h\nu_{kl} = E_k - E_l \qquad (40)$$

and we observe an emission line of frequency ν_{kl}. If E_k is smaller than E_l, then the atom absorbs a photon of energy

$$h\nu_{kl} = E_l - E_k \qquad (41)$$

and we observe an absorption line of frequency ν_{kl}.

We have seen that the stationary states of an atom are obtained as the eigenvalues and eigenfunctions of the Schrödinger equation,

$$\mathcal{H}_{op}\Phi_n = E_n\Phi_n \qquad (42)$$

Here the atom is represented by the Hamiltonian $\mathcal{H}_{op}$, the energy of stationary state n is given by the eigenvalue E_n, and the behavior of the electrons in this stationary state is described by the eigenfunction Φ_n. It is obvious that Eq. (42) describes the properties of stationary states, but it does not account for transitions between different stationary states. If we wish to study such transitions, we should realize that the time-independent Schrödinger equation (42) is a special case of a more general time-dependent Schrödinger equation, which has the form

$$\mathcal{H}_{op}\Psi = i\hbar \frac{\partial \Psi}{\partial t} \tag{43}$$

Here the function Ψ depends both on the electron coordinates, which we denote symbolically by x, and on the time t.

The two Schrödinger equations (42) and (43) are equivalent if we consider a stationary state and if the Hamiltonian $\mathcal{H}_{op}$ is time-independent (otherwise we could not have stationary states). In the time-independent formalism the stationary state n is described by the function Φ_n. In the time-dependent formalism, the same stationary state n is described by the function $\Psi_n(x; t)$, which is defined as

$$\Psi_n(x; t) = \Phi_n(x) \exp\left(-\frac{iE_n t}{\hbar}\right) \tag{44}$$

It is easily verified that substitution of the time-dependent function $\Psi_n(x; t)$ into the time-dependent Schrödinger equation (43) leads to the time-independent equation (42), because the energy E_n is a constant for the stationary state n.

The advantage of using the time-dependent Schrödinger equation is that it also describes the situation where our system is not in a stationary state. Supposedly, our system can be in any one of an infinite number of stationary states n. Each of these states is described by a time-dependent function $\Psi_n(x; t)$, defined by Eqs. (42) and (44), which satisfied the equation

$$\left(\mathcal{H}_{op} + \frac{\hbar}{i}\frac{\partial}{\partial t}\right)\Psi_n(x; t) = 0 \tag{45}$$

Obviously, an arbitrary linear combination,

$$\Psi(x; t) = \sum_n b_n \Psi_n(x; t) = \sum_n b_n \Phi_n(x) \exp\left(-\frac{iE_n t}{\hbar}\right) \tag{46}$$

also satisfies the time-dependent Schrödinger equations (43) and (45). It follows that the function (46) is the most general solution of the time-dependent Schrödinger equation and that the general behavior of our system, when it is not in any given stationary state, is described by the function (46).

If our system is described by the time-dependent wave function (46), we can ask what the probability P_m is of finding the system in a given stationary state m. By definition, this probability is given by

$$P_m = b_m b_m^* \tag{47}$$

Obviously, we must have

$$\sum_m P_m = \sum_m b_m b_m^* = 1 \tag{48}$$

Let us now proceed to the problem in which we are interested, namely how to describe spectroscopic transitions. The quantum mechanical analysis of this problem is quite complicated, because there are a number of different approaches involving various approximations and it is not easy to appreciate the consequences of these approximations. In addition, even the simplest approach is quite involved mathematically. Our discussion is therefore limited to a very broad outline of the problem, but we shall give the results.

In order to study spectroscopic transitions, we consider the case of a molecule in the presence of a radiation field. The molecule is described by a time-independent Hamiltonian $\mathcal{H}_{mol}$, and the interaction between the molecule and the radiation field is described by a time-dependent Hamiltonian $\mathcal{H}_{int}$. The total system is thus described by a Hamiltonian

$$\mathcal{H}_{op}(x;t) = \mathcal{H}_{mol}(x) + \mathcal{H}_{int}(x;t) \tag{49}$$

Strictly speaking, this Hamiltonian does not have stationary states because the Hamiltonian contains the time, but if its time-dependent part is much smaller than its time-independent part,

$$\mathcal{H}_{int}(x;t) \ll \mathcal{H}_{mol}(x) \tag{50}$$

then we may express the behavior of the system in terms of the stationary states of the operator $\mathcal{H}_{mol}$.

In mathematical terms, the behavior of the total system is described by a time dependent function $\Psi(x;t)$, which must be a solution of the Schrödinger equation

$$\mathcal{H}_{op}(x;t)\Psi(x;t) = -\frac{\hbar}{i}\frac{\partial \Psi}{\partial t} \tag{51}$$

This function may be expressed in the form

$$\Psi(x;t) = \sum_n b_n(t) \exp\left(-\frac{i\epsilon_n t}{\hbar}\right) \Phi_n(x) \qquad (52)$$

where ϵ_n and Φ_n are the eigenvalues and eigenfunctions of the molecular Hamiltonian $\mathcal{H}_{\text{mol}}$.

It may be shown that the above expansion (52) is permissible as long as the coefficients $b_n(t)$ are dependent on the time t. On the other hand, the expansion makes sense only if the time-dependent part of the operator is small and satisfies the condition (50). In that case, the coefficients $b_n(t)$ are slowly varying functions of the time, and they may be calculated by means of perturbation theory. Also, the behavior of the total system may be described in terms of the coefficients $b_n(t)$. The probability of finding the molecule in one of its stationary states n is given by

$$P_n(t) = b_n(t) b_n^*(t) \qquad (53)$$

just as in Eq. (47). The time dependence of the probabilities $P_n(t)$ describes how the probability distribution over the various molecular stationary states changes as a function of time, and it may be related to the three Einstein coefficients of absorption and emission.

We have seen in Eq. (39) of the previous section that there are two relations among the three Einstein coefficients. This is fortunate, because the simplest approach for calculating the probabilities $P_n(t)$ by means of perturbation theory gives a result only for the Einstein coefficient of absorption $B_{o \to n}$. The values of the other two coefficients of emission $B_{n \to o}$ and $A_{n \to o}$ may then be derived by using Eq. (39).

The theoretical result for the Einstein coefficient $B_{o \to n}$ is

$$B_{o \to n} = \left(\frac{2\pi}{3\hbar^2}\right)(\mathbf{u}_{n,o} \cdot \mathbf{u}_{o,n}) \qquad (54)$$

The vector $\mathbf{u}_{n,o}$ is known as the transition moment between the states n and o. It is the matrix element over the dipole moment operator

$$\mathbf{u} = e \sum_j \mathbf{r}_j \qquad (55)$$

between the two states

$$\begin{aligned} \mathbf{u}_{n,o} &= \langle \Phi_n | \mathbf{u} | \Phi_o \rangle \\ \mathbf{u}_{o,n} &= \langle \Phi_o | \mathbf{u} | \Phi_n \rangle \end{aligned} \qquad (56)$$

In most cases we are not so much interested in the exact value of a transition probability as in its order of magnitude. The experimental spectroscopists speak of allowed and forbidden transitions. An allowed transition has a large transition probability, and a forbidden transition has a small transition probability. The nature of the transition probabilities can often be derived from a set of rules known as selection rules. Ideally, these rules can tell us immediately which transitions are allowed and which are forbidden.

We like to define a forbidden transition as a transition for which the transition moment, as defined by Eqs. (55) and (56), is zero. We ought to realize that we have used a number of approximations in deriving our expression for the transition probability. Therefore, even if the transition moment $\mathbf{\mu}_{k,o}$ is zero, the transition may still be observable, but it is then caused by higher-order effects and its probability is generally much smaller than for an allowed transition.

If we use our definition of a forbidden transition, then a selection rule is simply a general rule that tells us when the transition moment $\mathbf{\mu}_{k,o}$ is zero and when it is nonzero. For example, the operator in Eq. (56) is spin-independent and the transition moment $\mathbf{\mu}_{k,o}$ is zero if the states k and o have different spin multiplicity, because then the spin functions of the states k and o are orthogonal to one another. This is known as the spin selection rule, $\Delta S = 0$. Other selection rules may be derived from the symmetry of the atom or molecule. For example, it is easily seen that the transition between the $(1s)$ and the $(2s)$ states of the hydrogen atom must be forbidden, because the transition moment of that transition is zero. We should realize that forbidden transitions may still be observed, because they are caused by mechanisms that were neglected in the derivation of the result (54) for the transition probability. Some symmetry-forbidden transitions can be observed because of these mechanisms. In a similar fashion, spin-forbidden transitions are nonzero if we take the interactions between the electron spins into account.

The probability of an ordinary allowed transition is of the order of 10^{10} sec^{-1}. This means that a molecule in an excited state returns to the ground state in a time of about 10^{-10} sec if the transition is allowed. Forbidden transition probabilities range all the way from 10^{-1} to 10^8 sec^{-1}. Some transitions are both spin-forbidden and symmetry forbidden, and they have very low probabilities. In such cases the lifetimes of molecular excited states can be as high as 10 sec. We postpone a more detailed discussion of the selection rules until later.

4. The Maser and the Laser

The word "maser" is an acronym for "*m*icrowave *a*mplification by *s*timulated *e*mission of *r*adiation," and the word "laser" is derived from "*l*ight *a*mplification by *s*timulated *e*mission of *r*adiation." The maser is a device

140 LIGHT AND SPECTROSCOPY

that amplifies microwaves. It was invented around 1950 and was used primarily as a noise-free amplifier. The first lasers were constructed around 1960, and they were used to produce beams of light with very high intensities in a very narrow frequency range. The invention of the laser has led to some interesting experimental discoveries, which are known as nonlinear optical phenomena or multiple-photon phenomena. The result (54) of the transition probability represents processes where only one photon at a time is absorbed or emitted. In principle, there are multiple-photon absorption or emission processes, but these are negligible in ordinary light. However, if we have light beams of very high intensity, then these higher-order effects may become observable. Recent experiments have uncovered a large variety of such multiple-photon processes, which might have useful practical applications.

The first maser that was constructed made use of ammonia. We know that the molecule NH_3 has a pyramidal structure (see Fig. 4.3), but it should be realized that the molecule has two equilibrium structures with the N atom either above or below the plane. The molecule can rock back and forth between these two structures. This motion can be represented by a harmonic oscillator with a frequency ν that corresponds to a wavelength of 1.26 cm. Under ordinary circumstances a sample of NH_3 molecules has N_0 molecules in the vibrational ground state and N_1 in the first excited state, and the ratio between N_0 and N_1 is given by the Maxwell distribution

$$\frac{N_1}{N_0} = e^{-h\nu/kT} \qquad (57)$$

Obviously N_1 is smaller than N_0.

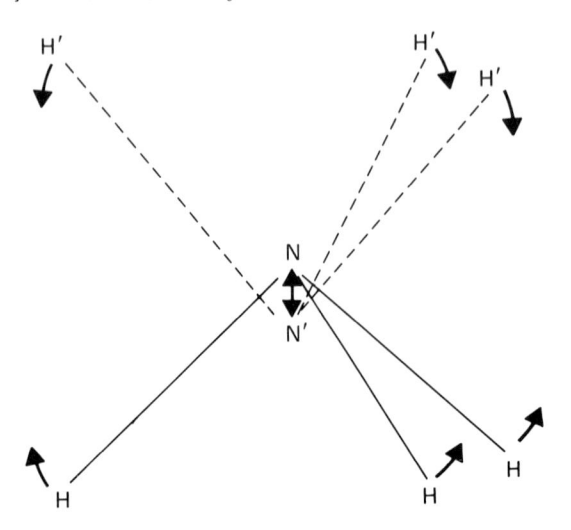

FIGURE 4.3 • The inversion motion of the ammonia molecule.

The operation of the maser is based on population inversion. If some way can be found to make N_1 larger than N_o, that is, to produce an excess of molecules in the excited state, then we can use stimulated emission to liberate the excess energy and in this way we produce an intense flash of radiation. It is possible to produce such a population inversion in the case of ammonia. If we let a beam of ammonia molecules pass through an inhomogeneous electric field, then the beam splits into two because the electric fields interacts differently with the molecules in the excited vibrational state and the molecules in the ground state. One of the two beams has a large fraction N_1 of molecules in the excited state, and in this way we can obtain a sample of ammonia molecules with N_1 larger than N_0. We can see what happens when a photon of energy $h\nu$ enters this sample. As soon as it encounters an excited molecule, it gives rise to stimulated emission; the molecule returns to its ground state and, as a result, we have two photons. These two photons produce more photons, and so on. In the most efficient device the ammonia sample is in a resonance cavity where the photons move back and forth through the sample and where they keep producing more and more photons. Such an ammonia maser can amplify an incoming signal by about a factor 100 to 1 in energy. What is more important, it is a practically noise-free amplifier. The ammonia maser is obviously too expensive for use in ordinary radios or televisions, but is was used in astronomy for the detection of microwave radiation from outer space.

It became clear that a maser or a laser can operate effectively if two conditions are satisfied. First, we must have a population inversion where there are more molecules in a higher excited state than there are in a lower state. Second, we must have a resonance condition where the radiation can travel back and forth through the sample and where it is amplified with each passage through. In order to construct a laser we must find a system in which the above two conditions can be satisfied, and it is clear that the first condition is the crucial one. The first two lasers that were constructed in 1960 were the ruby laser and the helium-neon laser. Other kinds of lasers were constructed later, but we shall limit our discussion to the two lasers mentioned above.

Ruby consists of aluminum oxide with small amounts of chromium added into it, the chromium atoms replacing some of the aluminum atoms in the lattice. In the ruby laser we use a cylinder of about 1 cm in diameter and about 5 cm long, which contains 0.05% chromium. In Fig. 4.4 we give the energy level diagram for the chromium atoms embedded in the lattice. Level 3 is not a single energy level, but consists of a very large number of closely spaced near-degenerate levels; in solid state physics it is called an energy band. If we radiate the ruby with green light, then we obtain excitation into the energy band. Some of the excitation drops back to the

142 LIGHT AND SPECTROSCOPY

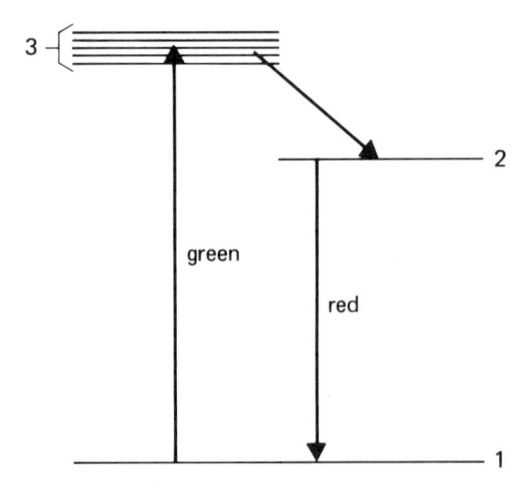

FIGURE 4.4 • Energy levels of the chromium ions in ruby.

nondegenerate level 2 through spontaneous emission. In ruby, level 2 is metastable, which means that the probability of the transition from level 2 to level 1 is quite small. Consequently, the chromium atoms in the excited state 2 have a relatively long lifetime and it is possible in this way to obtain a fairly large number of chromium atoms in the excited state 2. This procedure is known as optical pumping. We see that we can obtain a significant population inversion between levels 1 and 2 by means of optical pumping via the energy band 3.

Resonance is obtained in a ruby laser by placing a mirror at one end of the cylinder and a partially transparent mirror at the other end. In this way the bulk of the light travels back and forth between the two mirrors and acquires a greater and greater intensity. Eventually all of this light passes through the partially transparent mirror in the form of the laser beam. In the ruby laser all the emitted photons have a frequency that is very close to the energy difference between levels 2 and 1; in fact the line width of the frequency distribution is equal to the natural line width of the emission line, which is of the order of 0.01 Å. In this fashion the energy of the emitted light is concentrated within a very narrow frequency range, and its energy density is very high. The laser can be operated so that it emits pulses of light, and it can also be constructed so that it has a continuous light output. In the latter case the optical pumping must also be continuous.

A completely different type of lasers are the gas lasers. The helium-neon laser is the best-known example of this group. A gas laser consists of a mixture of two gases, and population inversion in one gas is obtained by means of energy transfer from the second gas. In the He-Ne laser the He atoms are excited to a higher state, and they transfer their excess energy to the Ne atoms through collisions. The Ne atoms lose their energy through a series of downward transitions from a higher excited state via the intermediate lower excited states to the ground state. It is clear that if one or more of these intermediate levels are metastable, the concentration of Ne atoms in this metastable excited state will increase. In this way we can obtain a significant population inversion over certain energy levels in the Ne atoms. The optimum efficiency for the He-Ne level occurs when we take five parts of He to one part of Ne. In general, the lasing material in the gas mixture is usually the smaller fraction in the mixture. Gas lasers are usually built in the form of a cylindrical glass tube. In the beginning two plane mirrors were usually placed within the glass tube at both ends, but it soon became apparent that it is more convenient to use spherical reflectors placed outside the tube. The final construction of the laser makes use of more sophisticated engineering devices but we do not wish to go into these. Actually, the engineering skills needed to construct a gas laser are much less sophisticated than those needed for solid state lasers. The difficult problem is to know the energy levels and their lifetimes for the gases involved and to have a good understanding of the energy transfer processes.

Problems

1. What is the representation for a plane light wave that propagates in the X direction and is polarized in the Y direction?
2. If an electromagnetic wave has a direction of propagation that bisects the positive X and Y axes, what are its possible directions of polarization? Give the corresponding equations.
3. The work function W is the potential barrier that an electron must pass in order to be able to leave a metal. If W is 2 eV, what is the value of the frequency ν_0 and wavelength λ of the light that we shine on the metal so that we begin to observe the photoelectric effect?
4. If the work function of a metal is 1 eV and we illuminate with light of 5,000 Å, what is the velocity v of the electrons that are ejected from the metal?
5. What is the ratio between the coefficients of spontaneous and stimulated emission for light with a wavelength of 6,000 Å and for microwave radiation with a wavelength of 3 cm?
6. What is the transition moment between a $(1s)$ and a $(2p_z)$ state of the hydrogen atom? Determine the corresponding Einstein coefficient of absorption.
7. Derive the value of the Einstein coefficient of spontaneous emission from the $(2p_z)$ state to the $(1s)$ state of the hydrogen atom.
8. Is the transition from a $(1s)$ state to a $(3d)$ state in the hydrogen atom allowed or forbidden? Why?
9. If we want to construct a laser, which two conditions must be satisfied?
10. How does the He–Ne laser operate?

144 LIGHT AND SPECTROSCOPY

Bibliography

The historical development of the various theories of light and radiation is treated very extensively in reference 1, which makes very interesting reading. The quantum theory of the radiation field is discussed in a number of books written for physicists, but we find that the majority of the modern books are very abstract and sophisticated and do not discuss any applications. The only exception is Heitler's book (2). My own book (3) was written specifically to deal with molecular applications of quantum electrodynamics. The classical theory of the radiation field is discussed in most physics textbooks, but we recommend particularly Feynman's book (4) because it gives such an excellent account of the subject. Time-dependent perturbation theory is discussed in (2) and (3) and in most of the quantum mechanics texts that we listed at the end of Chapter II. Finally, we list some books that deal with lasers.

1. E. Whittaker. *A History of the Theories of Aether and Electricity*. Harper & Row, New York (1960).
2. W. Heitler. *The Quantum Theory of Radiation*. Oxford University Press, London (1957).
3. H. F. Hameka. *Advanced Quantum Chemistry*. Addison-Wesley, Reading, Mass. (1965).
4. R. P. Feynman, R. B. Leighton, and M. Sands. *The Feynman Lectures on Physics*. Addison-Wesley, Reading, Mass. (1963).
5. B. A. Lengyel. *Introduction to Laser Physics*. Wiley, New York (1966).
6. A. L. Bloom. *Gas Lasers*. Wiley, New York (1968).

CHAPTER 5

THE SPECTRA OF DIATOMIC MOLECULES

1. Experimental Information

The bulk of the available experimental information on diatomic molecules has been derived from the molecular spectra. These are not just the spectra in the visible and ultraviolet region (with wavelengths ranging between 2,000 Å and 7,000 Å and wave numbers between 15,000 cm^{-1} and 50,000 cm^{-1}) but also the spectra in the far infrared region (wave numbers between 100 cm^{-1} and 5,000 cm^{-1}) and in the microwave region (wave numbers between 0.1 cm^{-1} and 100 cm^{-1}). It is necessary to consider all these spectra, because a diatomic molecule has a very large number of energy levels and the spectra in the visible region do not always give us enough information to identify and classify all the energy levels.

Naturally, the classification of the energy levels was originally derived from the experimental data, but in our discussion we shall reverse the historical order of events. We feel that it is easier if we outline first some of the general features of molecular spectra and then use these to discuss some of the experimental resutls.

It has been well established that the energy eigenvalues of a diatomic molecule can be represented as a sum of three terms,

$$E_{\text{mol}} = E_{\text{el}} + E_{\text{vib}} + E_{\text{rot}} \tag{1}$$

The first term is an energy eigenvalue of the electronic motion for a specific fixed, nuclear configuration, the second term is the vibrational energy where the two nuclei move relative to one another so that the internuclear distance R oscillates around an equilibrium distance R_0, and the third term represents the rotational motion of the whole molecule. It should be noted that Eq. (1) is only an approximation, but it is a fairly good approximation and agrees quite well with the experiments.

The rotational energy levels may be obtained by assuming that the molecule behaves like a rigid rotor with a distance R, which is equal to the internuclear distance, and with a mass μ_{AB}, which is the reduced mass of the two nuclear masses M_A and M_B,

$$\frac{1}{\mu_{AB}} = \frac{1}{M_A} + \frac{1}{M_B} \tag{2}$$

We have seen in Chapter II that the eigenvalues of the rigid rotor are given by

$$E_J = \frac{\hbar^2}{2\mu R^2} J(J+1) \tag{3}$$

Consequently, we represent the rotational energy levels of a diatomic molecule as

$$E_J = BJ(J+1) \tag{4}$$

where B is known as the rotational constant. It should be equal to

$$B = \frac{\hbar^2}{2\mu_{AB} R^2} \tag{5}$$

In Table 5.1 we list some experimental values for rotational constants. It should be noted that the energy differences between the rotational levels fall in the microwave region.

The vibrational energy levels are roughly approximated by the energy levels of a harmonic oscillator,

$$E_v = (v + \tfrac{1}{2})h\nu \tag{6}$$

This approximation is quite satisfactory for the lower vibrational levels, but it becomes less satisfactory for the higher states. In Table 5.2 we list the values of ν for the electronic ground states of some molecules. It may be seen that these energies correspond to photons in the far infrared.

Finally, the electronic eigenvalues, which we denote by ϵ_n, have the same order of magnitude as atomic energy eigenvalues. The differences between different electronic energies ϵ_n and ϵ_m are of the order of 1 to 10 eV, and the corresponding spectral transitions range from the infrared to the far ultraviolet region.

EXPERIMENTAL INFORMATION   147

TABLE 5.1 ● Rotational constants $B_{0,0}$ for the electronic and vibrational ground states ($n = 0, v = 0$) of some diatomic molecules (in terms of cm^{-1}). We also list the equilibrium distances R_0 for those states (in terms of Å).

Molecule	$B_{0,0}$	R_0
Cl$_2$	0.2348	1.988
CO	1.9314	1.1282
H$_2$	60.809	0.7417
HD	45.655	0.7414
D$_2$	30.429	0.7416
HBr	8.473	1.414
HCl	5.445	1.275
HF	20.939	0.9171
HI	6.551	1.604
I$_2$	0.03736	2.667
ICl	0.11416	2.3207
Li$_2$	0.6727	2.6725
LiH	7.5131	1.595
N$_2$	2.010	1.094
O$_2$	1.4457	1.2074
S$_2$	0.2956	1.889

It follows that each stationary state of a diatomic molecule may be represented by a set of three quantum numbers, n, v, and J. The first quantum number, n, refers to the electronic motion; the second quantum number, v, describes the vibrational state; and the third quantum number, J, represents the rotational state. We should realize that the vibrational frequency depends on the electronic structure of the molecule and, consequently, on the quantum number n. The same is true for the rotational constant B, because the equilibrium distance R is usually different for different values of n. This means that we ought to write the vibrational and rotational energies as

$$E_v^{(n)} = (v + \tfrac{1}{2})h\nu_n$$
$$E_J^{(n)} = B_n J(J + 1) \quad (7)$$

The total energy $E_{n,v,J}$ of the molecular eigenstate (n, v, J) is then approximated as

$$E_{n,v,J} = \epsilon_n + E_v^{(n)} + E_J^{(n)}$$
$$= \epsilon_n + (v + \tfrac{1}{2})h\nu_n + B_n J(J + 1) \quad (8)$$

We shall use this expression for discussing the experimental spectra.

148 THE SPECTRA OF DIATOMIC MOLECULES

TABLE 5.2 • Vibrational frequencies for the electronic ground states of some diatomic molecules, obtained as the energy differences $\nu_{0,1}$ between the lowest and the first excited vibrational levels (in terms of cm^{-1}).

Molecule	$\nu_{0,1}$
Cl$_2$	564.9
CO	2,170.2
H$_2$	4,395.2
HD	3,817.1
D$_2$	3,118.5
HBr	2,649.7
HCl	2,090.8
HF	4,138.5
HI	2,309.5
I$_2$	214.6
ICl	384.2
Li$_2$	351.4
LiH	1,405.6
N$_2$	2,359.6
O$_2$	1,580.4
S$_2$	725.7

A spectroscopic transition between two stationary states (n, v, J) and (n', v', J') has a frequency ν that is given by

$$h\nu = E_{n',v',J'} - E_{n,v,J}$$
$$= \epsilon_{n'} - \epsilon_n + h\nu_{n'}(v' + \tfrac{1}{2}) - h\nu_n(v + \tfrac{1}{2}) + B_{n'}J'(J' + 1) - B_n J(J + 1) \quad (9)$$

if the state (n', v', J') has the higher energy. If the quantum numbers n' and n are different, then we observe a transition between different electronic states and we expect the corresponding light to be in the visible or ultraviolet region. We should realize that Eq. (9) describes a very large number of transitions, because the quantum numbers v and v' and the quantum numbers J and J' may have a large number of possible values. If n and n' have specific values then we observe a specific electronic transition, but this transition contains a large number of different vibrational and rotational transitions with frequencies that are very close together. In many instances the various lines are so close together that they overlap, and we observe a continuous area of absorption. The resulting spectrum is known as a band spectrum. An area of continuous absorption, extending over 100 to 1,000 cm^{-1}, is called an absorption or emission band. We mentioned in Chapter III that an atomic spectrum is usually a line spectrum because it consists of discrete lines.

It may be pertinent to discuss briefly the experimental width of spectral lines. Ordinarily they are determined by experimental factors, namely the width of the slit in the spectrograph. In the visible region the widths are of the order of about 5 cm^{-1}. If we use more sophisticated equipment—for example, a grating instead of a spectrograph—then the widths of the spectral lines are smaller and they are determined by the collisions between the molecules, the Doppler effect, and collisions with the wall; they are of the order of 0.1 to 1 cm^{-1}. It may be shown that even in ideal circumstances a spectral line of one isolated molecule has a finite line width. This is known as the natural line width and it is of the order of 0.001 to 0.01 cm^{-1}. We should emphasize that all the above numbers refer to transitions in the visible; in other spectral regions the orders of magnitude may be different.

It is now obvious that for heavy molecules, where the values of B are small, it is not possible to detect the rotational structure in the electronic spectra. For lighter molecules, where B is larger, it is in principle possible to find the rotational fine structure in the electronic absorption bands if the experiments are done carefully so that the spectrograph or the grating has a high enough resolution.

The molecular spectra in the far infrared are usually vibration spectra, in which case we observe transitions between different vibrational levels within the same electronic state. Theoretically, if the vibrational motion is exactly described by a harmonic oscillator, we should see only one absorption line at the vibrational frequency ν_n. In practice, we observe more than one line, because the vibrational motion is not exactly harmonic. The additional lines have frequencies $2\nu_n$, $3\nu_n$, and so on. Again, there is rotational fine structure in the spectrum. This is due to the fact that the rotational constant B_n depends slightly on the vibrational quantum number, so that the rotational constants for different vibrational levels are slightly different.

Transitions between different rotational levels within the same electronic and vibrational states fall in the microwave region, as may be seen from Table 5.1. The microwave spectra give us quite accurate values of the rotational constants B and also quite accurate values of the internuclear distance R. As a result, these distances are quite well known for diatomic molecules.

In the following section we wish to discuss how the approximations we have sketched above can be understood from theoretical arguments. We should mention once again that the validity of the approximations can be justified from the experimental results, because the use of the approximations leads to a very satisfactory interpretation of most of the experimental results. Hence, we do not have to prove the validity of Eq. (1); rather, we wish to explain how it is compatible with the quantum mechanical description of diatomic molecules.

2. The Born-Oppenheimer Approximation

We mentioned in the previous section that the energy levels of a diatomic molecule may be written as the sum of an electronic energy, a vibrational energy, and a rotational energy. Strictly speaking, there is a fourth contribution to the total molecular energy, namely the kinetic energy of the overall translational motion of the total molecule. However, this translational motion does not play any role in the molecular spectrum, and we shall not pay any attention to it, although it should be remembered that this motion is always present and that it is important in the kinetic theory of gases.

It may be recalled that the molecular energy levels E_{mol} were written in the form

$$E_{mol} = E_{el} + E_{vib} + E_{rot} + E_{tr} \tag{10}$$

For completeness, we have added the fourth energy contribution, E_{tr}, here, even though we are not particularly interested in this term.

The molecular wave function Ψ may be written as a product of four factors,

$$\Psi = \Psi_{el}\Psi_{vib}\Psi_{rot}\Psi_{tr} \tag{11}$$

The first factor represents the electronic motion, the second the vibrational motion, the third the rotational motion, and the fourth the translational motion of the molecule.

Equations (10) and (11) imply that the overall motion of the nuclei and the electrons in a molecule may be separated into four separate motions, namely the electronic motion, the vibrational motion, the rotational motion, and the translational motion. It is helpful to give the mathematical analysis of this separation from a quantum mechanical viewpoint, not so much because we wish to give a rigorous proof but because this mathematical analysis gives us some insight into the behavior of the wave functions associated with each type of motion.

We first prove a theorem in quantum mechanics that deals with the separability of the Hamiltonian. Let us consider a system described by a Hamiltonian $\mathcal{H}(X, Y)$, which depends on two sets of coordinates X and Y and which may be separated into two parts:

$$\mathcal{H}(X, Y) = \mathcal{H}_1(X) + \mathcal{H}_2(Y) \tag{12}$$

The first term depends only on the coordinates X and the second term depends only on the coordinates Y. We assume that the eigenvalues and eigenfunctions of the two Hamiltonians $\mathcal{H}_1$ and $\mathcal{H}_2$ are given by

$$\mathcal{H}_1(X)\Phi_n(X) = \lambda_n \Phi_n(X)$$
$$\mathcal{H}_2(Y)\chi_m(Y) = \omega_m \chi_m(Y) \tag{13}$$

It is now easily proved that the eigenvalues $E_{n,m}$ and eigenfunctions $\Psi_{n,m}(X, Y)$ of $\mathcal{H}$ are given by

$$E_{n,m} = \lambda_n + \omega_m$$
$$\Psi_{n,m}(X, Y) = \Phi_n(X)\chi_m(Y) \tag{14}$$

We have

$$\mathcal{H}(X, Y)\Psi_{n,m}(X, Y) = \mathcal{H}_1(X)\Phi_n(X)\chi_m(Y) + \mathcal{H}_2(Y)\Phi_n(X)\chi_m(Y)$$
$$= (\lambda_n + \omega_m)\Phi_n(X)\chi_m(Y) = E_{n,m}\Psi_{n,m}(X, Y) \tag{15}$$

We can use this theorem to separate the translational and the rotational motion of the molecule, but the separation of the vibrational and the electronic motion is more complicated. It involves an approximation known as the Born-Oppenheimer approximation.

The total Hamiltonian of a diatomic molecule AB is

$$\mathcal{H}_{AB} = T_A + T_B + \frac{e^2 Z_A Z_B}{R_{AB}} + \mathcal{H}_{el} \tag{16}$$

Here T_A and T_B are the kinetic energies of the two nuclei,

$$T_A = -\frac{\hbar^2}{2M_A}\left(\frac{\partial^2}{\partial X_A^2} + \frac{\partial^2}{\partial Y_A^2} + \frac{\partial^2}{\partial Z_A^2}\right)$$
$$T_B = -\frac{\hbar^2}{2M_B}\left(\frac{\partial^2}{\partial X_B^2} + \frac{\partial^2}{\partial Y_B^2} + \frac{\partial^2}{\partial Z_B^2}\right) \tag{17}$$

The next term in Eq. (16) represents the Coulomb repulsion between the two nuclei, and the last term is the electronic Hamiltonian, which is given by

$$\mathcal{H}_{el} = -\frac{\hbar^2}{2m}\sum_j\left(\frac{\partial^2}{\partial x_j^2} + \frac{\partial^2}{\partial y_j^2} + \frac{\partial^2}{\partial z_j^2}\right) - \sum_j \frac{Z_A e^2}{r_{Aj}} - \sum_j \frac{Z_B e^2}{r_{Bj}} + \sum_{j>k} \frac{e^2}{r_{jk}} \tag{18}$$

It may be seen that the molecular Hamiltonian as we have written it depends on six nuclear coordinates and $3N$ electron coordinates if the molecule contains N electrons. We should realize that the translational motion of the molecule depends on the coordinates of the center of gravity of the molecule, the rotational motion depends on the two angles that describe the orientation of the molecular axis, and the vibrational motion

152   THE SPECTRA OF DIATOMIC MOLECULES

depends on the internuclear distance $R = R_{AB}$. Therefore, if we want to identify these various motions, we must make the corresponding coordinate transformations.

First we introduce the coordinates X_c, Y_c, Z_c of the center of gravity of the two nuclei and the coordinates X, Y, Z, which describe the relative orientations of the two nuclei,

$$X_c = \frac{M_A X_A + M_B X_B}{M_A + M_B} \qquad X = X_B - X_A$$

$$Y_c = \frac{M_A Y_A + M_B Y_B}{M_A + M_B} \qquad Y = Y_B - Y_A \tag{19}$$

$$Z_c = \frac{M_A Z_A + M_B Z_B}{M_A + M_B} \qquad Z = Z_B - Z_A$$

It follows that

$$\frac{1}{M_A}\frac{\partial^2}{\partial X_A^2} = \frac{1}{M_A}\frac{\partial^2}{\partial X^2} - \frac{2}{M_A + M_B}\frac{\partial^2}{\partial X \, \partial X_c} + \frac{M_A}{(M_A + M_B)^2}\frac{\partial^2}{\partial X_c^2} \tag{20}$$

$$\frac{1}{M_B}\frac{\partial^2}{\partial X_B^2} = \frac{1}{M_B}\frac{\partial^2}{\partial X^2} + \frac{2}{M_A + M_B}\frac{\partial^2}{\partial X \, \partial X_c} + \frac{M_B}{(M_A + M_B)^2}\frac{\partial^2}{\partial X_c^2}$$

Consequently,

$$T_A + T_B = -\frac{\hbar^2}{\mu}\left(\frac{\partial^2}{\partial X^2} + \frac{\partial^2}{\partial Y^2} + \frac{\partial^2}{\partial Z^2}\right) - \frac{\hbar^2}{M_A + M_B}\left(\frac{\partial^2}{2X_c^2} + \frac{\partial^2}{\partial Y_c^2} + \frac{\partial^2}{\partial Z_c^2}\right) \tag{21}$$

The second term represents the kinetic energy of the motion of the center of gravity with total mass $M_A + M_B$. We write this as $\mathcal{H}_{\text{trans}}$, the Hamiltonian for the overall translational motion of the whole molecule. The rest of the Hamiltonian, $\mathcal{H}_{\text{int}}$, does not depend on the coordinates of the center of gravity if we take this as the origin of the coordinates. We have thus

$$\mathcal{H}_{AB} = \mathcal{H}_{\text{int}} + \mathcal{H}_{\text{trans}} \tag{22}$$

We may separate the translational motion according to Eq. (14), and we need consider only the remaining Hamiltonian,

$$\mathcal{H}_{\text{int}} = -\frac{\hbar^2}{2\mu}\left(\frac{\partial^2}{\partial X^2} + \frac{\partial^2}{\partial Y^2} + \frac{\partial^2}{\partial Z^2}\right) + \frac{e^2 Z_A Z_B}{R} + \mathcal{H}_{\text{el}} \tag{23}$$

where μ is the reduced mass of the nuclei,

$$\frac{1}{\mu} = \frac{1}{M_A} + \frac{1}{M_B} \tag{24}$$

and R is the internuclear distance,

$$R = (X^2 + Y^2 + Z^2)^{1/2} \tag{25}$$

Let us now proceed to the separation of the electronic motion and the nuclear motion. Since the origin of the electronic coordinates is taken as the nuclear center of gravity, the vector **R** determines the orientation of the nuclear axis. We now take the nuclear axis as one of the coordinate axes, namely the Z axis, for the electron coordinates. The electronic Hamiltonian of Eq. (18) contains only the internuclear distance R as a parameter, and we can take it as representing the electronic motion on the assumption that the position of the nuclei is fixed. We assume now that the electronic energies E_n are eigenfunctions of the electronic Hamiltonian $\mathcal{H}_{el}$. We write this as

$$\mathcal{H}_{el}(r, R) F_n(r, R) = E_n(R) F_n(r; R) \tag{26}$$

Here r represents symbolically all the electron coordinates. Since the electronic Hamiltonian depends indirectly on the internuclear distance R, the eigenfunctions $F_n(r, R)$ must also depend on R, in addition to r, and the eigenvalues $E_n(R)$ also contain R as a parameter.

The separation of the electronic and nuclear motion depends on two approximations, which are both known as the Born-Oppenheimer approximation. This approximation cannot be proved rigorously but it is known to be quite satisfactory because its results agree quite well with the experimental facts. The first part of the Born-Oppenheimer approximation states that if a molecule is in a given electronic stationary state n, then the molecular eigenfunction Ψ_n of the Hamiltonian $\mathcal{H}_{int}$ of Eq. (23) can be written in the form

$$\Psi_n(r, X, Y, Z) \approx f_n(X, Y, Z) F_n(r; R) \tag{27}$$

This approximation may be understood by observing that any eigenfunction Ψ of the operator $\mathcal{H}_{int}$ may be expanded in terms of the eigenfunctions F_n because the latter form a complete set. Therefore, the exact eigenfunction Ψ may be written as

$$\Psi(r, X, Y, Z) = \sum_n g_n(X, Y, Z) F_n(r; R) \tag{28}$$

If the separation between the rotational and vibrational energy levels is much smaller than the separation between the electronic energy levels, as is usually the case, then each eigenfunction Ψ may be identified with a particular electronic state n. In that case the function g_n is much larger than any of the other functions g_k, so that Ψ_n may be accurately described by means of the approximation (27).

In order to discuss the second part of the Born-Oppenheimer approximation, we consider the effect of the total Hamiltonian $\mathcal{H}_{int}$ of Eq. (23) on the function Ψ_n of Eq. (27). We have

$$\mathcal{H}_{int}\Psi_n = -\frac{\hbar^2}{2\mu}\left(\frac{\partial^2}{\partial X^2} + \frac{\partial^2}{\partial Y^2} + \frac{\partial^2}{\partial Z^2}\right)[f_n(X, Y, Z)F_n(r; R)]$$

$$+ \left(\frac{e^2 Z_A Z_B}{R} + \mathcal{H}_{el}\right)[f_n(X, Y, Z)F_n(r; R)]$$

$$= -\frac{\hbar^2}{2\mu}\left(\frac{\partial^2}{\partial X^2} + \frac{\partial^2}{\partial Y^2} + \frac{\partial^2}{\partial Z^2}\right)[f_n(X, Y, Z)F_n(r; R)]$$

$$+ \left[\frac{e^2 Z_A Z_B}{R} + E_n(R)\right][f_n(X, Y, Z)F_n(r; R)] \tag{29}$$

It is generally accepted that the vibrational motion of a molecule extends over a relatively small area, of the order of 0.1 Å or less (we calculated this in Chapter II, Section 3). The electronic wave function F_n varies only slightly over this range, and it seems reasonable to assume that the derivative of F_n with respect to an arbitrary nuclear coordinate X is much smaller than the corresponding derivative of f_n,

$$\frac{\partial F_n}{\partial X} \ll \frac{\partial f_n}{\partial X} \qquad \frac{\partial^2 F_n}{\partial X^2} \ll \frac{\partial^2 f_n}{\partial X^2} \tag{30}$$

If we use this approximation, then the various derivatives in Eq. (29) may be approximated as

$$\frac{\partial^2}{\partial X^2}[f_n(X, Y, Z)F_n(r; R)] \approx F_n(r; R)\frac{\partial^2 f_n}{\partial X^2} \tag{31}$$

and Eq. (29) reduces to

$$\mathcal{H}_{int}[F_n(r; R)f_n(X, Y, Z)]$$

$$= -\frac{\hbar^2}{2\mu}F_n(r; R)\left(\frac{\partial^2}{\partial X^2} + \frac{\partial^2}{\partial Y^2} + \frac{\partial^2}{\partial Z^2}\right)f_n(X, Y, Z) + F_n(r; R)$$

$$\times \left[\frac{e^2 Z_A Z_B}{R} + E_n(R)\right]f_n(X, Y, Z)$$

$$= F_n(r; R)\left[-\frac{\hbar^2}{2\mu}\left(\frac{\partial^2}{\partial X^2} + \frac{\partial^2}{\partial Y^2} + \frac{\partial^2}{\partial Z^2}\right) + \frac{e^2 Z_A Z_B}{R} + E_n(R)\right]f_n(X, Y, Z) \tag{32}$$

It is obvious that the Schrödinger equation

$$\mathcal{H}_{int}[F_n(r; R)f_n(X, Y, Z)] = EF_n(r; R)f_n(X, Y, Z) \qquad (33)$$

may be divided by the function F_n if we substitute the result (32). We obtain, a Schrödinger equation for the nuclear function f_n only, namely

$$-\frac{\hbar^2}{2\mu}\left(\frac{\partial^2}{\partial X^2} + \frac{\partial^2}{\partial Y^2} + \frac{\partial^2}{\partial Z^2}\right)f_n(X, Y, Z) + \left[\frac{e^2 Z_A Z_B}{R} + E_n(R)\right]f_n(X, Y, Z)$$
$$= Ef_n(X, Y, Z) \qquad (34)$$

We have achieved our goal in separating the electronic and nuclear motions by making use of the Born-Oppenheimer approximation. The electronic motion is described by the Schrödinger equation (26), and the nuclear motion is described by the Schrödinger equation (34).

The potential energy in Eq. (34) is a sum of two terms. The first is the Coulomb repulsion of the two nuclei, and the second is the electronic energy of the eigenstate n as a function of the internuclear distance R. It is customary to combine the two terms and to introduce a function $U(R)$ as

$$U_n(R) = \frac{e^2 Z_A Z_B}{R} + E_n(R) \qquad (35)$$

This function is known as the potential curve of the electronic state n; it is a function of the internuclear distance R only. We discuss its behavior in the following section.

Finally, we consider the separation between the rotational and the vibrational motions of the molecule. If we substitute the expression (35) for the potential curve into the Schrödinger equation (34), we obtain

$$\left[-\frac{\hbar^2}{2\mu}\Delta + U_n(R)\right]f(X, Y, Z) = Ef(X, Y, Z) \qquad (36)$$

This equation is similar to the equation for the hydrogen atom that we considered in the previous chapter. Both equations describe the motion of a particle in a central force field where the potential depends only on the variable R and not on the polar angles θ, φ. We have seen that in this situation the angular momentum is a constant of the motion, and that the eigenfunctions in the stationary states must be eigenfunctions of both the Hamiltonian operator and of the angular momentum operator $(M^2)_{op}$. According to Chapter II, Sections 5 and 6, the eigenfunctions of Eq. (36) may thus be written in the form

$$f(X, Y, Z) = u(R)F_J(X, Y, Z) \qquad (37)$$

where $u(R)$ is a function of the variable R only and $F_J(X, Y, Z)$ is an Euler polynomial of the Jth degree, which is an eigenfunction of $(M^2)_{\text{op}}$.

In the present case we wish to separate the vibrational motion, which depends on the variable R, and the rotational motion, which depends on the polar angles θ and φ. We note that the rotational eigenfunctions should depend only on the two polar angles. We have seen in Chapter II, Section 4, that the rotational eigenfunctions should therefore have the form

$$\psi_J(\theta, \varphi) = R^{-J} F_J(X, Y, Z) \tag{38}$$

Accordingly, we write the eigenfunctions of Eq. (36) in the form

$$f(X, Y, Z) = \chi(R)\psi_J(\theta, \varphi) \tag{39}$$

where the function χ represents the vibrational motion and the function Ψ_J represents the rotational motion.

Let us substitute this expression into Eq. (36) and find out what the equations are for the rotational and the vibrational motion. We have discussed the rigid rotor in Chapter II, Section 4, and we found that the function ψ_J satisfies the equation

$$\Delta \psi_J(\theta, \varphi) = -J(J+1) R^{-2} \psi_J(\theta, \varphi) \tag{40}$$

or

$$-\frac{\hbar^2}{2\mu} \Delta \psi_J(\theta, \varphi) = \frac{J(J+1)\hbar^2}{2\mu R^2} \psi_J(\theta, \varphi) \tag{41}$$

We see that Eq. (41) defines the rotational eigenvalues and eigenfunctions.

Before substituting Eq. (39) into Eq. (36), we note first that

$$\left(X \frac{\partial}{\partial X} + Y \frac{\partial}{\partial Y} + Z \frac{\partial}{\partial Z}\right)\psi_J = F_J\left(X \frac{\partial}{\partial X} + Y \frac{\partial}{\partial Y} + Z \frac{\partial}{\partial Z}\right)\left(\frac{1}{R^J}\right)$$

$$+ \frac{1}{R^J}\left(X \frac{\partial}{\partial X} + Y \frac{\partial}{\partial Y} + Z \frac{\partial}{\partial Z}\right)F_J$$

$$= F_J\left(\frac{X^2}{R} + \frac{Y^2}{R} + \frac{Z^2}{R}\right)\left(-\frac{J}{R^{J+1}}\right) + \frac{JF_J}{R^J} = 0 \tag{42}$$

By making use of Eq. (42) we find now that

THE BORN-OPPENHEIMER APPROXIMATION 157

$$\left(\frac{\partial^2}{\partial X^2} + \frac{\partial^2}{\partial Y^2} + \frac{\partial^2}{\partial Z^2}\right)\chi(R)\psi_J = \psi_J\left(\frac{\partial^2}{\partial X^2} + \frac{\partial^2}{\partial Y^2} + \frac{\partial^2}{\partial Z^2}\right)\chi(R)$$

$$+ \chi(R)\left(\frac{\partial^2}{\partial X^2} + \frac{\partial^2}{\partial Y^2} + \frac{\partial^2}{\partial Z^2}\right)\psi_J$$

$$= \psi_J\left(\frac{\partial^2\chi}{\partial R^2} + \frac{2}{R}\frac{\partial\chi}{\partial R}\right) - \frac{J(J+1)}{R^2}\chi(R)\psi_J$$

(43)

If we substitute this into Eq. (36), we obtain the following equation for the vibrational motion:

$$-\frac{\hbar^2}{2\mu}\left(\frac{d^2\chi}{dR^2} + \frac{2}{R}\frac{d\chi}{dR} - \frac{J(J+1)}{R^2}\chi\right) + U_n(R)\chi = E\chi \qquad (44)$$

It may be useful to summarize the results of this section and to illustrate the conclusions from a classical viewpoint. We made use of the Born-Oppenheimer approximation to separate the nuclear and the electronic motions. The result is that the electronic motion is described by the Schrödinger equation (26), where it is assumed that the nuclei are at rest at a fixed nuclear separation. This means that the electronic cordinate system is attached to the nuclear framework with the electronic Z axis as the molecular axis and the origin as the molecular center of gravity.

The nuclear motion may be separated into three motions, namely the overall translational motion of the molecule, the rotational motion, and the vibrational motion. These motions are described by three different Schrödinger equations,

$$\mathcal{H}_{trans}\Phi_{trans} = E_{trans}\Phi_{trans}$$
$$\mathcal{H}_{rot}\Phi_{rot} = E_{rot}\Phi_{rot} \qquad (45)$$
$$\mathcal{H}_{vib}\Phi_{vib} = E_{vib}\Phi_{vib}$$

We have sketched in Fig. 5.1 how in the classical approach the total motion of the nuclei may be decomposed into the above three motions. The first (Fig. 5.1a) is the translational motion of the total molecule, which may be described by the motion of a particle with mass $M_A + M_B$ located at the center of gravity of the molecule. This motion is represented by the Hamiltonian of Eq. (21). The internal motion of the nuclei is characterized by the reduced mass μ, defined by Eq. (24), and it may be decomposed into the rotational motion of the molecule (Fig. 5.1b), which is described by the orientation of the nuclear axis and into the vibrational motion (Fig. 5.1c), which corresponds to changes in the internuclear distance R. The rotational motion of the molecule is exactly described by the rigid

158 THE SPECTRA OF DIATOMIC MOLECULES

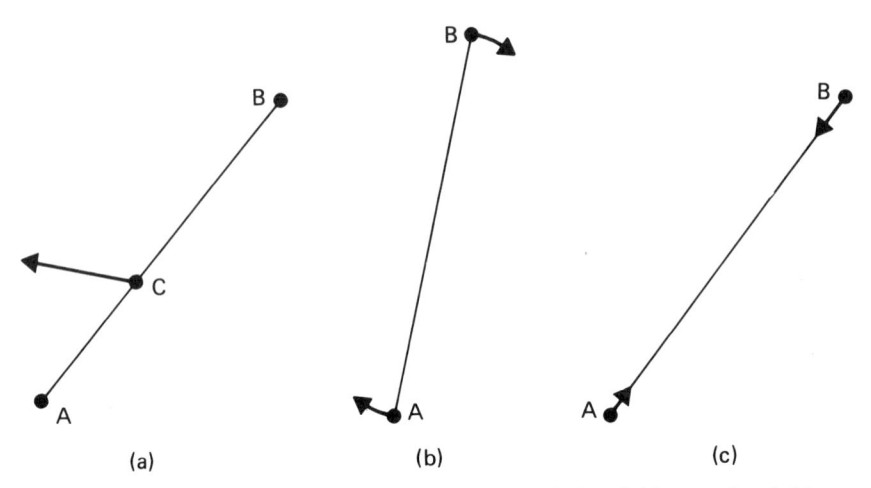

FIGURE 5.1 ● Schematic presentation of the translational (a), rotational (b), and vibrational (c) motion of a molecule.

rotor model. The vibrational motion is described by the Schrödinger equation (44), and we must know something about the behavior of the potential curves $U_n(R)$ before we can discuss its mathematical description. We do this in the following section.

3. The Vibrational Motion of a Diatomic Molecule

In the previous section we derived the Schrödinger equation for the vibrational motion of a diatomic molecule, namely Eq. (44). Clearly, we must know the potential curve $U_n(R)$ that describes this vibrational motion if we wish to solve the vibrational Schrödinger equation. The form of this curve depends on the electronic state the molecule is in. It can in principle be derived theoretically by solving the electronic Schrödinger equation and by determining the electronic eigenvalue $E_n(R)$ as a function of the internuclear distance R.

In general we do not know the detailed form of the potential curve, and it cannot be calculated theoretically. However, we can make fairly good predictions about the general behavior of the curve even though we do not know its exact form. First we note that every potential curve must tend to infinity when $R \to 0$, because in that case the Coulomb repulsion between the two nuclei tends to infinity. Also, the curve tends asymptotically to a constant value when R tends to infinity; this value is the sum of the energies of the isolated atoms. In Fig. 5.2 we show some typical potential curves for a diatomic molecule AB. If the molecule is in an electronic state where it has a stable configuration, then there must be a value R_0 for which

THE VIBRATIONAL MOTION OF A DIATOMIC MOLECULE 159

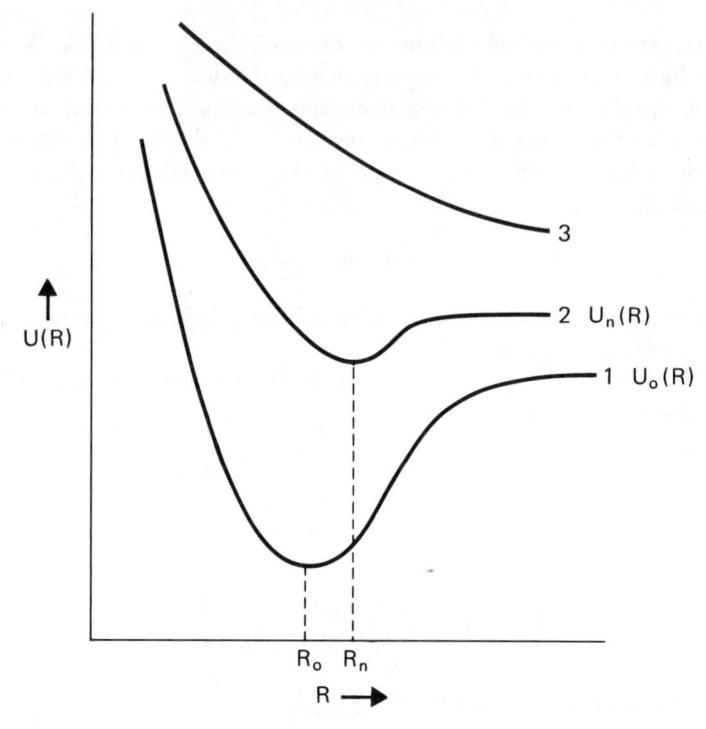

FIGURE 5.2 • Potential curves of a molecule AB.

the potential curve has a minimum. In Fig. 5.2 the potential curve 1 corresponds to the electronic ground state, and here the curve has a minimum so the molecule has a stable configuration with equilibrium distance R_0. Curve 2 corresponds to an excited electronic state, and the molecule also has a stable configuration in this excited state because the potential curve has again a minimum. The minimum of curve 2 is at the point R_n, and this is the equilibrium configuration of the molecule in this excited electronic state. In general, the equilibrium configuration changes going from one electronic state to another. Finally, potential curve 3 does not have a minimum. In this situation there is no value of R for which the molecule has a stable configuration, and the molecule will dissociate.

Let us first consider the case of an electronic ground state described by potential curve 1. Here it is customary to introduce the displacement coordinate q to describe the vibrational motion,

$$q = R - R_0 \qquad R = R_0 + q \qquad (46)$$

We should realize that the vibrational motion extends over a relatively small range; the average displacement from equilibrium is of the order of

0.05 Å, and the equilibrium distance R_0 ranges between 1 and 2 Å. We have seen in Chapter II that this information may be derived from the experimental data, R_0 may be derived from the rotational constant B of the molecule, and the range of q values may be derived from the vibrational frequency. Whatever the exact shape of the potential curve is, we may always assume that

$$q \ll R_0 \tag{47}$$

and we can use this assumption to simplify the Schrödinger equation (44) for the vibrational motion.

First, we replace the wave function $\chi(R)$ by another function $\Psi(R)$ by making the substitution

$$\chi(R) = \frac{\Psi(R)}{R} \tag{48}$$

We have

$$\left(\frac{d^2}{dR^2} + \frac{2}{R}\frac{d}{dR}\right)\frac{\Psi(R)}{R} = \frac{1}{R}\frac{d^2\Psi}{dR^2} \tag{49}$$

Substitution into Eq. (44) gives

$$-\frac{\hbar^2}{2\mu}\frac{d^2\Psi}{dR^2} + \frac{\hbar^2 J(J+1)}{2\mu R^2}\Psi + U_0(R)\Psi = E\Psi \tag{50}$$

Let us now imagine that the potential curve $U_0(R)$ has a minimum at the point R_0. We then make the substitution

$$V(q) = U(R) - U(R_0) \tag{51}$$

and we also introduce the displacement coordinate

$$q = R - R_0 \tag{52}$$

The Schrödinger equation (50) then becomes

$$-\frac{\hbar^2}{2\mu}\frac{d^2\Psi}{dq^2} + V(q)\Psi = \left(E - U_0(R_0) - \frac{\hbar^2 J(J+1)}{2\mu R^2}\right)\Psi \tag{53}$$

Since q is much smaller than R, we may approximate the rotational energy as

$$\frac{\hbar^2 J(J+1)}{2\mu R^2} \approx \frac{\hbar^2 J(J+1)}{2\mu R_0^2} \tag{54}$$

so that it becomes a constant term, independent of q. Both terms on the right side of Eq. (53) are then constants; the only effect they have is to change the zero point of the energy, and we may make the substitution

$$E - U_0(R_0) - \frac{\hbar^2 J(J+1)}{2\mu R^2} = \epsilon \tag{55}$$

The final form of the vibrational Schrödinger equation is thus

$$-\frac{\hbar^2}{2\mu}\frac{d^2\Psi}{dq^2} + V(q)\Psi = \epsilon\Psi \tag{56}$$

This begins to resemble the Schrödinger equation for the harmonic oscillator, which we discussed in Chapter II.

It is customary to approximate the potential curve $U(R)$ or $V(q)$ in order to find solutions to the Schrödinger equation (56). The crudest approximation is the harmonic approximation, where we approximate $V(q)$ by a parabola; we have sketched this in Fig. 5.3. It may be seen that this is not a bad approximation in the vicinity of the equilibrium distance R_0, but of course the parabola has the wrong asymptotic behavior and for larger values of R the harmonic approximation is none too satisfactory.

We can justify the harmonic approximation by expanding the function $U(R)$ in a power series of $R - R_0$ around the point R_0:

$$U_0(R) = U_0(R_0) + \frac{1}{2}\left(\frac{\partial^2 U_0}{\partial R^2}\right)_0 (R - R_0)^2 + \frac{1}{6}\left(\frac{\partial^3 U_0}{\partial R^3}\right)_0 (R - R_0)^3 + \cdots \tag{57}$$

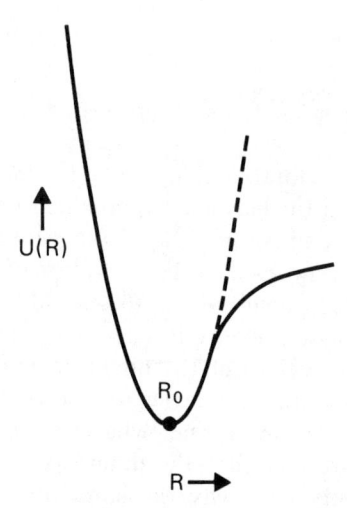

FIGURE 5.3 ● A potential function $U(R)$ approximated by a parabola near the minimum.

THE SPECTRA OF DIATOMIC MOLECULES

For small values of $R - R_0$ we may neglect the higher powers of $(R - R_0)$, and we take

$$U_0(R) = U_0(R_0) + \frac{1}{2}\left(\frac{\partial^2 U_0}{\partial R^2}\right)_{R=R_0}(R - R_0)^2 \tag{58}$$

Obviously the approximation breaks down if $(R - R_0)$ becomes too large. It follows from Eq. (51) that the function $V(q)$ becomes now

$$V(q) = \tfrac{1}{2}kq^2 \tag{59}$$

where the force constant k is given by

$$k = \left(\frac{\partial^2 U}{\partial R^2}\right)_{R=R_0} \tag{60}$$

If we substitute this into the Schrödinger equation (56), this reduces to

$$\left(-\frac{\hbar^2}{2\mu}\frac{d^2}{dq^2} + \tfrac{1}{2}kq^2\right)\Psi(q) = \epsilon\Psi(q) \tag{61}$$

which is exactly the equation for the harmonic oscillator.

We derived the eigenfunctions and eigenvalues of the harmonic oscillator in Chapter II, Section 2. It may be recalled that the eigenvalues are given by

$$E_v = (v + \tfrac{1}{2})\hbar\omega = (v + \tfrac{1}{2})\hbar\left(\frac{k}{\mu}\right)^{1/2} \quad v = 0, 1, 2, 3, \ldots \tag{62}$$

The corresponding eigenfunctions $\Psi_v(q)$ are defined and discussed in Chapter II, Section 2.

Before we consider the vibrational motion of the excited electronic states, we want to see how good the harmonic approximation is. We have already mentioned that the approximation of Eq. (59) is reasonable for small values of q, but that it is not very good for larger values of q. Consequently, the harmonic oscillator energy levels of Eq. (62) should be in good agreement with the lower experimental energy levels, but we expect deviations that increase with increasing values of the quantum number v. In Table 5.3 are listed the observed vibrational levels for the HCl molecule. It may be seen that these levels are lower than what we would predict from the harmonic approximation and that the differences become larger for the higher levels. It is easily seen why the harmonic approximation breaks down for the higher vibrational states: The harmonic oscillator has an infinite number of discrete energy levels, whereas in a diatomic molecule

THE VIBRATIONAL MOTION OF A DIATOMIC MOLECULE 163

TABLE 5.3 ● Vibrational energy levels of HCl in its electronic ground state. The first column represents the experimental levels, the second column the energy levels that are predicted from the harmonic approximation, and the third the differences. All energies are expressed in terms of cm^{-1}.

v	ν_{exp}	ν_{harm}	$\Delta\nu$
0	0	0	0
1	2,885.9	2,885.9	0
2	5,668.0	5,771.8	−103.8
3	8,347.0	8,657.7	−310.7
4	10,923.1	11,543.6	−620.5

we would expect a finite number of energy levels which are all below the dissociation energy of the molecule.

In order to give a more precise description of the higher vibrational levels, other potential energy functions have been proposed. The best known of these functions is the Morse potential,

$$V(q) = V_0(1 - e^{-aq})^2 \qquad q = R - R_0 \tag{63}$$

This function has at least the right asymptotic behavior. If $q \to \infty$ or $R \to \infty$, it approaches asymptotically the value V_0, which is equal to the dissociation energy of the molecule. Even though $V(q)$ does not become infinite for $R = 0$ or $q = -R_0$, it assumes a very large value and for all practical purposes it has the correct asymptotic behavior for $R \to 0$ and for $R \to \infty$. In addition, the function has a minimum at zero for $q = 0$. The force constant k at the minimum may be calculated from Eq. (60); it is

$$k = \left(\frac{\partial^2 V}{\partial q^2}\right)_0 = 2a^2 V_0 \tag{64}$$

The advantage of using the Morse potential is that the corresponding Schrödinger equation can be solved exactly if we substitute the potential function (63) into the vibrational equation (56). The mathematical analysis is somewhat complicated, and we give just the results for the energy levels:

$$\epsilon_v = (v + \tfrac{1}{2})\hbar\omega_0 - \delta(v + \tfrac{1}{2})^2 \hbar\omega_0 \tag{65}$$

Here ω_0 is given by

$$\omega_0 = \frac{a\sqrt{2V_0}}{\sqrt{\mu}} = \sqrt{\frac{k}{\mu}} \tag{66}$$

and the constant δ is

$$\delta = \frac{\hbar\omega}{4V_0} \qquad (67)$$

It may be seen that expression (65) for the energy levels agrees quantitatively with the trend in Table 5.3, where the higher energy levels are lower than would follow from the harmonic approximation. The Morse potential gives a finite number of discrete levels, determined by the condition

$$\epsilon_v < V_0 \qquad (68)$$

The experimental energy levels of the vibrational motion may be derived from the molecular spectra. It has been found that they can be described quite accurately by a power-series expansion in terms of the quantum number v,

$$E_v = \hbar\omega[(v + \tfrac{1}{2}) - x(v + \tfrac{1}{2})^2 + y(v + \tfrac{1}{2})^3 + \cdots] \qquad (69)$$

Even the Morse potential is not a very accurate representation of the potential curve, because its energy levels according to Eq. (65) do not agree too well with the experimental levels for higher values of v. A large number of other potentials have been proposed and calculated in order to get more precise agreement with the experimental data, but these potentials have fairly complex mathematical forms and we do not wish to discuss them here. However, for sufficiently small quantum numbers, all models, including the harmonic oscillator, give a fairly good representation of the lower vibrational levels. Since the harmonic approximation is so simple, we shall use it as a basis for discussing the vibrational motion of the excited electronic states.

In an excited electronic state n, the potential curve is different from the electronic ground state. We assume that we deal with a bound state where the potential curve has a minimum. The typical situation is represented by curve 2 of Fig. 5.2, where $U_n(R)$ is higher than the ground state curve $U_0(R)$ and where its minimum is located at a point R_n that is larger than the equilibrium distance R_0 for the ground state. If we use the harmonic approximation, then the function $U_n(R)$ may be approximated as

$$U_n(R) = U_n(R_n) + \tfrac{1}{2}k_n(R - R_n)^2 \qquad (70)$$

in the vicinity of the minimum. It is important to note that this curve has a different force constant and a different equilibrium distance as compared to the ground-state potential curve. In general, the equilibrium distance R_n for an excited state is larger than the equilibrium distance R_0 for the ground state by an amount that can vary between 0.1 Å and 0.5 Å. The force constant k_n is usually smaller than the force constant k_0 for the ground state. The difference is often a factor of two or more. The vibrational energy levels belonging to the electronic state n should thus be written as

$$\epsilon_v = (v + \tfrac{1}{2})\hbar\omega_n \qquad \omega_n = \left(\frac{k_n}{\mu}\right)^{1/2} \tag{71}$$

because the vibrational frequency depends on the force constant k_n. This frequency is thus different for a different electronic state. The vibrational eigenfunctions should also be written as $\Psi_v^n(q_n)$, because they also depend on the electronic state n. The displacement coordinate q_n is defined as $R - R_n$, and the excited-state harmonic oscillator is centered around a different point R_n than in the ground state. In addition, the unit of length of the harmonic oscillator functions depends on the force constant k_n, as we have seen in Chapter II, Section 2, so that the form of the vibrational eigenfunctions also depends on the electronic eigenstates.

Finally, it may be helpful to give the expressions for the total molecular energy and the total wave function. We are not interested in the overall translational motion of the molecule, so we shall omit it. Also, we use the harmonic approximation for the sake of simplicity. In Eq. (55) we related the total molecular energy E to the vibrational energy ϵ. It follows that

$$E(0, v, J) = U_0(R_0) + \frac{\hbar^2 J(J+1)}{2\mu R_0^2} + \left(v + \frac{1}{2}\right)\hbar\omega_0 \tag{72}$$

because Eq. (55) refers to the electronic ground state. We have substituted the equilibrium distance R_0 in the expression for the rotational energy. In the same way we obtain

$$E(n, v, j) = U_n(R_n) + \frac{\hbar^2 J(J+1)}{2\mu R_n^2} + \left(v + \frac{1}{2}\right)\hbar\omega_n \tag{73}$$

We have already mentioned that the vibrational frequency ω_n depends on the electronic state n. The rotational constant B also depends on n, because it contains R_n. It is less obvious why the rotational constant B also depends on the vibrational quantum number v, but we should realize that we make an approximation in substituting R_n for R in the rotational energy expression and the expectation value of $(1/R^2)$ that we should actually substitute instead depends on the quantum number v. Therefore, we should write Eq. (73) in the form

$$E(n, v, J) = U_n(R_n) + (v + \tfrac{1}{2})\hbar\omega_n + B_{n,v} J(J+1) \tag{74}$$

as the sum of an electronic, a vibrational, and a rotational energy.

The form of the molecular wave function follows from Eq. (26), which describes the separation of the electronic and nuclear motion, and from Eq. (39), which describes the separation of the vibrational and the rotational motion. We write it as

$$\Psi(n, v, J) = F_n(r; R_n)\Psi_v^n(q_n)\psi_J(\theta, \varphi) \tag{75}$$

The electronic wave function F_n is the eigenfunction of the electronic Hamiltonian for fixed positions of the nuclei and the equilibrium internuclear distance R_n. The vibrational wave function is an eigenfunction of the vibrational Schrödinger equation (56), and its form depends on the electronic quantum number n. In the harmonic approximation it reduces to a harmonic oscillator eigenfunction. The rotational wave function ψ_J is an eigenfunction of the rigid rotor.

It should be noted that we take the vibrational wave function as the function $\Psi(R)$ and not as the function $\chi = \Psi/R$ of Eq. (48). This is because the vibrational coordinate R was originally introduced as one of a set of polar coordinates (R, θ, φ). If we integrate over polar coordinates, then we must multiply by a factor R^2. This factor cancels out if we replace the function χ by the function Ψ, so that by using Ψ we may integrate over R as if it were a linear coordinate, consistent with the harmonic oscillator model.

4. Molecular Symmetry

In the previous section we considered only the frequencies of the various spectral lines. It is clear that additional information may be derived by considering the intensities as well. In general, it is not easy to calculate the absolute intensity of a given transition, but it is often possible to predict the relative intensities of a progression of lines within an absorption or emission band. The most useful tool for correlating spectroscopic intensities with molecular structure are the selection rules; these are rules that predict whether a given transition is allowed or forbidden. We have already mentioned that we define a forbidden transition as a transition for which the transition moment is zero.

Most selection rules are related to the symmetry properties of the eigenstates involved in the transition. We realize that a rigorous treatment of the symmetry properties of molecular eigenfunctions should be based on the branch of mathematics known as group theory. However, we feel that we can discuss the few theorems needed for our purpose without making use of group theory.

All diatomic molecules have cylindrical symmetry. In terms of mathematics, this means that the potential function for the electrons (or the electronic Hamiltonian) remains the same if we rotate the molecule around its axis. It may then be derived that if we rotate the molecule around its internuclear axis by an angle ϕ, then each electronic eigenfunction changes by an amount

$$e^{im\phi} \tag{76}$$

where m is an integer,

$$m = 0, \pm 1, \pm 2, \ldots \tag{77}$$

Each electronic eigenstate is thus characterized by a value of the quantum number m, and in describing the eigenstate the value of m is denoted by a symbol. For example, states with $m = 0$ are called Σ states, states with $m = \pm 1$ are Π states, states with $m = \pm 2$ are known as Δ states, and so on.

From a physical point of view, the projection M_z of the total electronic angular momentum is a constant of the motion, and its magnitude is described by the quantum number m. The magnitude of M_z is usually denoted by the symbol Λ. It is quantized, and it can have the values

$$\Lambda = 0, \hbar, 2\hbar, \ldots \tag{78}$$

In a Σ state $\Lambda = 0$, in a Π state $\Lambda = 1$, and so on. This nomenclature is similar to the hydrogen atom, where the letters s, p, d, and so on, were used to denote the magnitude of the angular momentum.

A homonuclear diatomic molecule A_2 has, in addition to cylindrical symmetry, a center of inversion, that is, the center of gravity or the midpoint of the internuclear axis. As a result, it may be shown that each molecular eigenfunction Ψ_n must be either symmetric or antisymmetric with respect to inversion. A symmetric eigenfunction obeys the relation

$$\Psi_n(\mathbf{r}) = \Psi_n(-\mathbf{r}) \tag{79}$$

Here $\Psi_n(-\mathbf{r})$ is obtained by replacing every electron coordinate $\mathbf{r}_i$, defined with respect to the center of inversion as origin, by its negative value $-\mathbf{r}_i$. Similarly, an antisymmetric eigenfunction must obey the relation

$$\Psi_n(\mathbf{r}) = -\Psi_n(-\mathbf{r}) \tag{80}$$

If a molecular has a center of inversion, its Hamiltonian must by symmetric with respect to inversion,

$$\mathcal{H}(\mathbf{r}) = \mathcal{H}(-\mathbf{r}) \tag{81}$$

168 THE SPECTRA OF DIATOMIC MOLECULES

Every eigenfunction $\Psi_n(\mathbf{r})$ must be a solution of the Schrödinger equation

$$\mathcal{H}(\mathbf{r})\Psi_n(\mathbf{r}) = E_n\Psi_n(\mathbf{r}) \tag{82}$$

and it must also obey the equation

$$\mathcal{H}(-\mathbf{r})\Psi_n(-\mathbf{r}) = E_n\Psi_n(-\mathbf{r}) \tag{83}$$

or

$$\mathcal{H}(\mathbf{r})\Psi_n(-\mathbf{r}) = E_n\Psi_n(-\mathbf{r}) \tag{84}$$

If the eigenvalue E_n is nondegenerate, then there is only one eigenfunction; the two functions $\Psi_n(\mathbf{r})$ and $\Psi_n(-\mathbf{r})$ must then be proportional to one another:

$$\Psi_n(\mathbf{r}) = \delta\Psi_n(-\mathbf{r}) \tag{85}$$

If we use this equation twice, we find that

$$\delta^2 = 1 \quad \delta = \pm 1 \tag{86}$$

This means that $\Psi_n(\mathbf{r})$ is either symmetric or antisymmetric with respect to inversion.

If the eigenvalue E_n is degenerate, then its eigenfunctions may be divided into a set of symmetric and a set of antisymmetric eigenfunctions and we may also say that every eigenfunction Ψ_n is either symmetric or antisymmetric.

If the eigenfunction of a molecular eigenstate is symmetric with respect to inversion, then we indicate this by a subscript g; this is the first letter of the German word *gerade*, meaning "even." If the eigenfunction is antisymmetric with respect to inversion, then we denote this by a subscript u; this is derived from the German word *ungerade*, meaning "odd."

Finally, the spin multiplicity of a molecular eigenstate is usually described by a superscript which has the value $2S + 1$. In this way, a superscript 1 refers to a singlet state, 2 is a doublet, 3 a triplet, and so on.

The lowest electronic eigenstates of the oxygen molecule O_2 are $^3\Sigma_g$, $^1\Delta_g$, $^1\Sigma_g$, and $^3\Sigma_u$. We see that the ground state is a triplet with spin $S = 1$, its eigenfunction is symmetric with respect to inversion, and it does not change if we rotate around the internuclear axis. The next two states are singlets, both *gerade*; the first is a Δ state so that its wave function changes by a factor $e^{2i\Phi}$ if we rotate around the molecular axis by an angle Φ, the second is a Σ state and its wave function is not affected by rotations. The next excited state is again a triplet, it is *ungerade* and its wave function has cylindrical symmetry.

In the subsequent discussion we shall use the symmetry properties of molecular eigenfunctions to describe some of the spectroscopic selection rules.

5. Selection Rules and Spectral Intensities in Electronic Bands

We have seen in Section 3 that an eigenstate in a diatomic molecule is characterized by three quantum numbers, n, which describes the electronic state, v, which describes the vibrational state, and J, which refers to the rotational state. In general, a spectroscopic transition may occur between two states (n, v, J) and (n', v', J'), but in calculating transition probabilities it is not practical to try and derive a general expression for the transition probabilities that applies to all situations. Instead we shall divide the general problem into a number of special cases, and we shall discuss some selection rules and make some predictions about relative intensities.

In this section we consider the situation where the two quantum numbers n and n' differ from one another. This is an electronic transition, and in most cases the light that is absorbed or emitted will be in the visible or the ultraviolet region of the spectrum. Experimentally we observe an absorption or an emission band that is composed of a large number of separate lines due to the various values that the quantum numbers v, v', J, and J' may assume. We discussed in Chapter IV how the intensity of each individual line is derived from the square of the transition moment, which is defined as

$$\mathbf{P}(n, v, J; n', v', J') = \langle \Psi_{n,v,J} | \mathbf{\mu}_{op} | \Psi_{n',v',J'} \rangle \tag{87}$$

Here $\mathbf{\mu}_{op}$ is defined as the electric dipole moment operator,

$$\mathbf{\mu}_{op} = e \sum_j \mathbf{r}_j \tag{88}$$

where we have to sum over all electrons in the molecule. For the eigenfunctions Ψ we substitute the approximate expression of Eq. (75),

$$\Psi_{n,v,J}(r; R, \theta, \varphi) = F_n(r; R)\Psi_v^n(q_n)\psi_J(\theta, \varphi) \tag{89}$$

where r stands symbolically for all the electron coordinates.

Here, the first term represents the electronic part of the wave function, the second part represents the vibrational part, and the third part represents the rotational part. If we substitute this function (89) into Eq. (87) for the transition moment, we obtain

$$\mathbf{P}(n, v, J; n', v', J')$$
$$= \langle F_n(r; R)\Psi_v^n(R)\psi_J(\theta, \varphi) | \mathbf{\mu}_{op}(r) | F_{n'}(r; R)\Psi_{v'}^{n'}(R)\psi_{J'}(\theta, \varphi) \rangle \tag{90}$$

170   THE SPECTRA OF DIATOMIC MOLECULES

for the transition moment of the transition $(n, v, J) \rightarrow (n', v', J')$. Here we must integrate over the electron coordinates, the vibrational coordinates, and the rotational coordinates.

In order to understand the general properties of the transition moment, it may be helpful to look at the situation sketched in Fig. 5.4. Here we have assumed that we have polarized light; this may be assumed without loss of generality, and it makes the situation a lot easier. We take the Z axis along the direction of polarization of the light and the origin as the center of gravity of the molecule. The molecular axis makes an angle θ with the Z axis and the angle θ, together with the second polar angle φ, describes the orientation of the molecular axis; the two angles θ and φ also describe the rotational motion of the molecule. The vibrational motion of the molecule is described by the distance AB, which we denote by R.

The electronic transition moment $\mathbf{u}_{n,n'}(R)$ is now defined as

$$\mathbf{u}_{n,n'}(R) = \int F_n^*(r, R) \mathbf{u}_{op} F_{n'}(r, R) \, dr \tag{91}$$

where we integrate over the electron coordinates r. These electron coordinates are defined with respect to a coordinate system that is attached to the molecular axis. It is important to realize that this coordinate system is

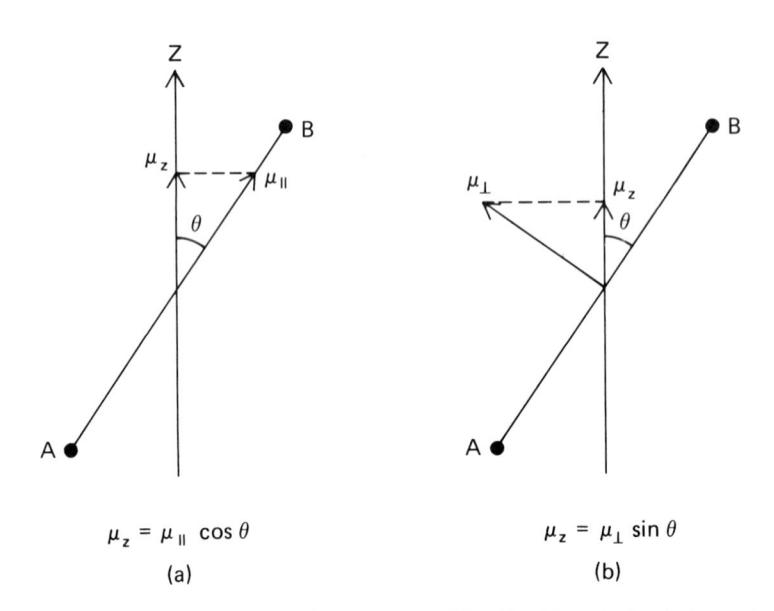

FIGURE 5.4 ● Electronic transition moments. The light is polarized along the Z axis. In (a) the electronic transition moment $\mathbf{u}$ is along the molecular axis and $\mu_z = \mu_\parallel \cos \theta$. In (b) the electronic transition moment $\mu_\perp$ is perpendicular to the molecular axis and $\mu_z = \mu_\perp \sin \theta$.

RULES AND SPECTRAL INTENSITIES IN ELECTRONIC BANDS 171

different from the space-fixed coordinate system that we use to describe the light.

There are three sets of selection rules for electronic transitions. The first selection rule states that the electronic transition moment is different from zero only if one of the two states n or n' is *gerade* and the other is *ungerade*. According to this selection rule, the only allowed transitions are

$$g \to u \qquad u \to g \qquad (92)$$

The other transitions,

$$g \to g \qquad u \to u \qquad (93)$$

are forbidden.

The second selection rule,

$$\Delta S = 0 \qquad (94)$$

requires that the two electronic states have the same spin functions. If the spins of the two states are different from each other, then the molecular eigenfunctions have orthogonal spin parts and the corresponding transition moment is zero.

The third selection rule is related to the quantum number Λ, which we described in Eq. (78). It may be shown that the electronic transition moment $\mathbf{\mu}_{n,n'}$ of Eq. (91) must be either directed along the molecular axis or directed perpendicular to the molecular axis. In the first case we call it $\mu_{\parallel}$ (see Fig. 5.4a), and this transition is allowed only if

$$\Delta \Lambda = 0 \qquad (95)$$

This means that the values of the quantum numbers Λ for the two states n and n' must be the same. In the second case it is called $\mu_{\perp}$ (see Fig. 5.4b), and this is nonzero only if the values of Λ for the two states differ by unity,

$$\Delta \Lambda = \pm 1 \qquad (96)$$

It follows that the selection rule for the quantum numbers Λ is given by

$$\Delta \Lambda = 0, \pm 1 \qquad (97)$$

To be more specific, the selection rule is

$$\begin{array}{lll} \Delta\Lambda = 0 & \mu_{\parallel} \neq 0 & \mu_{\perp} = 0 \\ \Delta\Lambda = \pm 1 & \mu_{\parallel} = 0 & \mu_{\perp} \neq 0 \end{array} \qquad (98)$$

As an example, let us apply the above selection rules to some specific cases. A $^1\Sigma_g$ state combines with a $^1\Sigma_u$ state ($\Delta\Lambda = 0$) to give $\mu_\parallel$ and with a $^1\Pi_u$ state ($\Delta\Lambda = 1$) to give $\mu_\perp$. A $^1\Pi_u$ state combines with a $^1\Pi_g$ state to give $\mu_\parallel$ and with a $^1\Sigma_u$ and a $^1\Delta_u$ state to give $\mu_\perp$. Other transitions from a $^1\Sigma_g$ or a $^1\Pi_u$ state are not allowed.

In the following we want to consider the intensities or transition moments of transitions between different electronic states. We should realize that absolute values of these transition probabilities are hard to determine, both theoretically and experimentally, and for that reason we are not too interested in them. The only question we want to answer is whether a certain transition is allowed or forbidden, and that answer is derived from the above selection rules.

Even though we may not know the absolute value of a certain electronic transition, we can make quite a few predictions about the relative intensities of the many separate transitions that occur in the same electronic band. These separate transitions occur between the various vibrational and rotational levels that belong to the same two electronic states, n and n'. For all these separate transitions the electronic transition moment $\mu_{n,n'}$ has the same value, and it may be considered as a parameter that describes the net intensity of each line. We shall see that we can predict the relative intensities of the various lines in a specific electronic band quite accurately, even though we do not know too much about the absolute intensity. Our discussion is therefore concerned only with relative intensities.

Let us now return to the situation sketched in Fig. 5.4, where we have a beam of light that is polarized in the Z direction. According to Eq. (90), the transition moment between the two states (n, v, J) and (n', v', J') is then given by

$$P(n, v, J; n', v', J') = \langle F_n(r; R)\Psi_v^n(R)\psi_J(\theta, \varphi) \mid \mu_z(r) \mid F_{n'}(r; R)\Psi_{v'}^{n'}(R)\psi_{J'}(\theta, \varphi) \rangle \quad (99)$$

This expression must be integrated successively over the electronic coordinate r, the vibrational coordinate R, and the rotational coordinates (θ, ϕ). We may substitute the expression (91) for the electronic transition moment, but we should take its projection along the Z axis. As we show in Fig. 5.4, in the first case where the electronic transition moment is parallel to the molecular axis we should substitute

$$\mu_z(R) = \cos\theta \, \mu_\parallel(R) = \cos\theta \, \mu_{n,n'}(R) \quad (100)$$

into Eq. (99). The result is

$$P(n, v, J; n', v', J') = \langle \Psi_v^n(R)\psi_J(\theta, \varphi) \mid \cos\theta \, \mu_\parallel(R) \mid \Psi_{v'}^{n'}(R)\psi_{J'}(\theta, \varphi) \rangle$$
$$= \langle \Psi_v^n(R) \mid \mu_\parallel(R) \mid \Psi_{v'}^{n'}(R) \rangle \langle \psi_J(\theta, \varphi) \mid \cos\theta \mid \psi_{J'}(\theta, \varphi) \rangle$$
$$(101)$$

RULES AND SPECTRAL INTENSITIES IN ELECTRONIC BANDS

In the second case we should substitute

$$\mu_z(R) = \sin\theta\,\mu_\perp(R) = \sin\theta\,\mu_{n,n'}(R) \qquad (102)$$

and the result is

$$P(n, v, J; n', v', J') = \langle \Psi_v^n(R) \mid \mu_\perp(R) \mid \Psi_{v'}^{n'}(R) \rangle \langle \psi_J(\theta, \varphi) \mid \sin\theta \mid \psi_{J'}(\theta, \varphi) \rangle \qquad (103)$$

It follows that the transition moment is the product of a vibrational and a rotational part and that these parts may be considered separately. The rotational part depends on the orientation of the electronic transition moment. We consider the vibrational part first. This part behaves differently for absorption and emission, so we shall consider these two situations separately.

In an electronic absorption band the electronic quantum number n of the initial state must be 0, representing the electronic ground state. In addition, the vibrational quantum number in the initial state must also be zero. It may be seen from Table 5.2 that the vibrational frequencies of diatomic molecules are fairly large, so that at ordinary temperatures only a very small fraction of molecules are in excited vibrational states due to the thermal distribution over these states. We may assume that in a typical sample practically all molecules will be in their vibrational ground state. In an absorption band we measure transitions from an initial state $(0, 0, J)$ to a set of excited states (k, v, J').

According to either of the Eqs. (101) or (103), the vibrational transition moments in an electronic absorption band are given by

$$P(0, 0; k, v) = \langle \Psi_0^0(R) \mid \mu_{0,k}(R) \mid \Psi_v^k(R) \rangle_R$$
$$\mu_{0,k}(R) = \langle F_0(r) \mid \mu_\alpha(r) \mid F_k(r) \rangle_r \qquad (104)$$

Here the subscripts R and r indicate that in the first integral we integrate over the vibrational coordinate R and in the second integral we integrate over the electronic coordinates r. The subscript α is either $\parallel$ or $\perp$, but the two situations are treated in the same way.

The electronic transition moment $\mu_{0,k}(R)$ of Eq. (104) is a function of the internuclear distance R, but it may be assumed that it varies very slowly over the range covered by the molecular vibrations. This means that in the integral (104) it may be approximated as

$$\mu_{0,k}(R) \approx \mu_{0,k}(R_0) \qquad (105)$$

where $\mu_{0,k}(R_0)$ is its value at the equilibrium distance of the electronic

174 THE SPECTRA OF DIATOMIC MOLECULES

ground state. If we substitute the approximation (105) into the integral (104), then $\mu_{0,k}(R_0)$ is a constant, so that we obtain

$$P(0, 0; k, v) = \mu_{0,k}(R_0)\langle \Psi_0^0(R) \mid \Psi_v^k(R) \rangle = \mu_{0,k}(R_0) I_{0,0;k,v} \quad (106)$$

The integrals $I_{0,0;k,v}$ are called overlap integrals. They contain the product of the vibrational eigenfunctions of the electronic ground state and of the excited electronic state k. They depend very strongly on the relative positions of the potential curves $U_0(R)$ and $U_k(R)$ for the two electronic states.

In Fig. 5.5 we have drawn the curves $U_0(R)$ and $U_k(R)$ for a typical situation in a diatomic molecule where the electronic state k has a stable configuration. It is known experimentally that the equilibrium distances R_k for excited states are generally larger than the equilibrium distance R_0 for the corresponding ground state. The differences are of the order of 0.1 Å to 1 Å. Also, the curvature of the upper curve is usually smaller than for the lower curve. This means that the force constant and the frequency ω for the higher electronic state are smaller than the corresponding quantities for the electronic ground state. This difference may be quite large; the frequencies may differ by a factor of two or more.

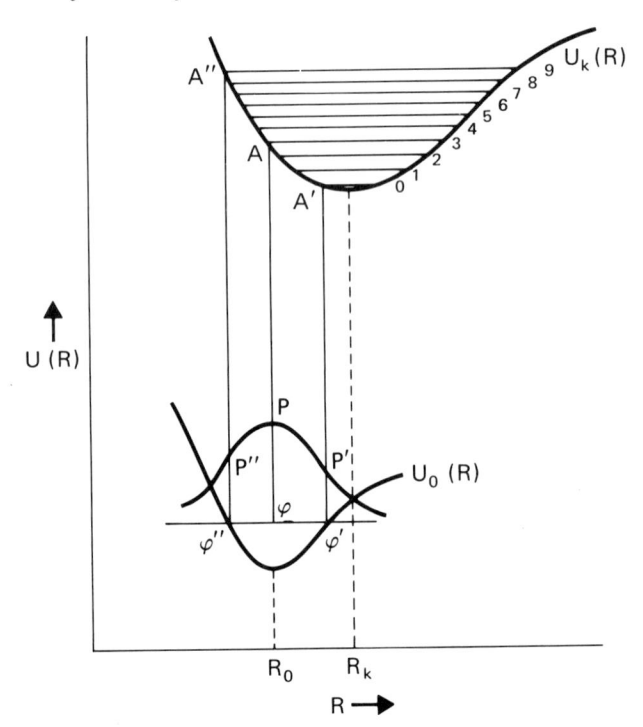

FIGURE 5.5 ● Vibrational fine structure of an electronic absorption band.

We consider the various transitions between the vibrational levels $v = 0, 1, 2, 3, \ldots$, of the electronic state k and the lowest vibrational level $v = 0$ of the electronic ground state 0 because these are the transitions that are observed in the spectrum. The term "vibronic" is used to denote an electronic-vibronic energy level; accordingly, we consider transitions between the vibronic states (k, v) and the vibronic state $(0, 0)$. The situation is sketched in Fig. 5.5. It may be seen here that the energy differences between the vibronic states are given by

$$\Delta \epsilon_{k,v} = U_k(R_k) + (v + \tfrac{1}{2})h\nu_k - U_0(R_0) - \tfrac{1}{2}h\nu_0$$
$$= [U_k(R_k) - U_0(R_0) + \tfrac{1}{2}h(\nu_k - \nu_0)] + vh\nu_k \qquad v = 0, 1, 2, 3, \ldots \quad (107)$$

Here ν_k is the vibrational frequency of the electronic state k and ν_0 is the vibrational frequency of the electronic ground state 0. We use the harmonic approximation for simplicity, even though its accuracy may leave something to be desired for the higher vibrational levels.

According to Eq. (107), the spectrum consists of a number of equidistant lines, separated by the vibrational frequency ν_k. It is important to note that in an absorption band the vibrational separation of the lines is the frequency ν_k belonging to the excited state curve $U_k(R)$.

Let us now consider the relative intensities of these lines. According to Eq. (106), the transition moments are the overlap integrals $I_{0,0;k,v}$, so that the intensity of the transition $(0, 0) \to (k, v)$ is proportional to the square of this overlap integral,

$$I_v = (\langle \Psi_0^0(R) \mid \Psi_v^k(R) \rangle)^2 \cdot \Delta E_{k,v} \qquad (108)$$

The relative intensities of the various lines is thus given by Eq. (108).

The overlap integrals of Eq. (108) are called Franck-Condon factors. They may be calculated exactly from the harmonic oscillator functions of the two electronic states, but their magnitudes may also be estimated by making use of the Franck-Condon approximation. We show this also in Fig. 5.5. Here we have sketched the ground-state vibrational function of the vibronic state $(0, 0)$; this is a Gaussian of the form

$$\exp -\lambda (R - R_0)^2 \qquad (109)$$

Let us now consider, for example, the transition $(k, 9) \to (0, 0)$. Here we draw a line for the energy level $v = 9$ and look where this intersects with the curve $U_k(R_k)$ on the left. If we then draw a vertical line from this intersection point A'', the overlap integral is about equal to the line $P''Q''$, which is the value of the ground-state eigenfunction (109) at the intersection point. The relative intensity is then the square of the line $P''Q''$.

176   THE SPECTRA OF DIATOMIC MOLECULES

In Fig. 5.6 we show the relative intensity curves for a few typical situations. The overall intensity distribution always has the form of a Gaussian, because the square of a Gaussian function (109) is also a Gaussian. The first curve (a) represents the situation of Fig. 5.5. Here it may be seen that the $0 \rightarrow 3$ transition is the one with the highest intensity. Toward the left, with decreasing intensity we find the transitions $0 \rightarrow 2$, $0 \rightarrow 1$, and $0 \rightarrow 0$. The latter is the cutoff, because there are no transitions to the left of it. The result is a Gaussian that is truncated on the left. Curve (b) represents the situation where $R_k - R_0$ is larger, that is, the two potential curves are shifted further with respect to one another. Here the transition $0 \rightarrow 6$ has the highest intensity and there is practically no cutoff on the left; this is a situation where $R_k - R_0$ is larger than in Fig. 5.5. Finally, in curve (c), $R_k - R_0$ is small and the cutoff point is closer to the maximum. It follows that we can derive the relative shift of the two potential curves from the cutoff point of the vibrational intensity distribution curve.

The Franck-Condon approximation may be explained either from classical or from quantum mechanics. In the classical explanation we note that the nuclei move much slower than the electrons, so that during an electronic transition the nuclei remain in the same configuration. The quantum mechanical explanation follows from the properties of the harmonic oscillator wave functions and leads to the same results.

We must recognize that the potential curve is not really a parabola but that it has the form of Fig. 5.7; for large values of R it approaches a constant value $U_k(\infty)$ and the harmonic approximation is no longer valid. There are only discrete vibrational levels below the energy $U_k(\infty)$; for higher energies the energy levels form a continuum. Therefore, if the frequency of the absorbed light becomes larger than $U_k(\infty) - U_0(R_0)$, the molecule is excited into a continuum of states and there is no structure left in the absorption band. In addition, we excite to a molecular state

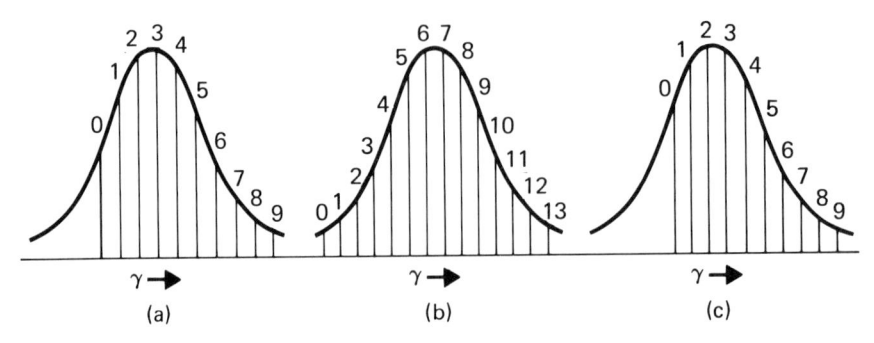

FIGURE 5.6 ● Relative intensities of vibrational lines. Figure (a) corresponds to Fig. 5, in (b) the difference $R_k - R_0$ is larger than in Fig. 5, and in (c) the difference $R_k - R_0$ is smaller.

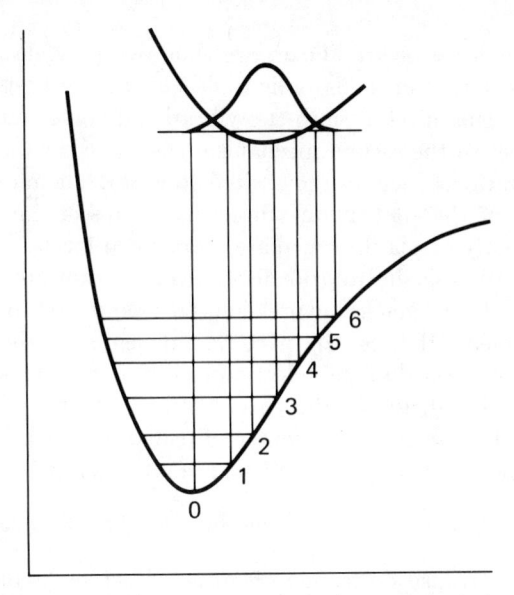

RULES AND SPECTRAL INTENSITIES IN ELECTRONIC BANDS 177

FIGURE 5.7 ● Vibrational fine structure of an electronic emission band.

that is not stable and the molecule will dissociate. We see that for this particular electronic transition the dissociation energy is given by

$$E_{\text{diss}} = U_k(\infty) - U_0(R_0) - \tfrac{1}{2}h\nu_0 \qquad (110)$$

An electronic emission band may be treated the same way as an absorption band, but the result is different. In the emission the molecules are initially in the state $(k, 0)$ and we observe transitions to the various vibronic levels $(0, v)$ of the electronic ground state. The energy differences for these transitions are given by

$$\begin{aligned}\Delta\epsilon_{0,v} &= U_k(R_k) + \tfrac{1}{2}h\nu_k - U_0(R_0) - (v + \tfrac{1}{2})h\nu_0 \\ &= [U_k(R_k) - U_0(R_0) + \tfrac{1}{2}h(\nu_k - \nu_0)] - vh\nu_0 \qquad v = 0, 1, 2, 3, \ldots\end{aligned} \qquad (111)$$

If we compare this with Eq. (107) for an absorption band, we see that the two expressions are equal for $v = 0$. In emission we have in addition a set of equidistant lines at lower frequencies separated by an amount ν_0. In absorption we have a set of equidistant lines at higher frequencies separated by an amount ν_k. We show both spectra in Fig. 5.8. At the right side we have the absorption spectrum and at the left an emission spectrum; both spectra have the $0 \rightarrow 0$ line in common.

In some cases the probability for transitions into the continuum of the ground-state potential curve become significant. If such a transition occurs,

178 THE SPECTRA OF DIATOMIC MOLECULES

the molecule will dissociate. There are thus two possible mechanisms for the photodissociation of a diatomic molecule. In the first mechanism we excite the molecule directly into the vibrational continuum of an excited electronic state. In the second mechanism we excite the molecule first into a discrete vibrational level of the excited state (k). The molecule then falls back into one of the continuum vibrational states of the electronic state (0), either directly or via intermediate vibrational levels.

In order to discuss the rotational structure within an electronic band, we must return to the general expressions (101) and (103) for the rotational transition moment, that is, the part that depends on the polar angles θ and φ. It may be seen that the selection rules for the rotational transitions depend on the symmetry of the two electronic states that are involved. If the electronic transition moment is directed along the molecular axis, then the rotational transition moments are given by

$$P_{J,J'} = \iint \psi_J^*(\theta, \varphi) \cos\theta \, \psi_{J'}(\theta, \varphi) \sin\theta \, d\theta \, d\varphi \tag{112}$$

If the electronic transition moment is directed perpendicular to the molecular axis, then it follows from Eq. (103) that the rotational transition moment is

$$P_{J,J'} = \iint \psi_J^*(\theta, \varphi) \sin\theta \, \psi_{J'}(\theta, \varphi) \sin\theta \, d\theta \, d\varphi \tag{113}$$

The first integral (112) is always zero unless $\Delta J = J - J' = \pm 1$. The second integral (113) is nonzero if $\Delta J = 0$ or $\Delta J = \pm 1$. The result is thus that we have the selection rule

$$\Delta J = \pm 1 \qquad \text{if } \Delta\Lambda = 0 \tag{114}$$

and

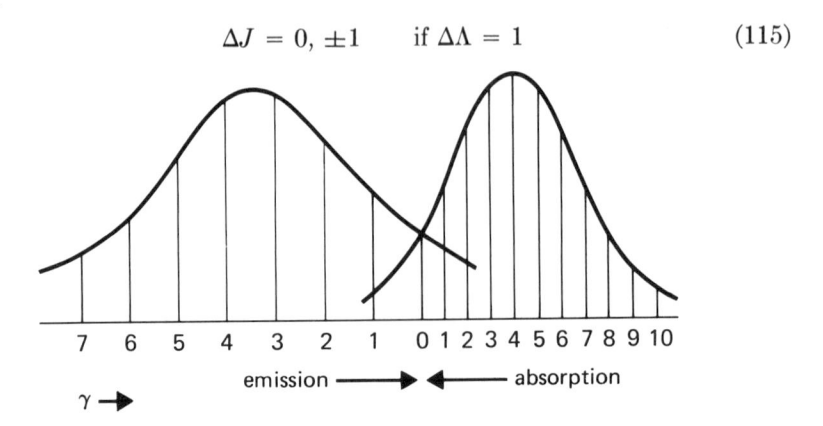

$$\Delta J = 0, \pm 1 \qquad \text{if } \Delta\Lambda = 1 \tag{115}$$

FIGURE 5.8 • Combined absorption and emission spectra; the vibrational fine structure of both.

The specific values of the transition moments $P_{J,J'}$ of Eqs. (101) and (103) may also be derived from the properties of the rotational eigenfunctions, but these expressions are fairly complicated and we shall not mention them here.

Let us now consider the frequencies of the rotational transitions that are allowed for given values of the vibrational and electronic quantum numbers of the upper and the lower states. It follows from Eq. (74) that the energies of the upper and the lower states are given by

$$E_{k,w,J'} = U_k(R_k) + (w + \tfrac{1}{2})h\nu_k + B_{k,w}J'(J' + 1)$$
$$E_{0,v,J} = U_0(R_0) + (v + \tfrac{1}{2})h\nu_0 + B_{0,v}J(J + 1) \qquad (116)$$

The difference between the upper and the lower states may be written as

$$\Delta E_{J,J'} = h\nu_{J,J'} = \Delta\epsilon_0 + B'J'(J' + 1) - BJ(J + 1) \qquad (117)$$

where the definition of the various quantities is self-explanatory. If we substitute the possible values $J' = J$ and $J' = J \pm 1$, we obtain three possible expressions for the rotational lines in the spectrum:

$$\begin{aligned}
R(J) &= \Delta E_{J,J'+1} = \Delta\epsilon_0 + B'(J + 1)(J + 2) - BJ(J + 1) \\
&= \Delta\epsilon_0 + 2B' + (3B' - B)J + (B' - B)J^2 \\
Q(J) &= \Delta E_{J,J} = \Delta\epsilon_0 + B'J(J + 1) - BJ(J + 1) \\
&= \Delta\epsilon_0 + (B' - B)J + (B' - B)J^2 \\
P(J) &= \Delta E_{J,J-1} = \Delta\epsilon_0 + B'J(J - 1) - BJ(J + 1) \\
&= \Delta\epsilon_0 - (B' + B)J + (B' - B)J^2
\end{aligned} \qquad (118)$$

These expressions give the frequencies of the spectral lines as a function of an integer J, which can in principle have the values $0, 1, 2, \ldots$. The distribution depends on the selection rule. It is customary to denote the lines that belong to transitions $J' = J + 1$ by the R branch of the spectrum, the lines due to $J' = J$ by the Q branch, and the lines due to $J' = J - 1$ by the P branch. The P and the R branches may both be described by one equation:

$$\Delta E_n = \Delta\epsilon_0 + (B' + B)n + (B' - B)n^2 \qquad (119)$$

For negative values of n, $n = -J$, this represents the P branch; for positive values of n, $n = J + 1$, this describes the R branch.

We can plot the values of the frequencies ν_n of the possible transitions as a function of the integer n by using Eq. (119). The result is a parabola called the Fortrat parabola. We have sketched such a parabola in Fig. 5.9, where we have also drawn the spectral lines obtained by projecting the

180 THE SPECTRA OF DIATOMIC MOLECULES

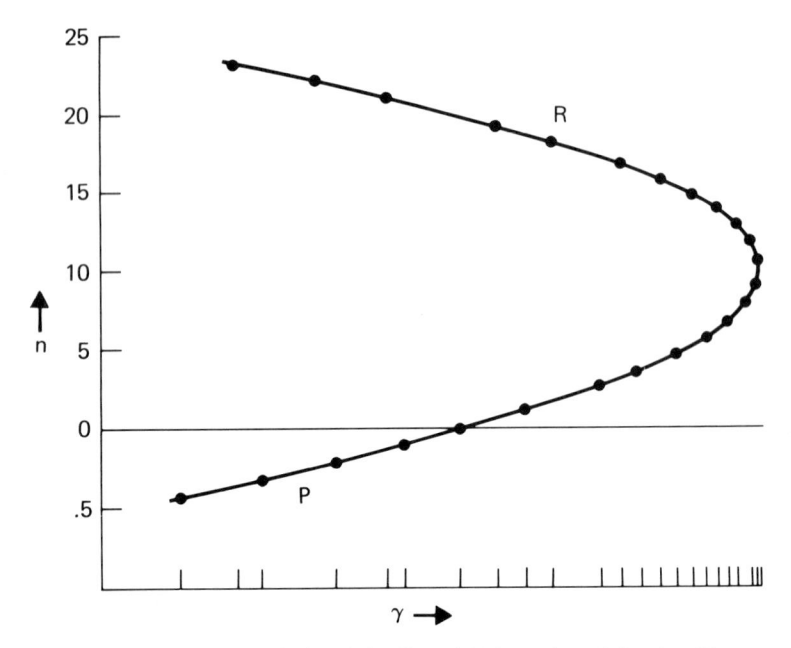

FIGURE 5.9 ● Fortrat parabola of the P and R branches, following Eq. (119).

points on the parabola on the horizontal axis. In drawing Fig. 5.9, we have assumed that $B' - B$ is negative, in which case the spectrum shows a band head toward the high-frequency side. The upper part of the parabola belongs to the R branch and the lower part to the P branch. Obviously, if we were to assume the opposite sign for $(B' - B)$, then the parabola would have a different direction and the band head would be toward the low-frequency side. The Q branch may also be represented by a parabola (or rather half a parabola) according to Eq. (118). Its representation is very similar to Fig. 5.9.

It is important to identify the P, Q, and R branches in the rotational fine structure, because this leads to some useful information about the symmetry of the electronic transitions involved. We have shown that the rotational selection rules are $\Delta J = \pm 1$ if $\Delta \Lambda = 0$. The two cases $\Delta J = \pm 1$ correspond to the P and R branches, and it follows that the Q branch is missing if $\Delta \Lambda = 0$, that is, if the electronic transition moment is along the molecular axis. If we observe a Q branch, then we must have $\Delta \Lambda = 1$ or the electronic transition moment is perpendicular to the molecular axis. For example, if we know that the electronic ground state has Σ symmetry, then we conclude that we have a Σ-Π transition if there is a Q branch in the rotational fine structure, whereas the electronic transition is Σ-Σ if the Q branch is absent.

So far we have discussed only the frequencies of the rotational lines. The relative intensities can also be calculated, but this is fairly complicated and we shall limit ourselves to an outline of the theory. We must first consider the thermal population over the various rotational levels, either in the lower electronic state in an absorption band or in the higher electronic state in an emission band. It may be seen that for most molecules the rotational constant is quite small, so that even the higher rotational levels are populated to an appreciable extent. For a heavy molecule such as N_2, the levels are almost equally populated at room temperature. In most cases the thermal distribution does not impose any limitation on the observation of the rotational lines, even for high values of the quantum number J. The exact values of the rotational transition moments may be derived from the properties of the rotational eigenfunctions. The relative intensities of the rotational lines can be calculated quite accurately. The results contain the exponential factors that represent the thermal distribution over the various levels, and from a comparison with the experimental intensities it is even possible to determine the temperature of the sample.

The above treatment of the rotational fine structure of the electronic energy levels is not entirely accurate except for $^1\Sigma$ states. In the general case we must consider the total angular momentum of the molecule. This is usually denoted by the symbol **J**, and it is the sum of three contributions: namely, the electronic orbital angular momentum, the angular momentum of the electron spin, and the angular momentum of the rotational motion of the molecule. The rotational energy levels depend, then, on the possible values of the total angular momentum **J** rather than on the angular momentum of just the rotational motion. Obviously, the electronic and spin angular momenta are both zero in a $^1\Sigma$ state, so that here it is permissible to consider only the rotational angular momentum.

In constructing the total angular momentum of a diatomic molecule from its three contributions, we must know the relative magnitudes of the interactions between these two contributions. The interaction between the orbital angular momentum and the spin angular momentum of the electrons is described by the spin-orbit coupling. It is necessary to distinguish between two cases, which are known as Hund's case (a) and Hund's case (b). In Hund's case (a) the spin-orbit coupling is large. Here we must first determine the sum of the spin and the orbital angular momenta of the electrons and then we must quantize this sum with respect to the rotational angular momentum. In Hund's case (b) the spin-orbit coupling is small. This usually applies to molecules with light nuclei. Here we consider first the quantization of the electronic angular momentum with respect to the rotational angular momentum where the spin angular momentum has a negligible effect on the rotational energy levels.

6. Intensities in the Infrared and Microwave Regions

The transitions that fall in the infrared usually occur between states that differ only in their vibrational and rotational quantum numbers. The theory of infrared intensities is based on the calculation of transition moments between states of the type $(0, 0, J)$ and $(0, 0, J')$. We consider only absorption spectra, and we assume that in the initial state all molecules are in their electronic and vibrational ground states. The transition moment is again given by Eq. (90), which now has the form

$$P(0, 0, J; 0, v, J') = \langle F_0(r; R) f_0^0(R) \psi_J(\theta, \varphi) | \mathbf{\mu}_{op}(r) | F_0(r; R) f_v^0(R) \psi_{J'}(\theta, \varphi) \rangle \quad (120)$$

The electronic part of the transition moment is now equal to the electric dipole moment of the molecule as a function of the distance R,

$$\mathbf{\mu}_0(R) = \langle F_0(r; R) | \mathbf{\mu}_{op}(r) | F_0(r; R) \rangle_r \quad (121)$$

This dipole moment is rigorously zero for homonuclear molecules, and it is to be expected that the infrared transitions are forbidden for such molecules. For heteronuclear molecules the electric dipole moment is directed along the molecular axis, and if we make use of Eq. (101) we find that the transition moment reduces to

$$P(0, 0, J; 0, v, J') = \langle f_0^0(R) \psi_J(\theta, \varphi) | \cos\theta \, \mu_0(R) | f_v^0(R) \psi_{J'}(\theta, \varphi) \rangle \quad (122)$$

In order to calculate the transition moment, we again assume that the dipole moment $\mu_0(R)$ varies only slightly in the vicinity of the equilibrium distance R_0, so that we may expand it as

$$\mu_0(R) \approx \mu_0(R_0) + \left(\frac{\partial \mu_0}{\partial R}\right)_{R_0} (R - R_0) = \mu_0(R_0) + \mu_0'(R_0)(R - R_0) \quad (123)$$

We first consider a vibrational transition where the quantum number v of the second state is different from zero. The vibrational transition moment is then given by

$$P_{0,v} = \langle f_0^0(R) | \mu_0(R) | f_v^0(R) \rangle$$
$$= \mu_0(R_0) \langle f_0^0(R) | f_v^0(R) \rangle + \mu_0'(R_0) \langle f_0^0(R) | R - R_0 | f_v^0(R) \rangle \quad (124)$$

The first term of Eq. (124) is rigorously zero because the vibrational eigenfunctions are orthogonal. The second term is in principle nonzero and it may be derived from the properties of the harmonic oscillator eigenfunctions that in the harmonic approximation the term is nonzero only if $v = 1$.

If we use the harmonic approximation, then the theory predicts only one line in the vibrational spectrum: namely, the line from $v = 0$ to $v = 1$, which has the vibrational frequency ν_0. This prediction is in reasonable agreement with the experimental results. Even though vibrational transitions to states $v = 2, 3, 4$, and 5 have been observed, these lines are extremely weak. The finite intensity of these lines is due to anharmonicity effects or to higher terms in the expansion of Eq. (123) which we have neglected, and a measurement of the intensities gives us some information about the magnitude of the anharmonicity effects.

From the vibrational intensity we can derive the magnitude of the quantity $\mu_0'(R_0)$. This is the derivative of the electronic dipole moment at the point R_0 with respect to the internuclear distance R.

Each vibrational line has a rotational fine structure, which is easier to measure than the rotational fine structure in an electronic band because the spectral resolution is higher in the infrared. The theoretical description of the rotational fine structure is identical with the theory for electronic bands that we discussed in the previous section. The only difference is that the electronic dipole moment $\mathbf{\mu}(R)$ is always directed along the molecular axis, and according to Eq. (114) the rotational selection rule must be $\Delta J = \pm 1$. This means that in vibrational transitions there is no Q branch in the rotational fine structure. It is possible to derive the values of the rotational constants $B_{0,0}$ and $B_{0,v}$ from the rotational structure, and we can see how the rotational constants depend on the vibrational quantum numbers.

Pure rotational transitions between states $(0, 0, J)$ and $(0, 0, J')$ fall within the microwave region of the spectrum because the energies range between 0.01 and 50 cm^{-1}. They can be observed only if the molecule has a nonzero dipole moment, and even then they are difficult to measure because the experimental techniques are fairly difficult. The selection rules for these transitions are again $\Delta J = \pm 1$. The results lead to quite accurate values of the rotational constant B_0, but the microwave spectra have been measured only for a limited number of molecules because the measurements are much more difficult than in the optical regions of the spectrum.

Problems

1. Calculate the reduced nuclear mass for the molecules H_2, HCl, HI, and N_2.
2. In the electronic ground state of HCl the rotational constant $B_{0,0}$ for the lowest vibrational level is 10.4400 cm^{-1} and the rotational constant $B_{0,1}$ for the first excited vibrational level is 10.136 cm^{-1}. Calculate the equilibrium distance R_0 between the two nuclei for both cases from the rotational constants.
3. The electronic ground state of N_2 is $^1\Sigma_g$. The lowest excited states have $^1\Pi_g$, $^1\Sigma_u$, $^1\Pi_u$, $^1\Sigma_u$, and $^1\Pi_u$ symmetry. Which electronic transitions between the

184   THE SPECTRA OF DIATOMIC MOLECULES

ground state and these excited states are allowed? What can you say about the direction of the electronic transition moment?

4. In the ground state of CO (C = 12, O = 16), the rotational constant $B_{0,0}$ is 1.9314 cm^{-1}. What is the value of the equilibrium internuclear distance R_0? In the first excited electronic state, $B_{1,0}$ has the value 1.6116 cm^{-1}. What is the value of the equilibrium internuclear distance R_1?

5. According to the Maxwell distribution the relative population of an energy level E_k is given by $\exp(-E_k/kT)$. Calculate the relative populations for the levels $J = 0$, $J = 2$, and $J = 10$ for the molecules H_2 and N_2 at liquid helium temperature ($T = 4°K$) and at room temperature ($T = 300°K$). The rotational constants are given in Table 5.1.

6. The vibrational transition in HCl^{35} is at 2989.74 cm^{-1}. Calculate the force constant of the vibrational motion on the assumption that it is a harmonic oscillator. If we now assume that the vibrational motion in DCl^{35} has the same force constant as in HCl^{35}, where does the vibrational transition occur in DCl^{35}?

7. Explain the Born-Oppenheimer approximation in simple, nonmathematical terms.

8. Why does the rotational constant B of a diatomic molecule depend on the electronic state the molecules is in?

9. If we want to determine the electric dipole moment of a diatomic molecule from the molecular spectra, which spectroscopic experiment should we perform?

10. It may be shown that the rotational selection rules of Eqs. (114) and (115) should be supplemented by the selection rule that the rotational transition from the state $J = 0$ to the state $J' = 0$ is always forbidden. Prove this by means of Eqs. (114) and (115).

11. The rotational constant B for a specific electronic state of a diatomic molecule depends slightly on the vibrational quantum number v. Explain why.

12. The potential function $U_0(R)$ of a diatomic molecule is sometimes approximated by the Morse potential $U_0(R) = D\{1 - \exp[-b(R - R_0)]\}^2$, where R_0 is the equilibrium distance and D is the dissociation energy. If the harmonic force constant of the motion is k, what should we take as the value of the parameter b?

13. The rotational constants for the ground states of H_2 and D_2 are 60.809 cm^{-1} and 30.429 cm^{-1}, respectively. Are the equilibrium internuclear distances R_0 for the two molecules different from one another, or are they basically the same?

14. The vibrational frequency in an excited electronic state is usually much smaller than in the ground state. Explain why this is so from the behavior of the potential curves $U(R)$.

15. Explain why you would not expect to observe vibrational transitions in the infrared spectra of the N_2 molecule.

16. The electronic ground state of H_2 has $^1\Sigma_g$ symmetry, and the first excited state has $^1\Sigma_u$ symmetry. Do you expect the rotational fine structure of the electronic transition between these two states to have a P, a Q, and an R branch? Explain your answer.

17. In the infrared spectrum of a heteronuclear diatomic molecule, we observe the transition $v = 0 \rightarrow v = 1$ with a large intensity, but we also observe the transition $v = 0 \rightarrow v = 2$ with a much smaller intensity. Why can this second transition be observed?

18. Do you expect the rotational constant $B_{0,1}$ belonging to the first excited vibrational state $v = 1$ to be larger or smaller than the corresponding rotational constant $B_{0,0}$ belonging to the vibrational ground state $B_{0,0}$? Justify your answer.

Bibliography

The best books on molecular spectra are the three books written by Herzberg (1, 2, 3), which contain everything there is to be known on molecular spectra. As a textbook we recommend Herzberg's recent book on free radicals (4), which gives a simplified and more concise version of the theory of molecular spectra and is easier to read than the more complete books (1, 2, 3). In our discussion we have avoided the use of group theory. For those students who are interested, we list a few good books on the subject; Hall's book (5) is more concerned with chemical questions than is Tinkham's.

1. G. Herzberg. *Molecular Spectra and Molecular Structure, I. Spectra of Diatomic Molecules.* Van Nostrand Reinhold, New York (1950).
2. G. Herzberg. *Molecular Spectra and Molecular Structure, II. Infrared and Raman Spectra.* Van Nostrand Reinhold, New York (1945).
3. G. Herzberg. *Molecular Spectra and Molecular Structure, III. Electronic Spectra and Electronic Structure of Polyatomic Molecules.* Van Nostrand Reinhold, New York (1966).
4. G. Herzberg. *The Spectra and Structures of Simple Free Radicals.* Cornell University Press, Ithaca, N.Y. (1971).
5. L. H. Hall. *Group Theory and Symmetry in Chemistry.* McGraw-Hill, New York (1969).
6. M. Tinkham. *Group Theory and Quantum Mechanics.* McGraw-Hill, New York (1964).

CHAPTER 6

THE CHEMICAL BOND

1. Introduction

From a theoretical point of view, the chemical bond in a diatomic molecule is characterized by its potential curve, that is, the ground-state eigenvalue $E_0(R)$ as a function of the internuclear distance R, and by the eigenfunction Ψ_0 of the electronic ground state. From an empirical point of view, the bond is characterized by quantities we can measure, such as the bond length, the bond energy, the force constant of the vibration, the electric dipole moment, and so on. It should be realized that the Schrödinger equation cannot be solved exactly for any molecule other than the hydrogen molecular ion, and even in that case the solutions are too complicated to have much practical use. Consequently, the theory of valence is based on approximate methods, that is, on the variation principle.

In the early papers on valence theory, the variational functions were usually constructed by making use of chemical ideas and, sometimes, of experimental information that was available. The variational function could contain some parameters that were determined by minimizing the energy. Nowadays electronic computers are widely used in quantum chemical calculations. In this way it is possible to do calculations with variational functions that have a very large number of parameters. This has the advantage that the accuracy of the more recent calculations is much higher than the results that were obtained some 30 or 40 years ago. On the other hand, the early work was more closely related to chemical ideas, and in many cases it gave a good understanding of the general nature of the chemical bond. In this chapter the emphasis will be on the early work in quantum chemistry, because we are basically interested in the general ideas on the chemical bond.

We have seen in the previous chapter that for a diatomic molecule the translational and rotational motions of the nuclei may be considered separately. If we then introduce the Born-Oppenheimer approximation, we may calculate the electronic wave function on the assumption that the nuclei have fixed positions in space. The solutions contain the internuclear distance R as a parameter because the electronic potential function contains R implicitly, but the kinetic energy of the nuclear motion may be disregarded while solving the electronic Schrödinger equation. The situation is basically the same for polyatomic molecules, although it is more complicated. Again, the translational and rotational motions of the nuclei may be separated off, but the theory of rotational motion is more complicated than for diatomic molecules. The rotational motion is described by the moments of inertia with respect to three mutually orthogonal symmetry axes. We speak of a spherical top if the three moments of inertia are the same, of a symmetric top if two moments are equal and the third one is different, and of an asymmetric top if all three are different. The symmetric top has more or less the same eigenvalues as the rigid rotor, and the eigenvalues of the asymmetric top are derived from the symmetric top by means of perturbation theory.

It is somewhat more difficult to formulate the Born-Oppenheimer approximation for a polyatomic molecule than it is for a diatomic molecule, but it may be concluded from general considerations that the approximation should be equally valid in both cases. If we use the Born-Oppenheimer approximation, we find that the electronic eigenvalues depend on a number of parameters, which represent the relative distances of the nuclei, rather than on just one parameter as for a diatomic molecule. Within the harmonic approximation these parameters may be combined in certain linear combinations of displacement coordinates, which are known as normal coordinates. Each normal coordinate vibrates as a harmonic oscillator with a given frequency and its vibration, which is called a normal mode, does not interact with any of the other normal modes. In this model the vibrational motion of a polyatomic molecule is a superposition of noninteracting normal modes, each with its own frequency. The theoretical description of one normal mode is identical with the theoretical description of the vibration in a diatomic molecule. The electronic eigenfunctions of a polyatomic molecule are again derived on the assumption that the nuclei are at rest, but the eigenvalues and eigenfunctions now contain more than one parameter. For example, in the case of the water molecule these parameters are the O—H bond distances and the H—O—H bond angle. In large molecules the number of parameters may be quite large. Usually the electronic Schrödinger equation is solved only for the equilibrium configuration of the nuclei, but if we wish to make predictions about the nuclear configuration or about changes in nuclear configuration we must solve the Schrödinger equation for different bond lengths and bond angles.

It is clear that the accuracy that can be achieved in calculations on small molecules is much higher than for large molecules. The various methods for deriving electronic eigenfunctions may be divided into two categories: At one extreme, there are methods that are quite accurate but so laborious that they can be applied only to small molecules. At the other extreme there are methods that can be applied to a large number of molecules but that have only limited accuracy. Such methods are often semiempirical, because they make use of experimental information. The various methods are usually tested first on some small molecules, in particular the hydrogen molecule and the hydrogen molecular ion, because here we can verify the accuracy of every approximation we wish to introduce and we obtain some understanding for the reliability of the method. The literature contains many calculations on the hydrogen molecule, and some of these are very accurate—the results agree with the experimental data to within the experimental error. Unfortunately, many of these methods that work so well for the hydrogen molecule cannot be applied to larger molecules, and their practical use is therefore limited.

In this chapter we first discuss the hydrogen molecular ion and the hydrogen molecule and we then proceed to larger molecules.

2. The Hydrogen Molecular Ion

In valence theory the concept of resonance has played a useful role. This concept is nothing more than an application of the variation principle that we discussed in Section 7 of Chapter I.

The benzene molecule is a good example of applying the resonance principle. It is well known in organic chemistry that we can write two possible structures of benzene, namely the two Kekulé structures, I and II, of Fig. 6.1. However, it follows from the experiments that the true structure of benzene is neither I nor II, because the experimental bond lengths in the

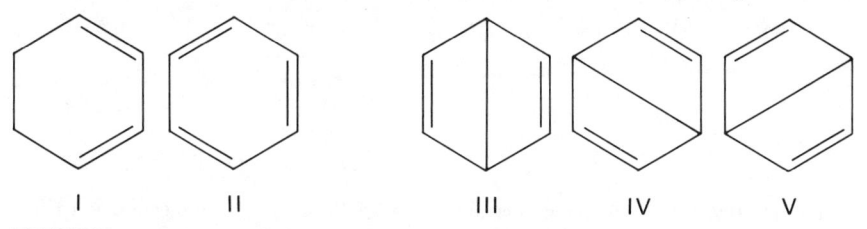

FIGURE 6.1 ● The two Kekulé structures, I and II, and the three Dewar structures, III, IV, and V, for the benzene molecule.

190 THE CHEMICAL BOND

benzene molecule are all equal to one another, whereas in either one of the structures I or II the double bonds should be shorter than the single bonds. It was proposed, therefore, that the benzene molecule oscillates between the structures I and II and that these oscillations are so rapid that we observe some kind of an average of the structures I and II; this is called resonance between the two structures I and II.

In quantum mechanics we may in principle represent each of the structures I or II by means of a ground-state eigenfunction Ψ_I and Ψ_{II}, respectively. The ground-state wave function Ψ of benzene is then represented as

$$\Psi = a_I \Psi_I + a_{II} \Psi_{II} \tag{1}$$

according to the resonance principle. It is easily seen that this is a special case of the variation principle in which we expand the molecular eigenfunction in terms of the wave functions of the separate chemical structures. In principle we obtain a more accurate approximation to the benzene eigenfunction if we include more terms in the expansion. In Fig. 6.1 we have sketched the three Dewar structures, III, IV, and V, in addition to the two Kekulé structures, I and II. If we represent the Dewar structures by wave functions Ψ_{III}, Ψ_{IV}, and Ψ_V, respectively, then we may also represent the benzene wave function as

$$\Psi = a_I \Psi_I + a_{II} \Psi_{II} + a_{III} \Psi_{III} + a_{IV} \Psi_{IV} + a_V \Psi_V \tag{2}$$

It is to be expected that the energy we derive from the function (1) is lower than the energy of structure I and that the energy we derive from the function (2) is lower yet. This decrease in energy is called the resonance energy, and it supplies additional stabilization to the benzene molecule.

Let us now apply the resonance principle to the hydrogen molecular ion. The two resonance structures, I and II, are sketched in Fig. 6.2. In structure I the electron is located in a (1s) orbital on atom a and in structure II

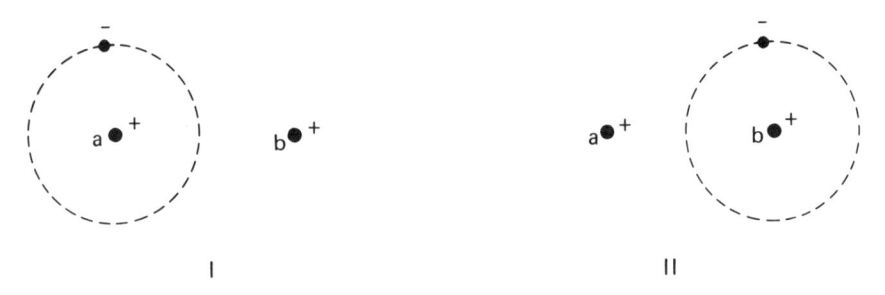

FIGURE 6.2 ● The two resonance structures of the hydrogen molecular ion H_2^+. In structure I the electron is on atom a and in structure II the electron is on atom b.

the electron occupies the (1s) orbital of atom b. The two resonance structures are described by the functions

$$\Psi_I = s_a = \frac{1}{\sqrt{\pi}} e^{-r_a}$$
$$\Psi_{II} = s_b = \frac{1}{\sqrt{\pi}} e^{-r_b} \tag{3}$$

The molecular wave function is then represented as

$$\Psi = A_I \Psi_I + A_{II} \Psi_{II} = A_I s_a + A_{II} s_b \tag{4}$$

It may be instructive to calculate the energy from this function. The molecular Hamiltonian is given by

$$\mathcal{H} = -\tfrac{1}{2}\Delta - \frac{1}{r_a} - \frac{1}{r_b} \tag{5}$$

in terms of atomic units. Since the molecule has a plane of symmetry perpendicular to the molecular axis, the eigenfunctions must be either symmetric or antisymmetric with respect to reflections. This means that $A_I = \pm A_{II}$, and the possible wave functions are

$$\Psi_1 = s_a + s_b$$
$$\Psi_2 = s_a - s_b \tag{6}$$

Because of symmetry we also have

$$\langle s_a | \mathcal{H} | s_a \rangle = \langle s_b | \mathcal{H} | s_b \rangle = \mathcal{H}_{aa}$$
$$\langle s_a | \mathcal{H} | s_b \rangle = \langle s_b | \mathcal{H} | s_a \rangle = \mathcal{H}_{ab} \tag{7}$$
$$\langle s_a | s_b \rangle = \langle s_b | s_a \rangle = S$$

The atomic functions are normalized,

$$\langle s_a | s_a \rangle = \langle s_b | s_b \rangle = 1 \tag{8}$$

The expectation values of the energies E_1 and E_2 with respect to the functions Ψ_1 and Ψ_2 are given by

$$E_1 = \frac{\langle \Psi_1 | \mathcal{H} | \Psi_1 \rangle}{\langle \Psi_1 | \Psi_1 \rangle} = \frac{\mathcal{H}_{aa} + \mathcal{H}_{ab}}{1 + S}$$
$$E_2 = \frac{\langle \Psi_2 | \mathcal{H} | \Psi_2 \rangle}{\langle \Psi_2 | \Psi_2 \rangle} = \frac{\mathcal{H}_{aa} - \mathcal{H}_{ab}}{1 - S} \tag{9}$$

The energies may all be expressed in terms of three integrals. Since s_a is an eigenfunction of the hydrogen atom with eigenvalue $-\tfrac{1}{2}$, we have

$$\mathcal{H} s_a = \left(-\frac{1}{2}\Delta - \frac{1}{r_a}\right)s_a - \frac{1}{r_b}s_a = -\frac{1}{2}s_a - \frac{1}{r_b}s_a \qquad (10)$$

Consequently,

$$\mathcal{H}_{aa} = \langle s_a | \mathcal{H} | s_a \rangle = -\frac{1}{2} - \left\langle s_a \left| \frac{1}{r_b} \right| s_a \right\rangle = -\frac{1}{2} - I$$

$$\mathcal{H}_{ab} = \langle s_b | \mathcal{H} | s_a \rangle = -\frac{1}{2} \langle s_b | s_a \rangle - \left\langle s_b \left| \frac{1}{r_b} \right| s_a \right\rangle = -\frac{1}{2} S - J \qquad (11)$$

and

$$E_1 = -\frac{1}{2} - \frac{I+J}{1+S} \qquad E_2 = -\frac{1}{2} - \frac{I-J}{1-S} \qquad (12)$$

The calculation of the three integrals S, I, and J is discussed in the appendix to this chapter. The results are

$$\begin{aligned}S &= e^{-R}\left(1 + R + \frac{R^2}{3}\right) \\ I &= R^{-1}[1 - e^{-2R}(1+R)] \\ J &= e^{-R}(1+R)\end{aligned} \qquad (13)$$

By substituting these results into Eq. (12), we find that the potential function $U_0(R)$ for the ground state is given by

$$U_0(R) = E_1(R) + \frac{1}{R} = -\frac{1}{2} - \frac{[(2R/3) - (1/R)] - e^{-R}[(1/R) + 1]}{e^R + [1 + R + (R^2/3)]} \qquad (14)$$

and the potential function $U'(R)$ for the excited state is

$$U'(R) = E_2(R) + \frac{1}{R} = -\frac{1}{2} + \frac{[(2R/3) - (1/R)] + e^{-R}[(1/R) + 1]}{e^R - [1 + R + (R^2/3)]} \qquad (15)$$

We have sketched the behavior of the two potential functions $U_0(R)$ and $U'(R)$ in Fig. 6.3. We have used atomic units in our calculations, and we find that the lower curve has a minimum for $R = 2.5$ and that the energy value at the minimum is -0.5648. The equilibrium distance is $R_0 = 2.5 \times 0.529$ Å $= 1.32$ Å, and the dissociation energy is 0.0648 a.u. $= 1.76$ eV. The experimental values are 1.06 Å for R_0 and 2.79 eV for the dissociation energy. The agreement with our calculation is not too bad.

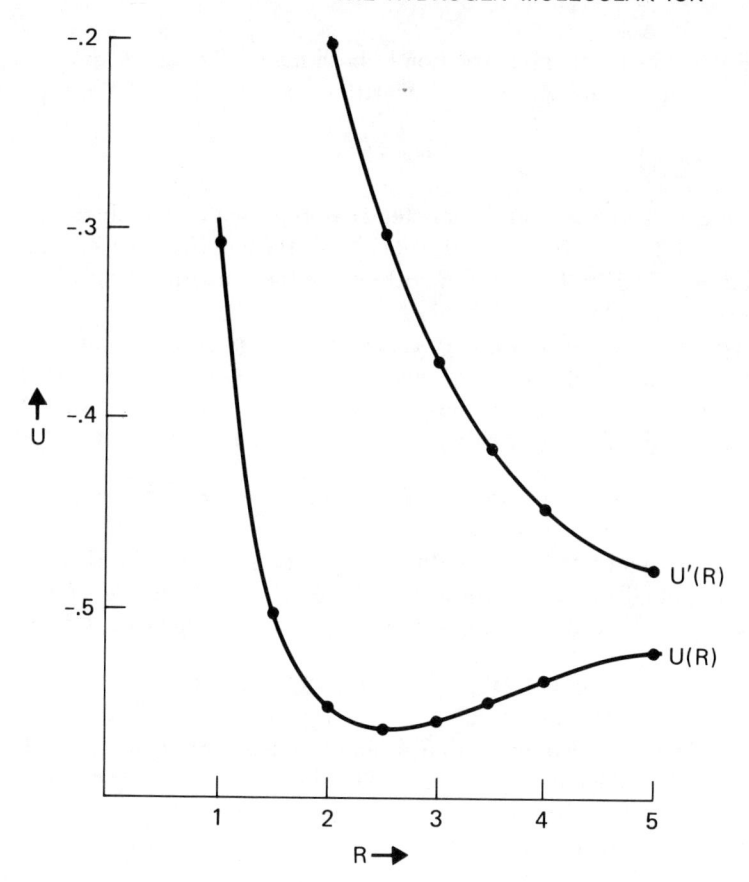

FIGURE 6.3 ● Potential curves for H_2^+ in atomic units.

It may be seen that $U'(R)$ does not have a minimum; there is no stable molecular configuration in this electronic state. Therefore, if the molecule is excited from the state E_1 to the state E_2, it dissociates.

By definition, a one-electron wave function is called an orbital. In an atomic orbital the electron moves in the vicinity of one particular atomic nucleus. The functions Ψ_1 and Ψ_2 are called molecular orbitals because they represent situations where the electron moves through the whole molecule. More specifically, they are molecular orbitals, which are linear combinations of atomic orbitals, or LCAO for short. It is customary to call the function Ψ_1 a bonding orbital because it represents a situation where there is a stable chemical bond between the two atoms. Similarly, the function Ψ_2 is known as an antibonding orbital.

It is possible to obtain better energies for the ground state of the hydrogen molecular ion by starting from more elaborate variational functions. We may argue that the electron, even when it is in the vicinity of atom a, is subject to the Coulomb attraction of both nuclei. To a first approximation we may account for this effect by writing the wave function as

$$\Psi = e^{-\rho r_a} + e^{-\rho r_b} \tag{16}$$

where ρ is a variational parameter that represents the effective nuclear charge. It follows from a variational calculation that the energy has a minimum for $\rho = 1.23$ and $R = 1.06$ Å; the corresponding dissociation energy is 2.25 eV.

The next effect that we may consider is the polarization of the atomic orbitals. If the electron is on nucleus a, then the Coulomb attraction of the other nucleus will pull the electron cloud slightly toward nucleus b. We can represent this effect by writing the wave function in the form

$$\Psi = e^{-\rho r_a} + \lambda z_a e^{-\rho' r_a} + e^{-\rho r_b} - \lambda z_b e^{-\rho' r_b} \tag{17}$$

The first two terms of this equation represent an atomic orbital on nucleus a that is pulled off-center toward nucleus b; the magnitude of the effect depends on the value of the parameter λ. A variational calculation gives

$$\rho = 1.247 \qquad \rho' = 0.934 \qquad \lambda = 0.145 \tag{18}$$

The equilibrium distance is 1.06 Å, and the dissociation energy is 2.71 eV.

The most efficient approximation to the ground-state eigenfunction of H_2^+ makes use of elliptical coordinates (see the appendix to this chapter). For example, the function

$$\Psi = e^{-\delta\mu} \tag{19}$$

gives a dissociation energy of 2.17 eV and an equilibrium distance R_0 of 1.06 Å. The improved function

$$\Psi = e^{-\delta\mu}(1 + a\nu^2) \tag{20}$$

with $\delta = 1.3540$ and $a = 0.4480$ gives $R_0 = 2$ a.u. $= 1.06$ Å and a dissociation energy of 0.102386 a.u. $= 2.78538$ eV. These values agree with the experimental data to within the possible experimental errors. We have combined the various results that are derived from the above-mentioned variational functions in Table 6.1.

It may be seen that the expansion of Ψ in terms of elliptical coordinates is the most efficient approach for calculating the dissociation energy and the equilibrium distance. However, this method is not easily extended to more complex molecules and also, it does not give a clear picture of the behavior of the electrons in relation to chemical ideas. Therefore, it is

TABLE 6.1 ● Values of dissociation energy and equilibrium distance for the hydrogen molecular ion, derived from various trial functions by means of the variational principle.

Wave function	R_0 (Å)	Diss. energy (eV)
Eq. (4)	1.32	1.76
Eq. (16)	1.06	2.25
Eq. (19)	1.06	2.17
Eq. (17)	1.06	2.71
Eq. (20)	1.06	2.785
experimental	1.06	2.79

generally preferred to represent molecular wave functions by expansion in terms of atomic orbitals even though the convergence of the expansion may be slower.

If we use atomic orbitals, we may try to relate the energy of a chemical bond to the charge distribution of the electron. Let us consider, for example, the simple molecular orbital of Eq. (6). The corresponding charge density function (normalized to unity) is given by

$$P = (2 + 2S)^{-1}[(s_a)^2 + 2(s_a)(s_b) + (s_b)^2] \qquad (21)$$

It may be said that the first term represents the probability that the electron is associated with nucleus a, the last term the probability that it is associated with nucleus b, and the middle term, which is called the overlap charge, is the probability that the electron may be found between the two nuclei. The numerical values of the three possibilities are obtained by integrating the corresponding terms in Eq. (21):

$$P_a = P_b = \frac{1}{2 + 2S}$$
$$P_{ab} = \frac{2S}{2 + 2S} \qquad (22)$$

It may be derived from Eq. (21) that for $R = 2$ we have $P_a = P_b = 31.5\%$ and $P_{ab} = 37\%$. It is interesting to note that the same calculation for the antibonding function Ψ_2 of Eq. (6) gives the strange results $P_a = P_b = 1.209$ and $P_{ab} = -1.418$. Here the overlap charge appears to be negative.

It seems to be an empirical fact that a chemical bond is stronger, that is, it has a higher dissociation energy, if the overlap charge is larger. Usually, in antibonding orbitals such as Ψ_2, the overlap charge is negative with respect to the electronic charge densities at the atoms. However, all semiempirical theories should be used with caution, and the various theories

3. The Hydrogen Molecule

A chemical bond usually contains a pair of electrons, and the properties of the bond are determined not only by the interaction between each of the electrons and the nuclei but also by the interaction between the electrons. The hydrogen molecule is the simplest molecule that contains at least one pair of electrons, and it is therefore a useful system to illustrate some of the general features of the chemical bond.

Just as in the previous section, we may approximate the molecular eigenfunction by writing it as a linear combination of functions that represent various resonance structures of the molecule. In Fig. 6.4 we have sketched the two most obvious resonance structures of the hydrogen molecule. In structure I, electron 1 is centered on nucleus a and electron 2 is centered on nucleus b; obviously this situation may be represented by a function Ψ_I, which is defined as

$$\Psi_I = s_a(1)s_b(2) \qquad (23)$$

where the atomic orbitals are defined in Eq. (3). In structure II, electron 1 is centered on b and electron 2 is centered on a; the corresponding function Ψ_{II} is given by

$$\Psi_{II} = s_b(1)s_a(2) \qquad (24)$$

The molecular wave function is then a linear combination of the functions Ψ_I and Ψ_{II}.

It is easily seen that we may use only the sum and the difference of the two functions because of the exclusion principle,

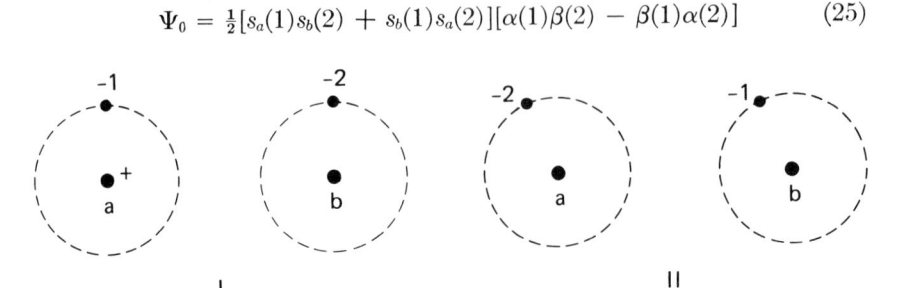

$$\Psi_0 = \tfrac{1}{2}[s_a(1)s_b(2) + s_b(1)s_a(2)][\alpha(1)\beta(2) - \beta(1)\alpha(2)] \qquad (25)$$

FIGURE 6.4 • Resonance structures I and II for the hydrogen molecule.

and
$$^3\Psi_1 = \tfrac{1}{2}\sqrt{2}[s_a(1)s_b(2) - s_b(1)s_a(2)]\,^3\zeta(1,2) \tag{26}$$

The total molecular wave function must be antisymmetric with respect to permutations of the electrons. It is easily verified that the sum of Ψ_I and Ψ_{II} is symmetric with respect to permutations; it must therefore be combined with an antisymmetric spin function, that is, a singlet spin function. By the same argument, the difference of Ψ_I and Ψ_{II} is antisymmetric with respect to permutations, and it must be combined with a triplet spin function in order that the total wave function is antisymmetric. The function Ψ_0 belongs to the ground state of the hydrogen molecule and the function $^3\Psi_1$ belongs to an excited state.

The Hamiltonian of the hydrogen molecule is

$$\begin{aligned}\mathcal{H}(1,2) &= -\tfrac{1}{2}\Delta_1 - \frac{1}{r_{a1}} - \frac{1}{r_{b1}} - \tfrac{1}{2}\Delta_2 - \frac{1}{r_{a2}} - \frac{1}{r_{b2}} + \frac{1}{r_{12}} \\ &= G(1) + G(2) + \frac{1}{r_{12}}\end{aligned} \tag{27}$$

The expectation value of this Hamiltonian with respect to the function Ψ_0 is obtained as

$$\begin{aligned}E_0 = \frac{\langle \Psi_0 \mid \mathcal{H}(1,2) \mid \Psi_0 \rangle}{\langle \Psi_0 \mid \Psi_0 \rangle} &= (1+S^2)^{-1}[2\langle s_a \mid G \mid s_a \rangle + 2S\langle s_a \mid G \mid s_b \rangle \\ &\quad + \langle s_a(1)s_b(2) \mid r_{12}^{-1} \mid s_a(1)s_b(2)\rangle \\ &\quad + \langle s_b(1)s_a(2) \mid r_{12}^{-1} \mid s_a(1)s_b(2)\rangle]\end{aligned} \tag{28}$$

with
$$S = \langle s_a \mid s_b \rangle \tag{29}$$

The first two integrals in Eq. (28) are identical with the integrals we encountered in the previous section, and they may again be calculated by means of the method that we discussed in the appendix to this chapter. It is much more difficult to calculate the other two integrals, namely

$$J = \langle s_a(1)s_b(2) \mid r_{12}^{-1} \mid s_a(1)s_b(2)\rangle \tag{30}$$

and
$$K = \langle s_b(1)s_a(2) \mid r_{12}^{-1} \mid s_a(1)s_b(2)\rangle \tag{31}$$

The integral J represents the Coulomb interaction between a $(1s)$ electron centered on atom a and a $(1s)$ electron centered on atom b; it is called a two-center Coulomb integral. It is possible to calculate this type of integral analytically, but the calculation is fairly tedious. Nowadays, there are programs available to calculate two-center Coulomb integrals by using electronic computers. The integral K is called a two-center exchange-type integral; it represents the Coulomb interaction between two identical overlap charges between atoms a and b. In the 1930s these integrals were calculated by hand. The customary procedure was to expand r_{12}^{-1} as an infinite power series in terms of the elliptical coordinates of electrons 1 and 2; this is called the Neumann expansion. The series can be integrated term by term, and in this way we can obtain a numerical result for the integral K. Again, this procedure has been programmed for use of an electronic computer. The computer programs for calculating both types of two-center integrals are quite efficient, and the calculations can be performed in a millisecond or even less. It may be derived that the function (25) predicts a dissociation energy of 3.14 eV and an equilibrium distance of 0.869 Å. The experimental values are 4.747 eV and 0.741 Å.

The above procedure, where the molecular wave function is approximated as a superposition of resonance structures, is called the valence-bond method. The function (25) is called a valence bond function or a Heitler-London function, after the two scientists who first proposed this method. We have seen that the valence bond method is closely related to chemical ideas, and its use leads to a general understanding of how the various resonance structures contribute to the chemical bond. However, the valence bond method is not appropriate for making numerical predictions for large molecules, because the calculations become very tedious. For that reason an alternative method has been developed. This approach is known as the molecular orbital theory. Here, the molecular wave function is written as an antisymmetrized product of one-electron functions (we have mentioned already that these functions are called molecular orbitals). It is customary to make the additional assumption that each molecular orbital may be approximated as a linear combination of atomic orbitals. This procedure is abbreviated in the literature as the LCAO MO method.

We illustrate the LCAO MO method for the H_2 molecule. To a first approximation we may expand the molecular orbitals in terms of the atomic ($1s$) orbitals s_a and s_b only. It follows from symmetry considerations that the possible linear combinations are

$$\phi_1 = s_a + s_b$$
$$\phi_2 = s_a - s_b \tag{32}$$

where ϕ_1 has the lower energy. The molecular ground state is thus obtained by placing two electrons in the orbital ϕ_1. This state must be a singlet state because of the Pauli exclusion principle,

$$\Psi = \phi_1(\mathbf{r}_1)\phi_1(\mathbf{r}_2)[\alpha(1)\beta(2) - \beta(1)\alpha(2)] \tag{33}$$

THE HYDROGEN MOLECULE

This function gives a dissociation energy of 2.70 eV and an equilibrium distance of 0.85 Å, which is slightly worse than the results from the valence bond method (we have collected the various results in Table 6.2).

It is interesting to substitute the expansion (32) into Eq. (33) and write it all out. The result is

$$\Psi_{MO} = s_a(1)s_a(2) + s_a(1)s_b(2) + s_b(1)s_a(2) + s_b(1)s_b(2) \quad (34)$$

where we have omitted the spin function. It is easily seen that this function contains two more terms than the corresponding valence bond function,

$$\Psi_{VB} = s_a(1)s_b(2) + s_b(1)s_a(2) \quad (35)$$

We may write Eq. (34) as a superposition of the four resonance structures sketched in Fig. 6.5. The first structures in Fig. 6.5 are identical with those in Fig. 6.4, but in addition we have structure III, where both electrons are centered on nucleus a, and structure IV, where both electrons are on nucleus b. These two resonance structures are known as ionic structures. It follows that in the valence bond method the ionic structures are completely neglected, whereas in the MO method they are treated on the same basis as the nonionic structures. It may be seen that a more realistic representation would be

$$\Psi = \rho[s_a(1)s_b(2) + s_b(1)s_a(2)] + \sigma[s_a(1)s_a(2) + s_b(1)s_b(2)] \quad (36)$$

where ρ and σ indicate the relative weights of the nonionic and the ionic structures, respectively. The parameters may be determined from the variational principle, and it turns out that the ionic character of the bond σ^2 is about 5%.

TABLE 6.2 ● Values of dissociation energy and equilibrium distance for the hydrogen molecule, derived from different trial functions by means of the variational principle.

Wave function	R_0 (Å)	Diss. energy (eV)
(25)	0.869	3.14
(32)–(33)	0.85	2.70
(25)–(37)	0.743	3.78
(33)–(37)	0.732	3.49
(36)–(37)	0.749	4.02
Hartree-Fock	0.74	3.62
(38)	0.71	4.11
James-Coolidge	0.740	4.72
Kolos-Roothaan	0.741	4.7467
experimental	0.741	4.747

200   THE CHEMICAL BOND

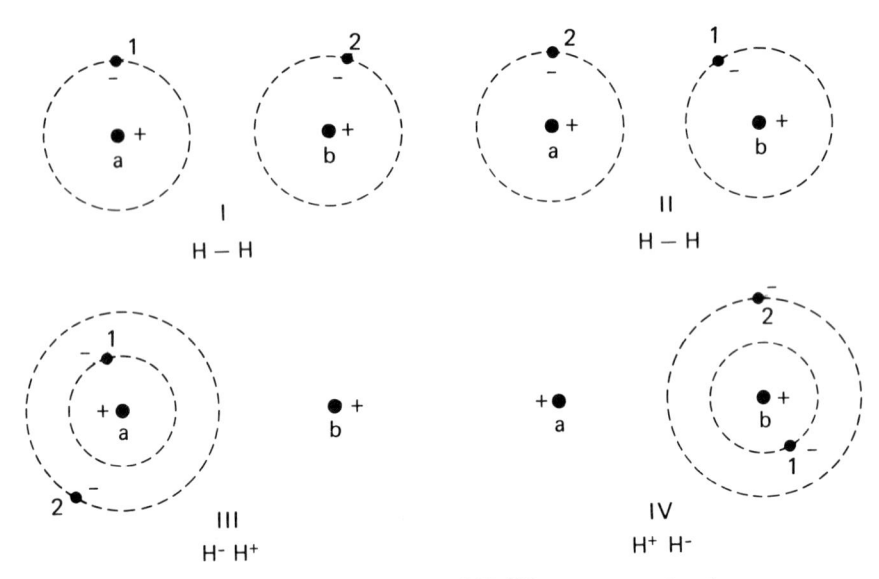

FIGURE 6.5 • Covalent (I, II) and ionic (III, IV) resonance structures of the hydrogen molecule.

The variational calculations may also be performed for the three functions (34), (35), and (36) by substituting

$$s_a = e^{-Zr_a} \qquad s_b = e^{-Zr_b} \qquad (37)$$

for the (1s) orbitals. Here, the effective nuclear charge Z is determined variationally; it is also called a screening constant, and its value is around 1.17. The results of the three variational calculations with the atomic orbitals (37) are 3.78 eV for the VB function (35), 3.49 eV for the MO function (34), and 4.02 eV for the intermediate function (36).

There are a number of ways to improve the wave functions by adding parameters and by using the variational principle. Some of the results are particularly interesting because they give us some indication of the accuracy we might expect in calculations on larger molecules. Once we write the molecular wave function in the form of Eq. (33) as a product of molecular orbitals, we can determine the best possible form of the function ϕ_1 by expanding it in a very large number of terms. This is known as the Hartree-Fock limit, and it means that we do not make any approximations in the wave function beyond the approximation (33). This method leads to a dissociation energy of 3.65 eV. In this approach we do not consider the electron repulsion in constructing the wave function, so the wave function does not contain the electron correlation. The repulsion between the electrons is represented in the Hamiltonian, and the energy expression contains the expectation value of the Coulomb repulsion. We note that the discrepancy between the energy derived from the best possible MO function and the experimental energy is a little more than 1 eV. This difference is known as the correlation energy and is defined as the difference between the exact energy and the energy of the Hartree-Fock limit. We may conclude that for larger molecules, MO wave functions should give energies that have errors of 1 eV or more, so that we should not expect too great an accuracy of the LCAO MO calculations.

Let us now consider what methods we have available to improve upon the accuracy of the Hartree-Fock method. In quantum chemistry this is known as the problem of finding the correlation energy, because this energy is defined as the difference between the Hartree-Fock energy and the true energy. It is clear that we must somehow incorporate the electron correlation in the wave function to improve upon the Hartree-Fock method. This may be done either directly or indirectly.

The most efficient method is to put the electron correlation directly in the eigenfunction. For example, instead of the MO function (33), we take

$$\Psi(1, 2) = \phi(\mathbf{r}_1)\phi(\mathbf{r}_2)(1 + ar_{12}) \tag{38}$$

where the parameter a is determined from the variation principle. If we simply take the MO ϕ as a linear combination of (1s) functions, then the function Ψ of Eq. (38) gives an energy of 4.11 eV, which is already significantly better than the Hartree-Fock energy. In the extreme case the function Ψ is expanded in terms of the elliptical coordinates of the two electrons and the electron distance r_{12}. Such a calculation was performed in 1933 by James and Coolidge, and they obtained a dissociation energy of 4.72 eV, very close to the experimental value. More recently, Kolos and Roothaan did a similar calculation on an electronic computer. Since they included many more terms in their expansion (about 50), the result of this calculation, a dissociation energy of 4.7467, is quite accurate. It agrees with the experimental value to within the experimental error, and it is generally considered to be more accurate even than the experimental dissociation energy.

The disadvantage of the above calculations is that the methods cannot be applied to molecules other than hydrogen, because it is not practical to include the electron distances r_{ij} in the wave functions of systems with more than two electrons. For that reason we must look for other methods, even though these methods may be less effective for small molecules. We have already mentioned a wave function that gives a better result than the Hartree-Fock energy, namely the function (36), which is a valence bond function including ionic terms. It is worth noting that this function can also be represented within the framework of the MO method. We have seen in Eq. (32) that we can derive two molecular orbitals from the two (1s) orbitals, namely

$$\phi_1 = s_a + s_b \qquad \phi_2 = s_a - s_b \tag{39}$$

Here ϕ_1 has a lower energy than ϕ_2. In the MO description we place two electrons in the orbital ϕ_1 (see Eq. 33). We can construct a more elaborate variational function by considering also the excited states,

$$\Psi = \lambda\phi_1(1)\phi_1(2) + \mu\phi_2(1)\phi_2(2) \tag{40}$$

It should be noted that we may not mix the configuration ($\phi_1\phi_2$) because its function has a different symmetry than the ground-state wave function. It is easily seen that Eq. (40) may also be written as

$$\Psi = (\lambda - \mu)[s_a(1)s_b(2) + s_b(1)s_a(2)] + (\lambda + \mu)[s_a(1)s_a(2) + s_b(1)s_b(2)] \tag{41}$$

which is the same as the function (36). We say that the function (40) is constructed by using configuration interaction; in addition to the lowest configuration that may be derived from the molecular orbitals it also contains excited configurations. It may be seen that the configuration interaction method is an indirect way of including electron correlation into the wave function. It is not as efficient as the direct method of Eq. (38), but it is at least possible to use the configuration interaction method (CI for short) in larger molecules.

Finally, we should point out that we can construct only a finite number of molecular configurations from a finite number of atomic orbitals. These configurations may either be constructed by means of the valence bond method or by means of the molecular orbital method. In order to obtain all possible molecular configurations, we must include the ionic states in the VB method and we must use CI in the MO method. We see that the two methods may differ if we consider only the lowest possible MO configuration and if we consider only a few resonance structures in the VB method. Naturally, the two methods should give the same result if we were to use the complete function, containing all possible molecular configurations, but in practice this is never done. Initially, the two methods were about equally popular, but it became clear that the MO method is easier to use for numerical calculations on large molecules and this method has become much more popular than the VB method.

4. Diatomic Molecules

The customary approach for dealing with diatomic molecules is by means of the MO method. In the early 1930s the derivation of approximate molecular eigenfunctions relied heavily on experimental information, mainly derived from the molecular spectra, and on chemical concepts. At that time it was not possible to perform elaborate numerical calculations, and it was not really possible to derive reliable numerical results by means of *ab initio* calculations, that is, calculations starting from first principles. This state of affairs was changed drastically by the invention of the electronic computer. A group of scientists at the University of Chicago—Mulliken, Roothaan, Ruedenberg, and others—recognized during the 1950s how the electronic computer might be used for performing molecular calculations. Even so, it took them many years to develop a method of calculation that may be applied to all diatomic molecules and that gives reasonably accurate results. The method is known as the Roothaan SCF method. We shall begin our discussion of the quantum theory of diatomic molecules with the Roothaan method.

The Roothaan SCF method is a special case of the Hartree-Fock method that we mentioned briefly in Chapter III. Here it is assumed that an atomic eigenfunction, or a molecular eigenfunction, may be written as an antisymmetrized product of one-electron functions. The method is most easily applied to a closed shell ground state where we have placed a pair of electrons with antiparallel spins in the atomic or molecular orbitals with lowest energies. For a molecule, where we use the symbol χ to denote the molecular orbitals, the configuration is thus $(\chi_1)^2(\chi_2)^2\cdots(\chi_N)^2$. The corresponding antisymmetrized molecular wave function Ψ_0 is

$$\Psi_0 = [(2N)!]^{-1/2} \sum_p P\,\delta_p\,[\chi_1(1)\alpha(1)\chi_1(2)\beta(2)\chi_2(3)\alpha(3)\chi_2(4)\beta(4)\cdots$$
$$\chi_N(2N-1)\alpha(2N-1)\chi_N(2N)\beta(2N)] \quad (42)$$

In the Hartree-Fock method we can derive a set of differential equations

$$F_{\text{op}}\chi_i = \epsilon_i\chi_i \quad (43)$$

for the orbitals χ_i by applying the variation theorem. In the case of an atom these differential equations can be solved numerically, although this is a fairly tedious procedure.

In the case of a molecule it is just not feasible to solve the Hartree-Fock Equations (43), even if we use electronic computers. It is therefore necessary to introduce additional approximations. Roothaan proposed that the molecular orbitals χ may be written as linear combinations of atomic orbitals $\phi_{a,k}$, centered on nucleus a, and of atomic orbitals $\phi_{b,k}$, centered on nucleus b,

$$\chi_i = \sum_k (c_{i,k}\phi_{a,k} + c'_{i,k}\phi_{b,k}) \quad (44)$$

The first set of calculations by the Chicago group was limited to diatomic molecules that contain only first-row elements (up to and including F).

In that case it is assumed that the expansion (44) may be limited to (1s), (2s), and (2p) atomic orbitals, which are defined as

$$k = \left(\frac{\zeta_1^3}{\pi}\right)^{1/2} \exp(-\zeta_1 r)$$

$$s = \left(\frac{\zeta_2^5}{3\pi}\right)^{1/2} r \exp(-\zeta_2 r) \qquad (45)$$

$$p_\alpha = \left(\frac{\zeta_3^5}{\pi}\right)^{1/2} r_\alpha \exp(-\zeta_3 r) \qquad (r_\alpha = x, y, z)$$

It is customary to use Slater orbitals. The orbital exponents ζ_1, ζ_2, and ζ_3 are approximately equal to the quantities derived from Slater's rules (see Chapter III, Section 5), but in Roothaan's method they are treated as unknown parameters that are determined variationally. Usually, the orbital exponents ζ_2 and ζ_3 for the 2s and 2p orbitals are set equal to each other.

We see that each molecular orbital χ_i contains ten unknown coefficients c and two unknown orbital exponents ζ. The number of molecular orbitals can be ten at most, so there is a maximum of 120 variational parameters in the molecular eigenfunction. The basic idea of the SCF method is to vary one orbital at a time. In the Roothaan method we vary the 12 parameters of one molecular orbital χ_i while keeping the other molecular orbitals constant.

What makes the method time consuming is the very large number of integrals that must be calculated. This problem arises in particular for the integrals over the Coulomb repulsion $(1/r_{ij})$. These integrals may all be expressed in terms of atomic orbitals, and they can then be subdivided into three categories, namely

$$C_{k,l;m,n} = \langle \phi_{a,k}(1)\phi_{b,l}(2) \mid r_{12}^{-1} \mid \phi_{a,m}(1)\phi_{b,n}(2) \rangle$$
$$L_{k,l;m,n} = \langle \phi_{a,k}(1)\phi_{a,l}(2) \mid r_{12}^{-1} \mid \phi_{a,m}(1)\phi_{b,n}(2) \rangle \qquad (46)$$
$$A_{k,l;m,n} = \langle \phi_{a,k}(1)\phi_{a,l}(2) \mid r_{12}^{-1} \mid \phi_{b,m}(1)\phi_{b,n}(2) \rangle$$

The first of these integrals is called a Coulomb integral; it represents the Coulomb interaction between a charge centered on atom a and a charge centered on atom b. The second integral is an ionic integral; it stands for the interaction between an atomic charge cloud and an overlap charge. The third integral represents the interaction between two overlap charges; it is called an exchange integral. The three types of integrals are known as two-center integrals because they contain atomic orbitals that are centered on two different nuclei.

It is possible to evaluate one particular two-center integral by hand if we have a given set of orbital exponents and a given value for the internuclear distance R, but the calculation is fairly time consuming. The problem is that even a simple SCF calculation involves the evaluation of from 1,000 to 10,000 two-center integrals for each value of R and the orbital exponents. It may be seen that it is not feasible to do such a calculation by hand. Even if an electronic computer is used, it is necessary to devise a program in which each integral is evaluated in a relatively short time (a fraction of a second); otherwise the whole calculation would take up an excessive amount of computer time. The Chicago group spent many years writing programs in which the integrals could be evaluated rapidly and efficiently. Once this problem had been solved, it became feasible to perform SCF calculations on diatomic molecules. The results for the first-row diatomic molecules were published in 1960.[*] We have reproduced the results for the nitrogen molecule in Table 6.3.

It may be seen that three different calculations were performed. In the first and second calculation a fixed set of orbital exponents were used. In the first calculation they were taken equal to the Slater values, derived by means of the Slater rules that we discussed in Chapter III. In the second calculation the orbital exponents were allowed to vary in the molecular SCF calculation. It may be seen that the three sets of orbital exponents differ relatively little, and we may conclude that the Slater rules seem to predict the orbital exponents quite satisfactorily.

It may be helpful to explain the notation used in Table 6.3. The molecular orbitals are classified according to symmetry, and the notation is the same as we used in Chapter V for the symmetry of the total molecular wave functions. Accordingly, if we rotate around the molecular axis through an angle ϕ, then the molecular orbital must change by an amount $\exp(i\lambda\phi)$, where λ is the quantum number that describes the rotational symmetry of the orbital. If $\lambda = 0$, then the orbital is cylindrically symmetric and we call it a σ orbital, if $\lambda = \pm 1$, then we speak of a π orbital, and so on. In the case of N_2, a homonuclear molecule, the σ orbitals are linear combinations of the $1s$, $2s$, and $2p$ atomic orbitals. The π orbitals are linear combinations of the $2p_x$ and the $2p_y$ atomic orbitals, and each π orbital is doubly degenerate. In homonuclear diatomic molecules, each of the molecular orbitals must be either *gerade* or *ungerade*. It is easily verified that there are three σ_g and three σ_u molecular orbitals. The other symmetry species, π_{xg}, π_{yg}, π_{xu}, and π_{yu}, have one orbital each.

The atomic orbitals are numbered by following the molecular symmetry considerations. The first three atomic orbitals are the $1s$, $2s$, and $2p_z$ orbitals of the first atom, and the next three atomic orbitals are the $1s$, $2s$, and $2p_z$ orbitals of the second atom. These six atomic orbitals combine to form the σ molecular orbitals. The seventh and eighth atomic orbitals

---

[*] B. J. Ransil, *Rev. Mod. Phys.* **32**, 245 (1960).

206 THE CHEMICAL BOND

TABLE 6.3 • Results of the Roothaan SCF calculation for the N_2 molecule. The first calculation, called Slater LCAO MO, uses Slater orbitals with the orbital exponents derived from Slater's rules. In the second calculation the orbital exponents are derived from atomic SCF calculations. In the third calculation the orbital exponents are allowed to vary; this should be the most accurate of the three calculations. The molecular configuration is taken as $(1\sigma_g)^2(1\sigma_u)^2(2\sigma_g)^2(2\sigma_u)^2(1\pi_u)^4(3\sigma_g)^2$, and the internuclear distance is taken as 1.094 Å = 2.068 a.u.

		Slater LCAO MO		Best atom LCAO MO		Best limited LCAO MO	
$\zeta(1s)$		6.70		6.6652		6.6675	
$\zeta(2s)$		1.95		1.9236		1.9170	
$\zeta(2p\sigma)$		1.95		1.9170		2.2524	
$\zeta(2p\pi)$		1.95		1.9170		1.9090	
		C_{ip}	ϵ_i (a.u.)	C_{ip}	ϵ_i (a.u.)	C_{ip}	ϵ_i (a.u.)
$1\sigma_g$	$C_{11} = C_{14}$	0.70447	−15.72176	0.70493	−15.80452	0.70480	−15.64705
	$C_{12} = C_{15}$	0.00842		0.00688		0.00743	
	$C_{13} = C_{16}$	0.00182		0.00184		0.00203	
$1\sigma_u$	$C_{21} = -C_{24}$	0.70437	−15.71965	0.70494	−15.80219	0.70494	−15.64423
	$C_{22} = -C_{25}$	0.01972		0.01656		0.01502	
	$C_{23} = -C_{26}$	0.00857		0.00744		0.00567	
$2\sigma_g$	$C_{31} = C_{34}$	0.16890	−1.42541	−0.16618	−1.47922	−0.16328	−1.42106
	$C_{32} = C_{35}$	−0.48828		0.48390		0.49029	
	$C_{33} = C_{36}$	−0.23970		0.24276		0.24824	

$2\sigma_u$	$C_{41} = -C_{44}$	0.16148	-0.73066	-0.15709	-0.75409	-0.15963	-0.71730
	$C_{42} = -C_{45}$	-0.74124		0.74773		0.80527	
	$C_{43} = -C_{46}$	0.26578		-0.26068		-0.23244	
$3\sigma_g$	$C_{51} = C_{54}$	-0.06210	-0.54451	-0.05986	-0.56759	-0.06640	-0.55548
	$C_{52} = C_{55}$	0.40579		0.40879		0.38257	
	$C_{53} = C_{56}$	-0.60324		-0.60431		-0.58527	
$3\sigma_u$	$C_{61} = -C_{64}$	-0.10969	1.100865	0.10547	1.04956	0.09239	1.22618
	$C_{62} = -C_{65}$	1.20696		-1.25520		-0.96578	
	$C_{63} = -C_{66}$	1.21625		-1.24382		-1.07595	
$1\pi_u$	$C_{77} = C_{78}$	-0.62450	-0.57951	0.62200	-0.60486	-0.62139	-0.54540
$1\pi_g$	$C_{87} = -C_{88}$	0.83452	0.27290	0.84059	0.24041	0.84211	0.30021
E_M (a.u.)			-108.57362		-108.55808		-108.63359

208 THE CHEMICAL BOND

are the $2p_x$ atomic orbitals of the first and the second atom, respectively. The ninth and tenth atomic orbitals are the two $2p_y$ atomic orbitals. These latter four atomic orbitals combine to form the π molecular orbitals.

Let us now discuss the general behavior of the molecular orbitals listed in Table 6.3. The energy of the $1s$ atomic orbital is considerably less than for the other atomic orbitals. It may therefore be expected that the molecular orbitals with the lowest energies consist almost exclusively of the atomic $1s$ orbitals. This may be seen in Table 6.3: The molecular orbitals $1\sigma_g$ and $1\sigma_u$ are basically the sum and the difference of the two atomic $1s$ orbitals, and they have very low energies.

The nature of the other σ orbitals may be understood by introducing the concept of hybridized atomic orbitals. In Fig. 6.6 we have drawn a contour map of a $2s$ orbital and a $2p_z$ orbital. It may be seen that in a linear combination of the type

$$\phi_r = \sqrt{1-\rho^2}\,s + \rho p_z \tag{47}$$

for positive ρ the two functions s and p_z reinforce one another for positive values of z and partially cancel for negative values of z. The result is the type of function sketched in Fig. 6.6c. In the hybridized function the charge cloud is shifted along the positive Z axis, and we may even say that it points in the positive Z direction. Naturally, for negative values of ρ we

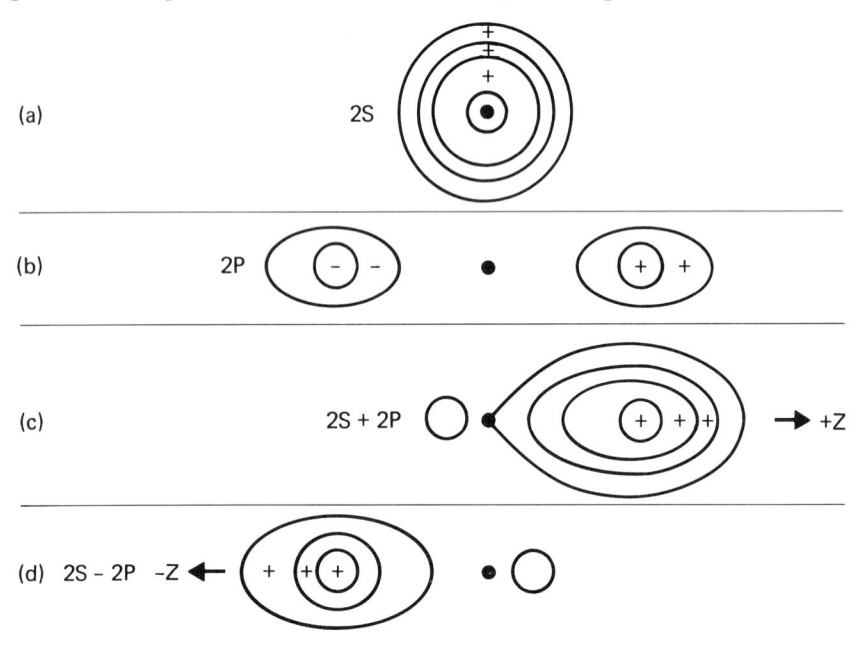

FIGURE 6.6 ● Hybridization of a $2s$ and a $2p$ orbital.

have a similar situation, but now the charge cloud is displaced in the negative Z direction and we have a hybridized orbital that points to the left, as shown in Fig. 6.6d.

If, in a molecule such as N_2, we have two hybridized orbitals, centered on the nuclei a and b and pointing toward each other, then it is clear that the overlap integral between these two orbitals may be quite large. The overlap integrals have been calculated for a variety of situations, and values as high as 0.6 or 0.7 are not uncommon. Since the degree of bonding may be associated with the magnitude of the overlap integral, it may be expected that hybridized orbitals give much stronger bonding than non-hybridized orbitals.

Let us now return to our discussion of the nitrogen molecule. On each atom we may combine the $2s$ and $2p_z$ function into two hybridized orbitals, ϕ_r, which points to the other atom, and ϕ_{lp}, which points away from the other atom. The two orbitals $\phi_{r,a}$ and $\phi_{r,b}$ may be combined into the symmetric and antisymmetric linear combinations:

$$\phi_{\text{bond}} = \phi_{r,a} + \phi_{r,b}$$
$$\phi_{\text{anti}} = \phi_{r,a} - \phi_{r,b} \tag{48}$$

We expect that the molecular orbital ϕ_{bond} has by far the lowest energy of the four remaining σ orbitals, and we can see that this orbital corresponds to the orbital $2\sigma_g$ in Table 6.3. Similarly, the antibonding orbital ϕ_{anti} must have the highest energy of the molecular orbitals, and it may be verified that this orbital corresponds to the molecular orbital $3\sigma_u$. The two orbitals ϕ_{lp} point away from the other atom and do not really participate in the bonding process. Such orbitals are known as lone-pair orbitals. Because of symmetry requirements, the two atomic lone-pair orbitals must be combined to give a symmetric and an antisymmetric σ orbital. These orbitals are neither strongly bonding nor strongly antibonding; they are known as nonbonding orbitals. It may be seen that the molecular orbitals $2\sigma_u$ and $3\sigma_g$ are nonbonding orbitals.

The two atomic $2p_x$ orbitals may be combined to form a bonding and an antibonding π orbital:

$$1\pi_u = 2p_{x,a} + 2p_{x,b}$$
$$1\pi_g = 2p_{x,a} - 2p_{x,b} \tag{49}$$

The $2p_y$ orbitals may be treated in the same way, and it is easily seen that the orbitals $1\pi_u$ and $1\pi'_u$ (constructed from $2p_y$ orbitals) are degenerate with one another. The overlap integral between the $2p_x$ orbitals is much smaller than for the hybridized σ orbitals, and we may conclude that the π orbitals are not as strongly bonding and antibonding as the σ orbitals.

We expect that the energy of the orbital $1\pi_u$ lies between the bonding and the nonbonding σ orbitals and that the energy of the orbital $1\pi_g$ lies between the nonbonding and the antibonding σ orbitals. It may be seen in Table 6.3 that these predictions are not quite correct, but that they are fairly close to the truth.

If we combine the above general arguments, we find that the rank-order of the molecular orbitals in terms of their energies is

$$(1s)_{\text{symm}} \to (1s)_{\text{antisymm}} \to \sigma_{\text{bonding}} \to \pi_{\text{bonding}} \to \text{lone pair 1}$$
$$\to \text{lone pair 2} \to \pi_{\text{antibond}} \to \sigma_{\text{antibond}} \quad (50)$$

or

$$1\sigma_g \to 1\sigma_u \to 2\sigma_g \to 1\pi_u \to 2\sigma_u \to 3\sigma_g \to 1\pi_g \to 3\sigma_u \quad (51)$$

A comparison with Table 6.3 shows that we made the wrong prediction about the order of $1\pi_u$ and $2\sigma_u$, but apart from that discrepancy our general arguments seem to be fairly satisfactory.

The ground-state configuration of the N_2 molecule is

$$(1\sigma_g)^2(1\sigma_u)^2(2\sigma_g)^2(2\sigma_u)^2(1\pi_u)^4(3\sigma_g)^2 \quad (52)$$

We must remember that the π orbitals are twofold degenerate, so that we can accommodate four electrons in them.

It is interesting to consider some other diatomic molecules. We should realize that the rank order (51) is approximately valid for all diatomic molecules. If we move from N_2 to O_2, we have two additional electrons that should be placed in the orbital $1\pi_g$ because this is the next orbital. However, this orbital is degenerate, because we have two antibonding π orbitals,

$$1\pi_g = 2p_{x,a} - 2p_{x,b}$$
$$1\pi_g' = 2p_{y,a} - 2p_{y,b} \quad (53)$$

There are three different ways to distribute two electrons over two orbitals, corresponding to the configurations

$$X^2 = (1\pi_g)^2 \quad Y^2 = (1\pi_g')^2 \quad XY = (1\pi_g)(1\pi_g') \quad (54)$$

The first two configurations must have antiparallel spins because of the exclusion principle, and they must therefore be singlet states. In the configuration XY the electron spins may be either parallel or antiparallel, and we expect to have a singlet XY state and a triplet XY state. It follows that the ground-state configuration of the oxygen molecule gives rise to four different eigenstates. The lowest state is the XY triplet state and the other three are the XY, X^2, and Y^2 singlet states. It may be shown that the singlet states give rise to two molecular states, one with Σ symmetry and the other with Δ symmetry. The oxygen molecule is one of the few molecules that has a triplet ground state, and we have seen now how this may be understood on the basis of molecular orbital theory.

If we proceed to the next homonuclear molecule, F_2, the situation becomes more straightforward. Here we have to accomodate four more electrons than in the N_2 molecule, and these can all be placed in the twofold degenerate $1\pi_g$ orbital. The ground-state configuration of F_2 is

$$(1\sigma_g)^2(1\sigma_u)^2(2\sigma_g)^2(2\sigma_u)^2(1\pi_u)^4(3\sigma_g)^2(1\pi_g)^4 \tag{55}$$

From the general arguments that we have outlined above, it is possible to predict the ground-state configuration of the molecule and to derive a crude approximation to the molecular orbitals. This is useful for two different reasons. First, it is convenient to have some general idea of the behavior of the molecular orbitals without having to resort to calculations. Second, in molecular SCF calculations just as in atomic SCF calculations, we must start with a rough approximation to the molecular orbitals because this is required for the construction of the zeroth-order Hartree-Fock operator. Through various iterations these approximate orbitals are refined until they converge to the self-consistent Roothaan-Hartree-Fock orbitals, which are the final result of the calculation.

So far we have considered only homonuclear diatomic molecules, where the molecular orbitals must be symmetric with respect to the coordinate z (the origin is taken at the midpoint of the internuclear axis). In the case of heteronuclear molecules, such as CO or HF, we cannot make use of this symmetry with respect to z and the molecular orbitals contain an additional unknown variable. For example, if we know the form of the hybridized atomic orbitals t_C and t_O on the carbon and on the oxygen atoms, then the molecular bonding orbital must be represented as

$$\chi_{\text{bond}} = t_C + \lambda t_O \tag{56}$$

This represents a charge cloud that is not symmetric. It is easily seen that the charge cloud is pulled toward the oxygen atom if λ is larger than unity and that it is displaced toward the carbon if λ is smaller than unity. If we want to make qualitative predictions about the asymmetry of the charges in heteronuclear chemical bonds, we should have some feeling for the relative power of the atoms in pulling electrons toward themselves. This question is related to the concept of electronegativity, which we discuss in the following section. We should point out that the electronegativity rules are useful only if we wish to make qualitative predictions about the general form of the charge distribution. From the Roothaan-type calculations we discussed above, we may derive much more precise information about the charge distribution in a molecule, and the whole concept of electronegativity has become somewhat obsolete since these more precise theoretical results have become available.

5. Electronegativity

The electronegativity of an atom may be defined as its capability to pull electrons toward itself in a chemical bond. This definition is somewhat vague, but then, the whole concept of electronegativity is somewhat imprecise. In a diatomic molecule AB it is possible to measure only the total charge distribution of all the electrons, and there is no unique way in which we can assign part of this charge cloud to atom A and part of it to atom B. It is difficult to define electronegativity in terms of molecular orbital theory, but it is possible to propose a definition by making use of valence bond theory. We can imagine that a molecule AB is a superposition of the three resonance structures AB, A^+B^-, and A^-B^+, and we can imagine also that each of these structures may be represented by a wave function, $\Psi(AB)$, $\Psi(A^+B^-)$, and $\Psi(A^-B^+)$, respectively. The total molecular wave function may then be written as a linear combination

$$\Psi = \Psi(AB) + \rho_1\Psi(A^+B^-) + \rho_2\Psi(A^-B^+) \tag{57}$$

If ρ_2 is larger than ρ_1, we may say that atom A is more electronegative than atom B and the difference $(\rho_2 - \rho_1)$ or the ratio (ρ_2/ρ_1) is a measure of the difference in electronegativity between the two atoms.

In polyatomic molecules there is ample evidence to support the idea that the various bonds in the molecule may be treated as separate, well-defined entities. For one thing, bond energies seem to be additive, and we can assign a fairly accurate value to the energy of a bond AB, independent of the rest of the molecule. As a result, the concept of electronegativity can also be used to describe chemical bonds in polyatomic molecules.

The goal of the various theories on electronegativity is the derivation of an electronegativity scale that applies to all atoms. This means that an electronegativity number x_A is assigned to each atom A in such a way that the relative electronegativity between two atoms A and B in a chemical bond AB is related to the difference $x_A - x_B$. The best-known electronegativity scales are the one by Pauling and the one by Mulliken. Both of these were derived from semiempirical arguments, and neither of them has been justified by rigorous theoretical derivations. However, the two electronegativity scales seem to be consistent with one another, and their predictions seem to be quite reasonable. Also, they have been useful in giving us a qualitative understanding of the chemical bond between different atoms.

Pauling noted that the dissociation energy of a bond AB between two different atoms is generally larger than the arithmetic average of the dissociation energies of the bonds AA and BB. Consequently, we can define a positive quantity Δ_{AB} as

$$\Delta_{AB} = D_{AB} - \tfrac{1}{2}(D_{AA} + D_{BB}) \qquad (58)$$

where D_{AB} stands for the dissociation energy. It is customary to express all quantities in the equation in terms of kilocalories. Pauling now defines the electronegativities x_A and x_B of the two atoms by means of

$$|x_A - x_B| = 0.208\sqrt{\Delta_{AB}} \qquad (59)$$

where Δ is expressed in terms of kilocalories/mole.

Unfortunately, the bond energies D_{AA} are available only for a limited number of elements, so Eq. (58) can be used only for a few atoms. In order to derive the electronegativity constants for the other elements, Pauling had to use more complicated thermochemical arguments involving the heats of formation of polyatomic molecules. In this way he derived the electronegativity values that we have listed in Table 6.4. It should be noted that the method gives only differences in negativity values. In order to choose a reference value, Pauling assigned the value $x_H = 2.1$ to the electronegativity value of the hydrogen atom. The electronegativities of the other elements are then defined uniquely.

The second electronegativity scale was derived by Mulliken. Here a quantity χ_M is defined as

$$\chi_M(A) = \tfrac{1}{2}(I_A + E_A) \qquad (60)$$

where I_A is the ionization potential and E_A is the electron affinity of atom A in its valence state. The valence state of an atom is its electron configuration when it participates in a chemical bond; this may be different from the configuration of the isolated atom. It is customary to express I_A, E_A, and χ_A in terms of electron volts. The relation between Mulliken's quantities χ_M and Pauling's electronegativities x is then given by

$$\chi_M(A) - \chi_M(B) = 2.78\,|x_A - x_B| \qquad (61)$$

It seems that the Mulliken electronegativity scale is more precisely defined than the Pauling scale because the ionization potentials and the electron affinities are usually better known than the bond energies used by Pauling. Also, the Mulliken scale is easier to understand, because I_A is

TABLE 6.4 • Electronegativity values. The first column contains Pauling's values. The second column contains Mulliken's figures, derived from Eq. (60) and converted to the Pauling scale by making use of Eq. (61). The third column contains the values currently used, which are considered the most suitable.

Atom	Pauling	Mulliken	Best
Ag	—	1.36	1.7
Al	1.5	1.81	1.5
As	2.0	1.75	2.0
B	2.0	2.01	2.0
Ba	0.9	—	0.9
Be	1.5	1.46	1.5
Br	2.8	2.76	2.8
C	2.5	2.63	2.6
Ca	1.0	—	1.0
Cl	3.0	3.00	3.0
Cs	0.7	—	0.7
F	4.0	3.91	4.0
H	2.1	2.28	2.1
I	2.5	2.56	2.5
K	0.8	0.80	0.8
Li	1.0	0.94	1.0
Mg	1.2	1.32	1.3
N	3.0	2.33	3.0
Na	0.9	0.93	0.9
O	3.5	3.17	3.5
Rb	0.8	—	0.8
S	2.5	2.41	2.5
Si	2.8	2.44	1.9

the energy of an electron in the highest filled molecular orbital and E_A is the energy of an electron in the lowest unfilled molecular orbital so that χ is roughly equal to the energy of a bonding electron when it is located on atom A. Clearly, the higher this energy is, the higher the chance that the electron is on atom A, so that χ_M should be a measure of the electronegativity of atom A. The Mulliken and Pauling scales are derived from different concepts, but the resulting set of electronegativity values are very similar if we use the conversion of Eq. (61). This may be verified from Table 6.4, where we have listed both sets of values.

Even though the electronegativity scale consists of a set of numbers, its practical use is more of a qualitative than of a quantitative nature. We should mention that several relations have been proposed between the electronegativity difference of the atoms in a bond and the parameters that occur in valence bond theory, but none of these relationships have been adequately justified. In addition, the valence bond parameters are poorly defined quantities that cannot be measured.

One quantity that is well defined and easily measured is the electric dipole moment $\boldsymbol{\mu}_0$ of a molecule. We have already mentioned that the dipole moment may be derived from the rotational spectrum of a molecule, but most experimental dipole moments are obtained by measuring the dielectric constant. In quantum mechanics the theoretical dipole moment is derived as

$$\boldsymbol{\mu}_0 = \langle \Psi_0 \mid \boldsymbol{\mu}_{op} \mid \Psi_0 \rangle$$
$$\boldsymbol{\mu}_{op} = e \sum_n Z_n \mathbf{R}_n - e \sum_j \mathbf{r}_j \tag{62}$$

Here Ψ_0 is the wave function of the molecular ground state. In the expression for the dipole moment operator, the first summation should be extended over all nuclei and the second summation over all electrons in the atom. A dipole moment has the dimension of charge multiplied by length. If we express the charge in terms of electrostatic units and the length in terms of centimeters, then the customary unit for dipole moment is 10^{-18} esu cm, which is known as a Debye unit (D). We should realize that the charge of an electron is 4.8×10^{-10} esu and that an angstrom unit is 10^{-8} cm, so that the dipole moment of two electronic charges with opposite signs, separated by 1 Å, is 4.8 D. Most experimental dipole moments range from 0.1 to 10 Debye units. It may be seen from Eq. (62) that the dipole moment is a vector, but experimentally we can determine only the magnitude of $\boldsymbol{\mu}_0$ and not its direction. In most cases we have a fairly good idea what the direction of the dipole moment should be, mainly by considering the difference in electronegativity and by making a comparison with similar molecules. However, for some molecules it is difficult to say what the direction of the dipole moment should be. A well-known example is CO, which has a small dipole moment of 0.11 D. Here it is still not quite certain whether the dipole moment is C^+—O^- or C^-—O^+.

It is remarkable that in the case of the hydrogen halides the dipole moments in terms of Debye units are almost equal to the electronegativity differences $x_X - x_H$, as may be seen from Table 6.5. However, this excellent

TABLE 6.5 ● A comparison between the electric dipole moments μ of hydrogen halides HX (in terms of Debye units) and the electronegativity differences $\Delta_{HX} = x_X - x_H$.

Molecule	μ	Δ
HF	1.9	1.9
HCl	1.07	0.9
HBr	0.8	0.7
HI	0.38	0.4

216 THE CHEMICAL BOND

agreement is nothing more than a fortuitous coincidence, because there are many reasons to believe that any simple relation between experimental dipole moments and electronegativity values cannot really be justified theoretically. From the various *ab initio* calculations, that is, the fairly accurate SCF calculations of the Roothaan group, and from subsequent more precise work, it may be seen that the theoretical dipole moments are quite sensitive to small variations in the eigenfunctions. Even these highly accurate calculations give relatively poor predictions about dipole moments. It also follows that the atomic charge clouds are not spherically symmetric, so that we must consider atomic dipole moments in addition to the effects due to differences in electronegativity. Altogether there are so many different effects that contribute to molecular dipole moments that simple rules for predicting their values generally do not work.

6. Hybridization

In Section 4, on diatomic molecules, we have discussed already some general features of the chemical bond. In this section we use these general ideas to treat the structures of some polyatomic molecules, in particular the hydrides water, ammonia, and methane. First we consider the methane molecule.

It has been shown experimentally that bond energies are additive. This means that we can assign a bond energy to a given bond in a saturated molecule. In Table 6.6 are listed some of these bond energies. The total

TABLE 6.6 ● Empirical bond energies.

Bond	Energy (kcal/mole)
H—H	104
H—C	99
H—F	135
H—Cl	103
H—Br	87
H—I	71
H—O	110
H—N	84
C—C	83
C=C	146
O—O	33
N—N	32
F—F	37
F—Cl	61
Br—Br	46
Br—I	42
I—I	36

dissociation energy of methane is approximately equal to four times the C—H bond energy (the difference with the experimental value is 4 kcal/mole). The dissociation energy of ethane is six C—H bond energies plus one C—C bond energy, and so on. For the higher alkanes the sums of the separate bond energies agree with the experimental energies of formation to within less than 1 kcal/mole. We may conclude, therefore, that the additivity rules for bond energies are surprisingly accurate for saturated molecules.

Let us now consider the quantum mechanical description of the methane molecule. We assume that the four C—H bonds may be treated as separate entities because of the additivity rules for the bond energies. This means that each C—H bond is represented by a molecular orbital χ_i and that the bond consists of two electrons with opposite spins which are both placed in the molecular orbital χ_i. The methane molecule has a total of ten electrons. Two of these are in the carbon (1s) orbital, which we denote by k_C, and the other eight electrons form four pairs, each of which occupies a bonding orbital χ_i. The total, antisymmetrized, molecular wave function is then obtained as

$$\Psi_0 = (10!)^{-1/2} \sum_p P \, \delta_p \, [k_C(1)\alpha(1) k_C(2)\beta(2) \chi_1(3)\alpha(3) \chi_1(4)\beta(4) \chi_2(5)\alpha(5)$$
$$\chi_2(6)\beta(6) \chi_3(7)\alpha(7) \chi_3(8)\beta(8) \chi_4(9)\alpha(9) \chi_4(10)\beta(10)] \quad (63)$$

The ground state of the carbon atom has the configuration $(1s)^2(2s)^2(2p)^2$. It seems reasonable to assume that the bonding orbitals χ_i may be represented as linear combinations of the (2s) and (2p) orbitals on the carbon and of the (1s) orbital of the hydrogen atom. We write this as

$$\chi_i = t_i + \rho s_H \quad (64)$$

where t_i is a linear combination of the carbon (2s) and (2p) orbitals and ρ is a parameter that depends on the electronegativities of carbon and hydrogen. It may be seen in Table 6.4 that $x_C = 2.5$ and $x_H = 2.1$; it seems therefore that ρ must be smaller than unity but that it is only slightly smaller than unity because the difference $x_C - x_H$ is only 0.4.

We have already mentioned in Section 4 that the strongest bonds are obtained if the overlap between the atomic orbitals that form the bond is large. The most favorable situation, that is, the largest amount of overlap, occurs when the atomic orbital is hybridized. This means that the atomic orbital should be a linear combination

$$t_i = \alpha s_C + \sqrt{1 - \alpha^2} \, p_{C,i} \quad (65)$$

where $p_{C,i}$ is the linear combination of (2p) orbitals that points in the direction of the ith hydrogen atom. Naturally, the magnitude of the overlap integral depends on the value of the parameter α.

218 THE CHEMICAL BOND

It is customary to impose the condition that the four orbitals t_i are orthogonal to one another, but it is not immediately obvious why this condition should be imposed, so it may be useful to say a few words about it. The total molecular wave function Ψ_0 is given by Eq. (63), and its right-hand side is the same as a mathematical quantity known as a determinant. A determinant remains unchanged under certain transformations; for example, its value remains the same if we add or subtract rows and columns. As a result, the individual molecular orbitals χ_i are not uniquely defined; they may be changed in certain ways without causing a change in the total wave function Ψ_0. Because of this freedom of choice, we may choose the molecular orbitals χ_i in such a way that they are orthogonal to one another. It is convenient to do this because it simplifies all subsequent calculations with the function Ψ_0 considerably. In our present discussion on methane, we choose the hybridized atomic orbitals t_i orthogonal to one another rather than the molecular orbitals χ_i, because this is more convenient in the present case.

In the case of methane, we know that the four C—H bonds are equivalent and that they form a tetrahedral structure. The same must be true for the hybridized atomic orbitals t_i. Furthermore, we impose the condition that the orbitals t_i are orthogonal to one another. The specific form of the orbitals is then easily derived from these conditions.

It is convenient to choose the Z axis along one of the bonds, for example, the bond corresponding to t_1. We have then

$$t_1 = as_C + bp_z \tag{66}$$

We choose the X axis in the plane of bonds 1 and 2 (see Fig. 6.7). The orbital t_2 is then given by

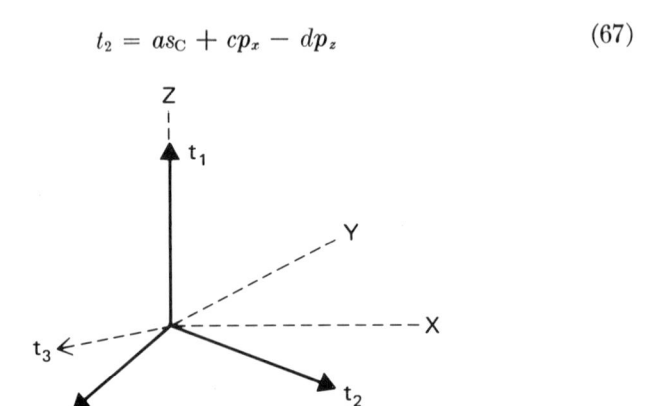

$$t_2 = as_C + cp_x - dp_z \tag{67}$$

FIGURE 6.7 ● Coordinate system for the hybridized orbitals in methane.

Since the orbitals t_1 and t_2 are equivalent, they differ only in their orientations and the coefficients of the orbitals s_C must be the same in t_1 and t_2. In Fig. 6.8 we have drawn the projections of the orbitals t_3 and t_4 in the XY plane. They have the form

$$t_3 = as - fp_x + gp_y - dp_z$$
$$t_4 = as - fp_x - gp_y - dp_z \qquad (68)$$

The three orbitals, t_2, t_3, and t_4, make the same angle with t_1 or with the Z axis, so that the coefficient of the orbital p_z must be the same for all three orbitals. In the same way it may be seen in Fig. 6.8 that the two orbitals t_3 and t_4 have the same coefficient for the orbital p_x and that the two coefficients of the orbital p_y must have the same magnitude but opposite signs.

The values of the coefficients may now be derived from the orthogonality and normalization conditions between the orbitals. Since the atomic orbitals s_C and p_α are all orthogonal to one another, we find that

$$\langle t_3 | t_3 \rangle = \langle t_4 | t_4 \rangle = a^2 + f^2 + g^2 + d^2 = 1 \qquad (69)$$

We also have

$$\langle t_3 | t_4 \rangle = a^2 + f^2 - g^2 + d^2 = 0 \qquad (70)$$

By subtracting Eq. (70) from Eq. (69) we obtain

$$2g^2 = 1 \qquad g = \tfrac{1}{2}\sqrt{2} \qquad (71)$$

and also

$$a^2 + f^2 + d^2 = \tfrac{1}{2} \qquad (72)$$

From the orthogonality between t_2 and t_3, we obtain the condition

$$\langle t_2 | t_3 \rangle = a^2 - fc + d^2 = 0 \qquad (73)$$

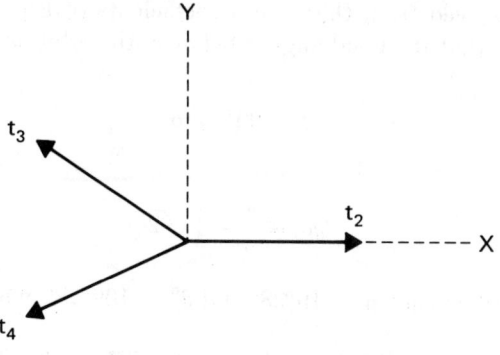

FIGURE 6.8 • Projections of methane orbitals in the XY plane.

and from the normalization condition for t_2 we find that

$$\langle t_2 | t_2 \rangle = a^2 + c^2 + d^2 = 1 \tag{74}$$

From the three equations (72), (73), and (74), it is easily derived that

$$c = \tfrac{1}{3}\sqrt{6} \qquad f = \tfrac{1}{6}\sqrt{6} \tag{75}$$

and also that

$$a^2 + d^2 = \tfrac{1}{3} \tag{76}$$

Finally, we have the equations

$$\begin{aligned} \langle t_1 | t_1 \rangle &= a^2 + b^2 = 1 \\ \langle t_1 | t_2 \rangle &= a^2 - bd = 0 \end{aligned} \tag{77}$$

The solution of Eqs. (76) and (77) is

$$a = \tfrac{1}{2} \qquad b = \tfrac{1}{2}\sqrt{3} \qquad d = \tfrac{1}{6}\sqrt{3} \tag{78}$$

By substituting the values of the coefficients into the expressions (66), (67), and (68), we find that the four orbitals are given by

$$\begin{aligned} t_1 &= \tfrac{1}{2}s_c + \tfrac{1}{2}\sqrt{3}\,p_z \\ t_2 &= \tfrac{1}{2}s_c + \tfrac{1}{3}\sqrt{6}\,p_x - \tfrac{1}{6}\sqrt{3}\,p_z \\ t_3 &= \tfrac{1}{2}s_c - \tfrac{1}{6}\sqrt{6}\,p_x + \tfrac{1}{2}\sqrt{2}\,p_y - \tfrac{1}{6}\sqrt{3}\,p_z \\ t_4 &= \tfrac{1}{2}s_c - \tfrac{1}{6}\sqrt{6}\,p_x - \tfrac{1}{2}\sqrt{2}\,p_y - \tfrac{1}{6}\sqrt{3}\,p_z \end{aligned} \tag{79}$$

The bond angle H—C—H in the methane molecule is known: It is equal to the tetrahedral angle. We want to show that the value of the bond angle may also be derived from the orbital coefficients of Eq. (79). It may be seen in Fig. 6.7 that the bond angle θ between the orbitals t_1 and t_2 may be written as

$$\theta = 90° + \alpha \tag{80}$$

and that

$$tg\alpha = \frac{d}{c} = \tfrac{1}{4}\sqrt{2} \tag{81}$$

It is easily verified that $\alpha = 19°28'$ and $\theta° = 109°28'$, which is the tetrahedral angle.

We may look upon the formation of the CH_4 molecule as a two-step procedure. First we "prepare" the carbon atom for bonding by rearranging its atomic structure so that we have four hybridized atomic orbitals, pointing in the four tetrahedron directions and each of them with one electron. This atomic configuration may be written as $(1s)^2(t_1)(t_2)(t_3)(t_4)$ and is known as the valence state of the carbon atom. The methane molecule is then formed by placing the four hydrogen atoms in the proper positions. Each hydrogen atom contributes one electron, which combines with the electron in the corresponding orbital t_i to form a two-electron bond.

In each of the one-electron states the probability of finding the electron in the $(2s)$ state is given by a^2, which is 0.25. In the valence state there is on the average $4 \times 0.25 = 1$ electron in the $(2s)$ state and $4 \times 0.75 = 3$ electrons in the $2p$ state, and we can write the atomic configuration of the atomic valence state also as $(1s)^2(2s)(2p)^3$. We have already mentioned that the configuration of the carbon atom ground state is $(1s)^2(2s)^2(2p)^2$, so that the valence state is derived from the ground state by transferring (or promoting) an electron from the $2s$ to the $2p$ state. It may be shown that the valence state has a higher energy than the atomic ground state; the energy difference, known as the promotion energy, is 96 kcal/mole.

We may explain the situation by assuming that the valence state of carbon is much more suitable for bond formation than the original atomic ground state. Apparently, it pays to excite the atom first to its valence state. This costs an amount of energy of 96 kcal/mole, which is 24 kcal/mole per bond. We have listed the C—H bond energy in Table 6.6. It is around 100 kcal/mole, and it seems that the increase in bond energy due to suitable hybridization more than compensates the excitation energy that is needed to bring the carbon atom to its valence state.

We can use the same arguments to construct the wave function of the ammonia molecule NH_3 (see Figure 6.9). Again we construct four hybridized

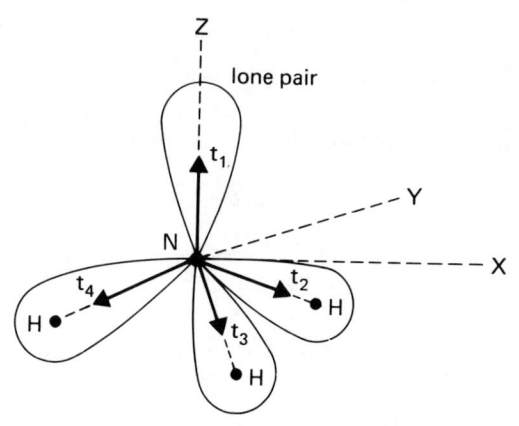

FIGURE 6.9 • Bonding and lone-pair orbitals in NH_3.

222 THE CHEMICAL BOND

orbitals t_i in order to represent the valence state of the nitrogen atom, but we should remember that now only three of the four orbitals, t_2, t_3, and t_4, are equivalent because there are only three NH bonds. We take the Z axis as the symmetry axis of the molecule. Then the three bonding orbitals may be written as

$$\begin{align} t_2 &= as_N + bp_z + cp_x \\ t_3 &= as_N + bp_z - dp_x + fp_y \\ t_4 &= as_N + bp_z - dp_x - fp_y \end{align} \tag{82}$$

There is a fourth orbital, t_1, which is not associated with any of the bonds. It follows from symmetry reasons (and from the orthogonality conditions) that this orbital must point in the negative Z direction, and we write it as

$$t_1 = gs_N - hp_z \tag{83}$$

We have

$$\begin{align} \langle t_3 | t_3 \rangle = \langle t_4 | t_4 \rangle &= a^2 + b^2 + d^2 + f^2 = 1 \\ \langle t_3 | t_4 \rangle &= a^2 + b^2 + d^2 - f^2 = 0 \end{align} \tag{84}$$

Hence,

$$f = \tfrac{1}{2}\sqrt{2} \qquad a^2 + b^2 + d^2 = \tfrac{1}{2} \tag{85}$$

We also have

$$\begin{align} a^2 + b^2 + c^2 &= 1 \\ a^2 + b^2 - cd &= 0 \end{align} \tag{86}$$

so that

$$c = \tfrac{1}{3}\sqrt{6} \qquad d = \tfrac{1}{6}\sqrt{6} \tag{87}$$

The remaining coefficients must satisfy the conditions

$$\begin{align} a^2 + b^2 &= \tfrac{1}{3} \\ g^2 + h^2 &= 1 \\ ag - bh &= 0 \end{align} \tag{88}$$

We cannot solve these equations, but if we set g equal to a parameter ρ, we find that

$$g = \rho \qquad h = \sqrt{1 - \rho^2} \qquad a = \tfrac{1}{3}\sqrt{3}\sqrt{1 - \rho^2} \qquad b = \tfrac{1}{3}\sqrt{3}\rho \tag{89}$$

The orbitals are then obtained as

$$\begin{align}
t_1 &= \rho s_N - \sqrt{1-\rho^2}\, p_z \\
t_2 &= \tfrac{1}{3}\sqrt{3}\sqrt{1-\rho^2}\, s_N + \tfrac{1}{3}\sqrt{3}\rho p_z + \tfrac{1}{3}\sqrt{6}\, p_x \\
t_3 &= \tfrac{1}{3}\sqrt{3}\sqrt{1-\rho^2}\, s_N + \tfrac{1}{3}\sqrt{3}\rho p_z - \tfrac{1}{6}\sqrt{6}\, p_x + \tfrac{1}{2}\sqrt{2}\, p_y \\
t_4 &= \tfrac{1}{3}\sqrt{3}\sqrt{1-\rho^2}\, s_N + \tfrac{1}{3}\sqrt{3}\rho p_z - \tfrac{1}{6}\sqrt{6}\, p_x - \tfrac{1}{2}\sqrt{2}\, p_y
\end{align} \quad (90)$$

The value of ρ may be derived from the experimental H—N—H bond angle, which is 108°. We note that the coefficients of the p orbitals represent vectors that point in the direction of the bond. The direction of t_2 is given by the vector $(\tfrac{1}{3}\sqrt{6}, 0, \tfrac{1}{3}\rho\sqrt{3})$, and the direction of t_3 is given by the vector $(-\tfrac{1}{6}\sqrt{6}, \tfrac{1}{2}\sqrt{2}, \tfrac{1}{3}\rho\sqrt{3})$. The angle θ between two vectors $\mathbf{u}$ and $\mathbf{v}$ may be derived from their inner product:

$$\mathbf{u} \cdot \mathbf{v} = uv \cos\theta \quad (91)$$

In our case this gives

$$-\tfrac{1}{3} + \tfrac{1}{3}\rho^2 = (\tfrac{2}{3} + \tfrac{1}{3}\rho^2) \cos 108° \quad (92)$$

or

$$\rho^2 = \frac{1 - 2\sin 18°}{1 + \sin 18°} \qquad \rho = 0.540 \quad (93)$$

The molecular wave function is now easily obtained from the hybridized orbitals t_i. Each of the bonding orbitals t_2, t_3, and t_4 combines with the $1s$ orbital s_i of the corresponding hydrogen atom to form a molecular orbital χ_i,

$$\chi_i = t_i + \lambda s_i \quad (94)$$

It may be seen in Table 6.4 that the electronegativity values of nitrogen and hydrogen are $x_N = 3.0$ and $x_H = 2.1$, so that λ is somewhat smaller than unity. The ammonia molecule has ten electrons, and its configuration is $(k_N)^2(\chi_2)^2(\chi_3)^2(\chi_4)^2(t_1)^2$. Two electrons are in the $(1s)$ orbital of the nitrogen atom, six electrons are in the three N—H bonds, and the remaining two electrons must be accomodated in the atomic orbital t_1. This orbital points away from the hydrogen atom, and it does not participate in the bonding process. It is known as a lone-pair orbital, and the electrons in it are called lone-pair electrons.

We may take the point of view that the ammonia molecule is a slightly perturbed tetrahedral structure. If we start off with four tetrahedral atomic orbitals (as in methane), then we place two electrons in the lone-pair orbital t_1 and use the other three to form the N—H bonds. It may be seen that the charge density of the lone pair in the vicinity of the nitrogen nucleus is somewhat higher than in the other orbitals, and this causes the Coulomb repulsion between the lone pair and a bond to be somewhat larger than the Coulomb repulsion between two bonds. The result is that the bond orbitals t_2, t_3, and t_4 are pushed away from the lone-pair orbital t_1, and the angle between a pair of bonding orbitals should be a little smaller than the tetrahedral angle of 109°28′. The experimental H—N—H bond angle is 108°, and this agrees with our model of the ammonia molecule as a perturbed tetrahedral structure.

It may be seen in Fig. 6.6 that for a hybridized atomic orbital the electrons are displaced in the direction of the hybridization, and this means that a lone-pair electron contributes to the molecular dipole moment. It is easily derived that the dipole moment of a pair of electrons in the orbital t_1 is given by

$$\mu_1 = 4\rho\sqrt{1 - \rho^2}\langle s_N \mid ez \mid p_z \rangle \tag{95}$$

If we evaluate this number we obtain a surprisingly large value, namely $\mu_1 = 3.42$ D. The experimental dipole moment of the ammonia molecule is only 1.47 D. If we assume that the total molecular dipole moment is the vector sum of the lone-pair moment $\mathbf{\mu}_1$ and of the three N—H bond moments, $\mathbf{\mu}_2$, $\mathbf{\mu}_3$, and $\mathbf{\mu}_4$, then we find that each N—H bond moment must be 1.9 D (N$^+$—H$^-$) in order to reproduce the experimental value of the total dipole moment. This may seem a large value, but it is not too different from what we calculate from the molecular orbitals χ_i. It may be seen that in the hybridized orbital t_i the electronic charge is shifted toward the proton (compare Eq. 95), whereas the hydrogen orbital s_i is spherically symmetric. It follows that the bond moment N$^+$—H$^-$ may be fairly large as a result of these atomic contributions even though the electronegativity scale predicts an excess of electronic charge on the nitrogen. We might add that we have little confidence in the various theoretical descriptions of electric dipole moments that are based on electronegativity differences only because they disregard the existence of the lone-pair moments and of the atomic dipole moments we mentioned above.

Finally, let us discuss the water molecule. Again, we treat the molecule as a perturbed tetrahedral structure. The four hybridized orbitals t_i consist of two equivalent pairs, the orbitals t_1 and t_2 participate in the two O—H bonds, and the orbitals t_3 and t_4 are the lone-pair orbitals. We have sketched the molecule in Fig. 6.10. It may be seen that we take the

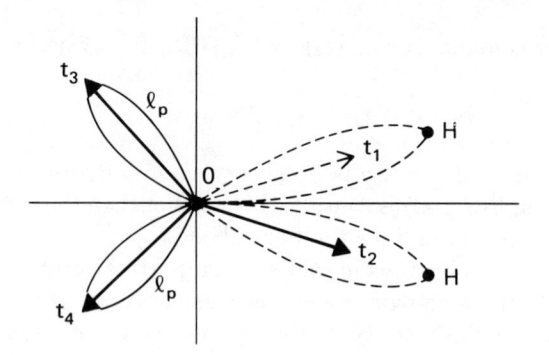

FIGURE 6.10 • Electronic structure of the water molecule.

bonding orbitals t_1 and t_2 in the XY plane, symmetric with respect to the X axis. They are then obtained as

$$t_1 = as_0 + bp_x + cp_y \\ t_2 = as_0 + bp_x - cp_y \tag{96}$$

Because of the molecular symmetry the lone-pair orbitals t_3 and t_4 must be in the XZ plane, symmetric with respect to the X axis. They may be written as

$$t_3 = ds_0 - fp_x + gp_z \\ t_4 = ds_0 - fp_x - gp_z \tag{97}$$

It is easily derived from the orthonormality relations between the orbitals that they must have the form

$$\begin{aligned} t_1 &= \tfrac{1}{2}\sqrt{2}(\rho s_0 + \sqrt{1-\rho^2}\, p_x + p_y) \\ t_2 &= \tfrac{1}{2}\sqrt{2}(\rho s_0 + \sqrt{1-\rho^2}\, p_x - p_y) \\ t_3 &= \tfrac{1}{2}\sqrt{2}(\sqrt{1-\rho^2}\, s_0 - \rho p_x + p_z) \\ t_4 &= \tfrac{1}{2}\sqrt{2}(\sqrt{1-\rho^2}\, s_0 - \rho p_x - p_z) \end{aligned} \tag{98}$$

If the H—O—H bond angle is θ then the parameter ρ is determined from the relation

$$\cot\frac{\theta}{2} = \sqrt{1-\rho^2} \tag{99}$$

The bond angle θ is 105°, and $\rho = 0.4865$.

The molecular configuration is $(k_0)^2(\chi_1)^2(\chi_2)^2(t_3)^4(t_4)^2$, where

$$\chi_i = t_i + \lambda s_i \tag{100}$$

The electronegativity difference between oxygen and hydrogen is 1.4, and we expect that λ in Eq. (100) is considerably smaller than the corresponding quantities for the N—H and for the C—H bonds.

Again, we have two electrons in each lone-pair orbital, and the electronic charge density in the lone-pair orbitals is considerably higher than in the bonding orbitals, in the vicinity of the oxygen molecule. Because of the Coulomb repulsions among the four orbitals (and some additional effects), the two bonding orbitals are slightly squeezed together as compared with the tetrahedral structure. This explains why the H—O—H bond angle of 105° is a bit smaller than the tetrahedral angle of $109\frac{1}{2}°$.

The experimental dipole moment of the water molecule is 1.82 D. This is due mainly to the dipole moment of the lone-pair electrons. If we calculate the lone-pair moments from the orbitals t_3 and t_4 of Eq. (98), we obtain a value that is considerably larger than the experimental moment 1.82 D. Again we are forced to conclude that the O—H bond moments are of the type O⁺—H⁻ in spite of the fact that the electronegativity values would predict just the opposite sign for the bond moment.

Today there are several theoretical methods available to derive approximate wave functions for polyatomic molecules. In Section 4 we mentioned the Roothaan SCF method, where the molecular orbitals for a diatomic molecule are expressed as linear combinations of atomic Slater orbitals. The coefficients and orbital exponents of the Slater orbital are derived by means of an electronic computer by using an SCF approach that is basically an application of the variation principle. It is clear that this method can in principle also be applied to polyatomic molecules and, in fact, it has been applied to some smaller polyatomic molecules. However, in applying the method to larger molecules we meet with some serious difficulties. In the first place we now encounter, in addition to the two-center integrals listed in Eq. (46), three-center or even four-center integrals over the Coulomb repulsion $(1/r_{ij})$. In the second place, the number of integrals that must be calculated increases very rapidly with the size of the molecule. This means that there is in practice a limit to the size of the molecules that can be calculated with the Roothaan method, because the amount of computer time required to calculate molecules of intermediate size such as ethyl alcohol, acetone, and so on, becomes prohibitive.

Because of these practical limitations on the possible applications of the Roothaan method, several attempts have been made to devise new methods that can be applied to larger molecules. Most of these methods are in principle based on the Roothaan method, but they contain additional approximations, which drastically reduce the amount of computer time required to get results. The best known of these methods are the CNDO method (complete neglect of differential overlap) and the INDO method (intermediate neglect of differential overlap), which were developed by Pople and his collaborators. Both of these methods contain some drastic approximations, but they seem to give fairly reliable results. They have become quite popular because of their simplicity, and they are widely used because the corresponding computer programs are readily available. We shall not discuss this method in detail; instead we refer the reader to the book by Pople and Beveridge listed in the bibliography at the end of this chapter. This book gives an excellent description of both the CNDO and INDO methods, and it even contains a FORTRAN IV computer program for doing the calculations.

7. Unsaturated Molecules

The chemical bonds that we discussed in the previous section are all known as single bonds. We have seen that each of them is formed by one pair of electrons with opposite spins, which occupy a localized molecular orbital. We know that there are molecules with double or even triple bonds, and it should be expected that these bonds differ in some respects from single bonds. We discuss the properties of multiple bonds by considering some of the hydrocarbons, namely ethane, ethylene, and acetylene. In Table 6.7 are listed the distances and energies of the various bonds in these molecules, and it may be seen that these quantities may vary considerably in going from one molecule to another. The structures of the three molecules are also quite different. The ethane molecule consists of two tetrahedral CH_3 groups, linked together by a C—C bond along two tetrahedral directions. The structure of ethane is thus quite similar to methane. The ethylene molecule is planar, with a H—C—C bond angle of 120°. Finally, the acetylene molecule is linear.

TABLE 6.7 ● Bond lengths and bond energies in some hydrocarbons.

Molecule	Bond	Length (Å)	Energy (kcal/mole)
CH_4	CH	1.090	103
C_2H_6	CH	1.10	99
C_2H_4	CH	1.069	106
C_2H_2	CH	1.060	121
C_2H_6	CC	1.54	83
C_2H_4	CC	1.34	146
C_2H_2	CC	1.205	201

228 THE CHEMICAL BOND

The electronic structure of the ethane molecule is readily understood by drawing an analogy with the methane molecule. We first construct two sets of tetrahedral hybridized orbitals u_i and t_i on the two carbon atoms, just as in methane, and we see to it that two of the orbitals on different carbon atoms, say t_1 and u_1, point toward each other. The C—C bond is then represented by an orbital

$$\chi_{CC} = t_1 + u_1 \tag{101}$$

and the C—H bonds by orbitals

$$\begin{aligned} \chi_i &= t_i + \lambda h_i \\ \chi_i' &= u_i + \lambda h_i' \end{aligned} \tag{102}$$

The molecular configuration is then $(k_C)^2(k_C')^2(\chi_1)^2(\chi_2)^2(\chi_3)^2(\chi_1')^2(\chi_2')^2(\chi_3')^2$ $(\chi_{CC})^2$, where the orbitals k_C and k_C' are $(1s)$ orbitals on the two carbons. The structures of propane and all the other saturated hydrocarbons may be derived in the same way, and all their bonds have the same orbitals as in ethane or methane.

Let us now consider the ethylene molecule. Its structure is quite different from ethane, because ethylene is planar. Also, the energies and distances of its bonds differ significantly from ethane, as may be seen from Table 6.7. If we wish to construct the various bonds from hybridized carbon orbitals, then these orbitals must all have directions that are in the plane of the molecule. We use the coordinate system of Fig. 6.11. The hybridized orbitals t_i on the first carbon atom are then given by

$$\begin{aligned} t_1 &= as_C + bp_x \\ t_2 &= as_C - cp_x + dp_y \\ t_3 &= as_C - cp_x - dp_y \end{aligned} \tag{103}$$

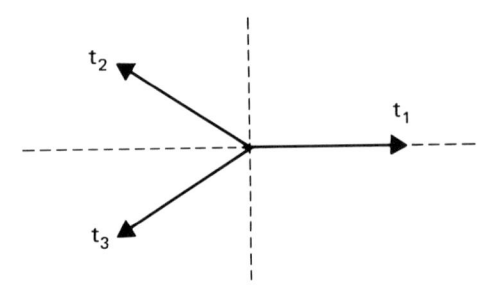

FIGURE 6.11 ● The three sp^2 hybridized orbitals on a carbon atom in ethylene.

The values of the coefficients are easily derived from the orthonormality relations between the orbitals, and we find that the result is given as

$$t_1 = \tfrac{1}{3}\sqrt{3}s_C + \tfrac{1}{3}\sqrt{6}p_x$$
$$t_2 = \tfrac{1}{3}\sqrt{3}s_C - \tfrac{1}{6}\sqrt{6}p_x + \tfrac{1}{2}\sqrt{2}p_y \qquad (104)$$
$$t_3 = \tfrac{1}{3}\sqrt{3}s_C - \tfrac{1}{6}\sqrt{6}p_x - \tfrac{1}{2}\sqrt{2}p_y$$

It is useful to note the differences between these orbitals and the methane orbitals of Eq. (79). In the present case we have only three hybridized orbitals, which are constructed from the $(2s)$ orbital and from two $(2p)$ orbitals; the $(2p_z)$ orbital has not been used in the construction of the hybridized orbitals. In the case of methane we have four hybridized orbitals, which are constructed from the $(2s)$ orbital and from all three $(2p)$ orbitals. The methane orbitals are known as sp^3 hybridized orbitals, and the ethylene orbitals are called sp^2 hybridized orbitals, for obvious reasons.

The various bonds in the ethylene molecule may now be constructed from the sp^2-type hybridized orbitals of Eq. (104) and from the corresponding set of hybridized orbitals u_i that belong to the other carbon atom. The C—C bond is represented by a molecular orbital

$$\chi_{CC} = t_1 + u_1 \qquad (105)$$

and the C—H bonds are represented by the orbitals

$$\chi_1 = t_2 + \lambda h_1 \qquad \chi_3 = u_2 + \lambda h_3$$
$$\chi_2 = t_3 + \lambda h_2 \qquad \chi_4 = u_3 + \lambda h_4 \qquad (106)$$

We may accomodate ten electrons in the above bonding orbitals and four electrons in the $(1s)$ orbitals of the two carbon atoms, for a total of 14 electrons. The ethylene molecule has 16 electrons, so we have two electrons left that are not accounted for. We should also remember that we have not yet considered the two $(2p_z)$ orbitals of the two carbon atoms because they do not occur in the bonding orbitals of Eqs. (105) and (106). The two p_z orbitals (p_z and p_z') may be combined to form a bonding and an antibonding molecular orbital,

$$\pi_1 = p_z + p_z'$$
$$\pi_2 = p_z - p_z' \qquad (107)$$

The molecular configuration is then obtained as $(k_C)^2(k_C')^2(\chi_1)^2(\chi_2)^2(\chi_3)^2(\chi_4)^2(\chi_{CC})^2(\pi_1)^2$.

If we compare the two molecules, C_2H_6 and C_2H_4, we see that the greatest difference between the molecules occurs in the C—C bond. In the case of ethane the C—C bond is formed by one electron pair in an orbital χ_{CC}; in the case of ethylene the C—C bond consists of two electron pairs, one in an orbital χ_{CC} and one in an orbital π_1, which is composed of the two p_z orbitals. It is customary in conjugated molecules to speak of σ orbitals, which are symmetric with respect to the plane of the molecule, and of π orbitals, which are antisymmetric with respect to the molecular plane. The double C—C bond in ethylene consists thus of a σ bond and a π bond. It may be seen in Table 6.7 that the C—C bond distances are 1.54 Å and 1.34 Å in ethane and in ethylene and the bond energies are 83 and 146 kcal/mole, respectively. The overlap integral between two $2p_z$ orbitals is much smaller than the overlap integral between two hybridized σ orbitals. The first integral is roughly 0.25 and the second integral is about 0.65. It should therefore be expected that the π bond energy is smaller than the σ bond energy, which is in agreement with the experimental values 83 and 146. It is not really permissible to say that the bond energy in ethylene is the sum of a 83 kcal/mole σ bond energy and a 63 kcal/mole π bond energy, because the C—C distance in the two molecules is also different, but it seems safe to conclude that the π bond energy is smaller than the σ bond energy.

It may be seen in Table 6.7 that there are also differences in the C—H bond lengths and energies between the two molecules. These differences may be ascribed to the different hybridization parameters in the carbon orbitals from which the bonds are constructed. We may recall that an sp^3 hybridized orbital has the form

$$t(sp^3) = \tfrac{1}{2}s_C + \tfrac{1}{2}\sqrt{3}\,p_\sigma \tag{108}$$

and that an sp^2 hybridized orbital is given by

$$t(sp^2) = \tfrac{1}{3}\sqrt{3}\,s_C + \tfrac{1}{3}\sqrt{6}\,p_\sigma \tag{109}$$

The two orbitals in Eqs. (108) and (109) should have somewhat different overlap integrals with the hydrogen $(1s)$ orbitals, and for that reason we expect that the C—H bonds that result from these orbitals should differ somewhat in their lengths and bond energies. It follows from Table 6.7 that the C—H bond is a little stronger in ethylene than it is in ethane, and this seems to indicate that the sp^2 hybridized orbitals have the larger overlap integrals. An exact calculation of the overlap integral agrees with this prediction.

Finally, let us consider the acetylene molecule, C_2H_2 (see Fig. 6.12). Since the molecule is linear, the bonds must all be formed by hybridized

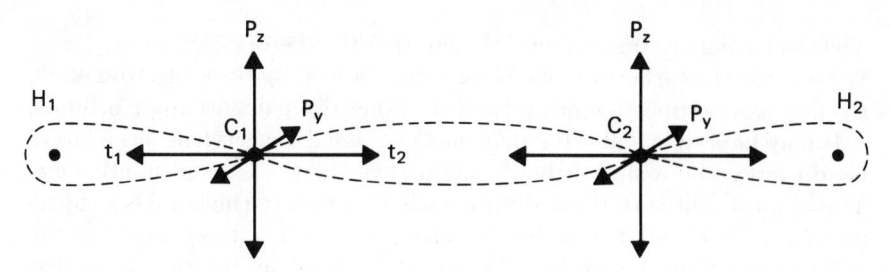

FIGURE 6.12 • Electronic structure of acetylene, C_2H_2.

orbitals that point in the X direction, hence they must be linear combinations of the carbon $2s$ and $2p_x$ orbitals only. It is easily derived that the proper hybridized orbitals are

$$t_1 = \tfrac{1}{2}\sqrt{2}(s_C + p_x)$$
$$t_2 = \tfrac{1}{2}\sqrt{2}(s_C - p_x) \qquad (110)$$

on the first carbon and

$$u_1 = \tfrac{1}{2}\sqrt{2}(s'_C + p'_x)$$
$$u_2 = \tfrac{1}{2}\sqrt{2}(s'_C - p'_x) \qquad (111)$$

on the second carbon. The two C—H bonds are then represented by the molecular orbitals

$$\chi_1 = t_2 + \lambda h_1$$
$$\chi_2 = u_2 + \lambda h_2 \qquad (112)$$

and the C—C bond by an orbital

$$\chi_{CC} = t_1 + u_1 \qquad (113)$$

(It should be noted that the coordinate systems on the individual atoms are always chosen in such a way that the positive coordinate direction points toward the other atom. In this case this means that the coordinate directions for the X axes on the two atoms have opposite directions.) The acetylene molecule has two π bonds. The first is represented by a linear combination of $2p_z$ orbitals,

$$\pi_1 = p_z + p'_z \qquad (114)$$

and the second by a linear combination of $2p_y$ orbitals,

$$\pi'_1 = p_y + p'_y \qquad (115)$$

The molecular configuration is then $(k_C)^2(k_C')^2(\chi_1)^2(\chi_2)^2(\chi_{CC})^2(\pi_1)^2(\pi_1')^2$. We see that the carbon–carbon bond is now formed by three electron pairs; the first pair occupies a σ orbital and the other two pairs occupy π orbitals.

It may be seen in Table 6.7 that the C—C bond in acetylene has a larger bond energy and a shorter bond length than the C—C bond in ethylene. This is consistent with the description where the acetylene bond is a superposition of a σ and two π bonds. The second π bond accounts for the difference of 55 kcal/mole between the C—C bonds in the two molecules. In acetylene the hybridized orbitals are linear combinations of one s and one p orbital, and they are called sp hybridized orbitals. These orbitals again have different overlap integrals, bond energies, and bond lengths from the corresponding quantities in ethylene and ethane. It may be seen in Table 6.7 that the experiments bear this out: The C—H bond energy in acetylene is 121 kcal/mole, which is quite a bit higher than the 106 kcal/mole in ethylene.

The concept of separate σ and π bonds and of σ and π orbitals is quite important in organic molecules. In general, if a molecule has a plane of symmetry, we may divide the molecular orbitals into σ orbitals, which are symmetric with respect to the molecular plane, and π orbitals, which are antisymmetric. It is customary, then, to make use of the Σ-Π approximation, where it is assumed that the σ and π orbitals may be treated separately, as if there is no interaction between them. In the examples we have discussed in this section, all the bonds and all the orbitals are localized, but there exist molecules where the π orbitals are not localized. We shall discuss these molecules in the following section.

8. Conjugated and Aromatic Molecules

We have already mentioned the benzene molecule in Section 2, and we used its valence bond description as an example in discussing the resonance principle. In Fig. 6.1 we sketched some of the possible valence bond structures of benzene, and we concluded that the benzene molecule is a superposition of these various structures. Experimentally, the benzene molecule is a regular hexagon with a C—C bond length of 1.397 Å. It should be noted that all the bond lengths are the same, and that the value lies somewhere between the value of 1.54 Å for the single C—C bond in ethane and the value of 1.34 Å for the double C—C bond in ethylene. It is easily seen that any theoretical description of benzene that is based on three localized single C—C bonds and three localized double C—C bonds is not consistent with the experimental information.

It was proposed in the early 1930s that in aromatic molecules the π electrons are basically delocalized. In the molecular orbital description we may treat the σ electrons in the same way as in the ethylene molecule. This means that we construct three sp^2 hybridized orbitals for each carbon atom as linear combinations of the $2s$, the $2p_x$ and the $2p_y$ orbitals on the atom. (We take the Z axis perpendicular to the plane of the molecule.) From the sp^2 hybridized orbitals we can then construct the bonding molecular orbitals that form the six C—H bonds and the six σ C—C bonds. At this point we have six electrons not accounted for (one for each carbon atom), and at the same time we have six p_z orbitals that we have not yet considered in deriving the total molecular wave function. It is assumed now that each of the six π electrons can move through the whole molecule. In the molecular orbital description, this means that a π electron is represented by a molecular orbital ϕ that is a linear combination of the six $2p_z$ orbitals,

$$\phi = \sum_j a_j p_{z,j} \tag{116}$$

The numbering of the carbon atoms is shown in Fig. 6.13; $p_{z,j}$ is the $2p_z$ orbital of the jth carbon atom.

The molecular orbitals for the π electrons are derived by means of the Hückel molecular orbital theory. This theory is based on a number of rather drastic approximations, and the method cannot really be justified from theoretical principles. On the other hand, the results of the Hückel theory have been quite useful, and they have been used to interpret a large number of experimental phenomena. In short, the method has worked much better than might have been expected. We shall briefly discuss it and illustrate it for the benzene molecule.

First we assume that the molecular orbitals for the π electrons are eigenfunctions of some sort of effective Hartree-Fock Hamiltonian $\mathcal{H}_{\text{eff}}$.

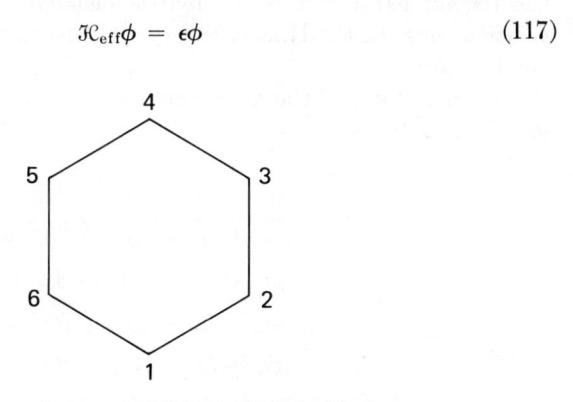

$$\mathcal{H}_{\text{eff}}\phi = \epsilon\phi \tag{117}$$

FIGURE 6.13 ● Numbering of atoms in MO calculation of benzene.

We substitute the expansion (116) into this equation,

$$\sum_j a_j \mathcal{H}_{\text{eff}} p_{z,j} = \epsilon \sum_j a_j p_{z,j} \tag{118}$$

If we multiply this equation on the left by one of the atomic orbitals $p_{z,l}$ and integrate, we obtain a set of linear equations for the coefficients a_j,

$$\sum_{j=1}^{6} (H_{l,j} - \epsilon S_{l,j}) a_j = 0 \qquad l = 1, 2, \ldots, 6 \tag{119}$$

where

$$\begin{aligned} H_{l,j} &= \langle p_{z,l} | \mathcal{H}_{\text{eff}} | p_{z,j} \rangle \\ S_{l,j} &= \langle p_{z,l} | p_{z,j} \rangle \end{aligned} \tag{120}$$

We can calculate the overlap integrals $S_{l,j}$, and we find that they are equal to unity if $l = j$, they are about 0.25 if l and j denote neighboring atoms, and they are 0.01 or less in all other cases. In the Hückel method it is assumed that they may be set equal to zero if $l \neq j$, and they are unity for $l = j$. It is difficult to calculate the matrix elements $H_{l,j}$ because we do not really know the form of the Hamiltonian. We do know that the matrix elements depend only on the distance between the atoms l and j and not on their absolute positions. We may therefore define two parameters α and β as

$$\begin{aligned} \alpha &= H_{l,l} \\ \beta &= H_{l,l+1} \end{aligned} \tag{121}$$

The second parameter is the matrix element between two neighboring carbon atoms. In the Hückel theory all other matrix elements $H_{l,j}$ are set equal to zero.

If we make use of the above approximations, we may write the set of equations (119) as

$$\begin{aligned} (\alpha - \epsilon) a_1 + \beta a_2 + \beta a_6 &= 0 \\ \beta a_1 + (\alpha - \epsilon) a_2 + \beta a_3 &= 0 \\ \beta a_2 + (\alpha - \epsilon) a_3 + \beta a_4 &= 0 \\ \beta a_3 + (\alpha - \epsilon) a_4 + \beta a_5 &= 0 \\ \beta a_4 + (\alpha - \epsilon) a_5 + \beta a_6 &= 0 \\ \beta a_1 + \beta a_5 + (\alpha - \epsilon) a_6 &= 0 \end{aligned} \tag{122}$$

We do not know the values of the parameters α and β, but they are treated as adjustable parameters in the Hückel theory. Besides, they may be eliminated from the equations (122). If we divide each equation by β and substitute

$$\frac{\alpha - \epsilon}{\beta} = -x \qquad \epsilon = \alpha + \beta x \tag{123}$$

we obtain

$$\begin{aligned}
-xa_1 + a_2 + a_6 &= 0 \\
a_1 - xa_2 + a_3 &= 0 \\
a_2 - xa_3 + a_4 &= 0 \\
a_3 - xa_4 + a_5 &= 0 \\
a_4 - xa_5 + a_6 &= 0 \\
a_1 + a_5 - xa_6 &= 0
\end{aligned} \tag{124}$$

The set of equations (124) form a set of six homogeneous equations in six unknowns. They have a solution only if the determinant of the coefficients is zero, and this condition gives an equation in the variable x from which the eigenvalues may be derived. This may be a fairly laborious procedure for large molecules, but today this is a routine problem for an electronic computer. Most Hückel calculations on organic molecules are now easily performed by using computers. In the specific case of benzene, we can solve the equations without having to resort to the theory of determinants.

We solve Eq. (124) by substituting

$$a_n = e^{in\rho} \tag{125}$$

All equations, except the first one and the last one, may be written as

$$a_{l-1} - xa_l + a_{l+1} = 0 \tag{126}$$

and if we substitute Eq. (125) into (126), we obtain

$$x = e^{i\rho} + e^{-i\rho} = 2\cos\rho \tag{127}$$

It follows that Eq. (125) in combination with Eq. (127) gives a solution for all equations except the first one and the last one. We now consider these two equations separately:

$$\begin{aligned}
-xe^{i\rho} + e^{2i\rho} + e^{iN\rho} &= 0 \\
e^{i\rho} + e^{(N-1)i\rho} - xe^{iN\rho} &= 0
\end{aligned} \tag{128}$$

Here we use N to show that our method is generally valid; in the present case $N = 6$. We now substitute for x the result of Eq. (127) and obtain

$$-e^{i\rho}(e^{i\rho} + e^{-i\rho}) + e^{2i\rho} + e^{iN\rho} = 0$$
$$e^{i\rho} + e^{i(N-1)\rho} - (e^{i\rho} + e^{-i\rho})e^{iN\rho} = 0 \qquad (129)$$

or

$$-1 + e^{iN\rho} = 0$$
$$e^{i\rho} - e^{i(N+1)\rho} = e^{i\rho}(1 - e^{iN\rho}) = 0 \qquad (130)$$

We see that both equations are solved if we take

$$e^{iN\rho} = 1 \qquad (131)$$

or

$$iN\rho = (k)\cdot 2\pi i \qquad k = 0, \pm 1, \pm 2, \ldots \qquad (132)$$

The corresponding eigenfunctions are given by

$$\phi_k = \sum_n \exp\left(\frac{2\pi i n k}{N}\right) p_{z,n} \qquad (133)$$

In Table 6.8 we have listed the eigenvalues and eigenfunctions for the benzene molecule, where $N = 6$. We obtain eigenvalues and eigenfunctions for $k = 0, \pm 1, \pm 2, 3$. For higher values of k the eigenfunctions are all identical with one of the previous eigenfunctions, so we should not consider

TABLE 6.8 ● Eigenvalues and eigenfunctions of the benzene molecule ($\gamma = 2\pi/6$).

k	ϵ	
0	$\alpha + 2\beta$	$p_{z,1} + p_{z,2} + p_{z,3} + p_{z,4} + p_{z,5} + p_{z,6}$
1	$\alpha + \beta$	$e^{i\gamma}p_{z,1} + e^{2i\gamma}p_{z,2} - p_{z,3} - e^{i\gamma}p_{z,4} - e^{2i\gamma}p_{z,5} + p_{z,6}$
-1	$\alpha + \beta$	$e^{-i\gamma}p_{z,1} + e^{-2i\gamma}p_{z,2} - p_{z,3} - e^{-i\gamma}p_{z,4} - e^{-2i\gamma}p_{z,5} + p_{z,6}$
2	$\alpha - \beta$	$e^{2i\gamma}p_{z,1} + e^{4i\gamma}p_{z,2} + p_{z,3} + e^{2i\gamma}p_{z,4} + e^{4i\gamma}p_{z,5} + p_{z,6}$
-2	$\alpha - \beta$	$e^{-2i\gamma}p_{z,1} + e^{-4i\gamma}p_{z,2} + p_{z,3} + e^{-2i\gamma}p_{z,4} + e^{-4i\gamma}p_{z,5} + p_{z,6}$
3	$\alpha - 2\beta$	$-p_{z,1} + p_{z,2} - p_{z,3} + p_{z,4} - p_{z,5} + p_{z,6}$
	$\alpha + 2\beta$	$(1/\sqrt{6})(p_{z,1} + p_{z,2} + p_{z,3} + p_{z,4} + p_{z,5} + p_{z,6})$
	$\alpha + \beta$	$(\sqrt{3}/6)(p_{z,1} - p_{z,2} - 2p_{z,3} - p_{z,4} + p_{z,5} + 2p_{z,6})$
		$(1/2)(p_{z,1} + p_{z,2} - p_{z,4} - p_{z,5})$
	$\alpha - \beta$	$(\sqrt{3}/6)(-p_{z,1} - p_{z,2} + 2p_{z,3} - p_{z,4} - p_{z,5} + 2p_{z,6})$
		$(1/2)(p_{z,1} - p_{z,2} + p_{z,4} - p_{z,5})$
	$\alpha - 2\beta$	$(1/\sqrt{6})(-p_{z,1} + p_{z,2} - p_{z,3} + p_{z,4} - p_{z,5} + p_{z,6})$

these higher k values. The eigenvalues $\alpha + \beta$ and $\alpha - \beta$ are twofold degenerate, and the eigenvalue that belongs to $k = 3$ is nondegenerate because $k = -3$ gives the same eigenfunction as $k = 3$. We may choose different basis sets for the eigenfunctions that belong to degenerate eigenvalues, and it is thus possible to select real eigenfunctions. These are the functions listed in the lower half of Table VIII; they are sums and differences of the complex functions.

We should realize that α and β must be negative quantities. The lowest eigenvalue is therefore $\alpha + 2\beta$, and the next lowest is the twofold degenerate eigenvalue $\alpha + \beta$. In the ground state we have two electrons in the state $k = 0$ and four electrons in the states $k = \pm 1$. The ground state energy of the π electrons in benzene is $6\alpha + 8\beta$.

The value of the parameter β may be related to a quantity known as the resonance energy and defined as the difference between the energies of the actual structure of a conjugated molecule and of one of the resonance structures. The resonance energy can be both measured and calculated in terms of β, and by comparing the experimental and theoretical values we can derive the value of β.

The calculation of the resonance energy is quite simple. The energy of a localized bonding π orbital (as in ethylene) is $\alpha + \beta$, so the energy of an electron pair in a π bond is $2\alpha + 2\beta$. In the resonance structure (see Fig. 6.1) benzene has three π bonds, so the energy of the π electrons in the resonance structure is $6\alpha + 6\beta$. It follows, then, that the resonance energy of benzene is given by

$$E_{\text{res}} = (6\alpha + 8\beta) - (6\alpha + 6\beta) = 2\beta \qquad (134)$$

The resonance energy can be derived experimentally by measuring the heat of hydrogenation of the molecule and by making a comparison with the experimental bond energy of a localized π bond. In this way we obtain a value of 37 kcal/mole. In Table 6.9 we have listed the theoretical and experimental resonance energies for some other aromatic molecules. It follows that β is roughly equal to 20 kcal/mole.

It may be interesting to consider a second example of the Hückel theory,

TABLE 6.9 • Resonance energies for some aromatic molecules.

Molecule	Theoretical	Experimental (kcal/mole)	β
benzene	2.00β	37	18.5
naphthalene	3.68β	75	20.4
anthracene	5.32β	105	19.7
phenanthrene	5.45β	110	20.2

namely the butadiene molecule. Here the normal structure is given by Fig. 6.14, but there are various indications that the π electrons are not strictly localized in this molecule and we may attempt to settle this question by performing the Hückel calculation. We write the molecular orbitals for the π electrons again as

$$\phi = \sum a_n p_{z,n} \qquad (135)$$

The coefficients must satisfy the equations

$$\begin{aligned}(\alpha - \epsilon)a_1 + \beta a_2 &= 0 \\ \beta a_1 + (\alpha - \epsilon)a_2 + \beta a_3 &= 0 \\ \beta a_2 + (\alpha - \epsilon)a_3 + \beta a_4 &= 0 \\ \beta a_3 + (\alpha - \epsilon)a_4 &= 0\end{aligned} \qquad (136)$$

or

$$\begin{aligned}-xa_1 + a_2 &= 0 \\ a_1 - xa_2 + a_3 &= 0 \\ a_2 - xa_3 + a_4 &= 0 \\ a_3 - xa_4 &= 0\end{aligned} \qquad (137)$$

if we substitute Eq. (123).

These equations may be solved in a similar manner as in the case of benzene. Again, we consider all equations except the first one and the last one, and we note that these equations have a solution

$$a_n = e^{in\rho} \qquad (138)$$

if we take

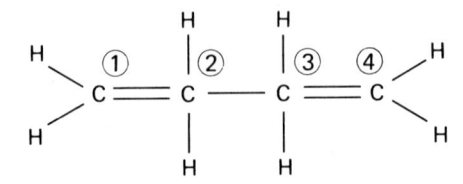

$$x = e^{i\rho} + e^{-i\rho} = 2\cos\rho \qquad (139)$$

FIGURE 6.14 ● Normal structure of butadiene and numbering of carbon atoms.

In the present case we note that there is a second solution,

$$a_n = e^{-in\rho} \tag{140}$$

if we take $x = 2\cos\rho$. The general solution of the equations is therefore given by

$$a_n = Ae^{in\rho} + Be^{-in\rho} \tag{141}$$

Let us now substitute equations (139) and (141) into the first and into the last equation (137),

$$\begin{aligned}-(e^{i\rho} + e^{-i\rho})(Ae^{i\rho} + Be^{-i\rho}) + Ae^{2i\rho} + Be^{-2i\rho} = 0 \\ (Ae^{i(N-1)\rho} + Be^{-i(N-1)\rho}) - (e^{i\rho} + e^{-i\rho})(Ae^{iN\rho} + Be^{-iN\rho}) = 0\end{aligned} \tag{142}$$

The two equations reduce to

$$\begin{aligned}A + B &= 0 \\ Ae^{i(N+1)\rho} + Be^{-i(N+1)\rho} &= 0\end{aligned} \tag{143}$$

and the condition for the eigenvalues is

$$e^{2i(N+1)\rho} = 1 \qquad (N+1)\rho = k\pi \tag{144}$$

The allowed values of k are

$$k = 1, 2, 3, \ldots, N \tag{145}$$

and the eigenfunctions are given by

$$\phi_k = \sum_n \sin\left(\frac{nk}{N+1}\right) p_{z,k} \tag{146}$$

In the case of butadiene, $N = 4$ and the eigenvalues are given by

$$\begin{aligned}E_1 &= \alpha + 2\beta \cos 36° = \alpha + 1.6180\beta \\ E_2 &= \alpha + 2\beta \cos 72° = \alpha + 0.6180\beta \\ E_3 &= \alpha + 2\beta \cos 108° = \alpha - 0.6180\beta \\ E_4 &= \alpha + 2\beta \cos 144° = \alpha - 1.6180\beta\end{aligned} \tag{147}$$

In the ground state there are two electrons in the state E_1 and two electrons in the state E_2, and the energy of these electrons is

$$E_{tot} = 4\alpha + 4.4720\beta \tag{148}$$

This energy is lower than the resonance structure with two localized π bonds because the resonance state has an energy $\alpha + 4\beta$. The resonance energy of the butadiene molecule is thus 0.47β.

The Hückel molecular orbital theory has also been used to make theoretical predictions about the C—C bond lengths in conjugated and aromatic molecules. The bond length in a molecule may be related to a quantity known as the bond order, which was introduced by Coulson. Before we give its formal definition, let us consider the case of the ethylene molecule. In the Hückel approximation the normalized bonding π orbital is given by

$$\phi = \tfrac{1}{2}\sqrt{2}p_{z,1} + \tfrac{1}{2}\sqrt{2}p_{z,2} \tag{149}$$

The bond order of the orbital is now defined as the product of the coefficients of the two orbitals, which is equal to one half. The total bond order is the sum of the bond orders for the individual electrons in their orbitals. Since we have two electrons in the orbital ϕ, the total bond order is two times one half, which is unity. The C—C bond order in ethylene is thus one, and this agrees with the fact that there is one π bond in the molecule.

In general, the π bond order between two adjacent carbon atoms k and l is defined as

$$P_{k,l} = 2 \sum_n c_{n,k} c_{n,l} \tag{150}$$

where the sum is to be extended over all occupied orbitals ϕ_n and where the orbitals are defined as

$$\phi_n = \sum_k c_{n,k} p_{z,k} \tag{151}$$

In Eq. (150) it has been assumed that the coefficients are all real, otherwise the bond order should be defined as

$$P_{k,l} = \sum_n (c_{n,k} c^*_{n,l} + c^*_{n,k} c_{n,l}) \tag{152}$$

In Table 6.10 we have calculated the bond orders in the benzene molecule from the real orbitals of Table 6.8 (obviously, the complex orbitals give the same result). We see that the π–π bond orders in benzene are all equal to $\tfrac{2}{3}$.

In butadiene the molecular orbitals that belong to the eigenvalues of Eq. (137) are easily derived from Eq. (146). The normalized orbitals are

$$\begin{aligned}
\phi_1 &= 0.3717 p_{z,1} + 0.6015 p_{z,2} + 0.6015 p_{z,3} + 0.3717 p_{z,4} \\
\phi_2 &= 0.6015 p_{z,1} + 0.3717 p_{z,2} - 0.3717 p_{z,3} - 0.6015 p_{z,4} \\
\phi_3 &= 0.6015 p_{z,1} - 0.3717 p_{z,2} - 0.3717 p_{z,3} + 0.6015 p_{z,4} \\
\phi_4 &= 0.3717 p_{z,1} - 0.6015 p_{z,2} + 0.6015 p_{z,3} - 0.3717 p_{z,4}
\end{aligned} \tag{153}$$

CONJUGATED AND AROMATIC MOLECULES 241

TABLE 6.10 ● Bond orders for benzene and butadiene.

Benzene

Bond	1–2	2–3	3–4	4–5	5–6	6–1
orb 1	$\frac{1}{6}$	$\frac{1}{6}$	$\frac{1}{6}$	$\frac{1}{6}$	$\frac{1}{6}$	$\frac{1}{6}$
orb 2	$-\frac{1}{12}$	$\frac{1}{6}$	$\frac{1}{6}$	$-\frac{1}{12}$	$\frac{1}{6}$	$\frac{1}{6}$
orb 3	$\frac{1}{4}$	0	0	$\frac{1}{4}$	0	0
total	$\frac{1}{3}$	$\frac{1}{3}$	$\frac{1}{3}$	$\frac{1}{3}$	$\frac{1}{3}$	$\frac{1}{3}$
bond order	$\frac{2}{3}$	$\frac{2}{3}$	$\frac{2}{3}$	$\frac{2}{3}$	$\frac{2}{3}$	$\frac{2}{3}$

Butadiene

Bond	1–2	2–3	3–4
orb 1	0.2236	0.3618	0.2236
orb 2	0.2236	−0.1382	0.2236
total	0.4472	0.2236	0.4472
bond order	0.8944	0.4472	0.8944

It has been found empirically that the theoretical bond orders may be related to the experimental bond lengths. We know the bond orders and the C—C bond lengths for ethane, ethylene, acetylene, benzene, and butadiene, and in Fig. 6.15 we have plotted the bond lengths versus the

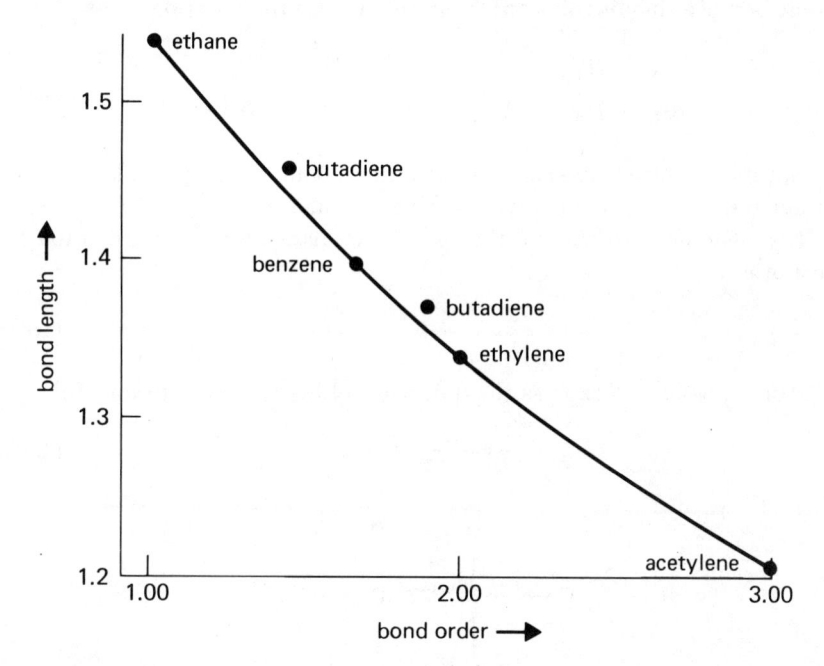

FIGURE 6.15 ● Relation between bond order and bond length in some conjugated molecules.

bond orders. We see that all the points that we have are situated on a curve, and we feel confident that we may use this curve for predicting the bond lengths of conjugated molecules from the calculated bond orders.

It should be noted that the 1–2 bond length in butadiene is 1.37 Å, which is somewhat longer than the bond length 1.34 in ethylene, and that the 2–3 bond length is 1.46, which is shorter than the bond length of 1.54 Å in ethane. We can see immediately that the 1–2 bond is not a pure π bond, so it must have a bond order smaller than unity. At the same time, we see that the 2–3 bond must have some π character. It follows thus from the bond lengths in butadiene that the simple valence bond structure of Fig. 6.14 does not give an adequate representation of the true butadiene structure. If we wish to use valence bond theory, we should consider resonance with the structure of Fig. 6.16 in order to explain the bond lengths in the molecule.

The Hückel molecular orbital method has been widely used in organic chemistry to make theoretical predictions about chemical reactivity, bond lengths, charge densities, and so on, of conjugated molecules. It is also possible to extend its applicability to heterocyclic molecules. As an example, we consider pyridine with the N atom in position 1 (see Fig. 6.13). We have to realize now that some of the Coulomb and exchange integrals are different because they involve the N atom. We use the notation

$$\alpha_N = H_{1,1} \qquad \alpha_C = H_{2,2} = H_{3,3} \cdots$$
$$\beta_{CN} = H_{6,1} = H_{1,2} \qquad \beta_{CC} = H_{2,3} = H_{3,4} \cdots \qquad (154)$$

Obviously we must have some information about the magnitude of these parameters, or we cannot solve the Hückel equations.

It is customary to assume that all the exchange integrals are equal to each other,

$$\beta_{CN} = \beta_{CC} = \beta \qquad (155)$$

It is then possible to express the difference of α_N and α_C in terms of β,

$$\alpha_N = \alpha_C + \gamma_N \beta \qquad (156)$$

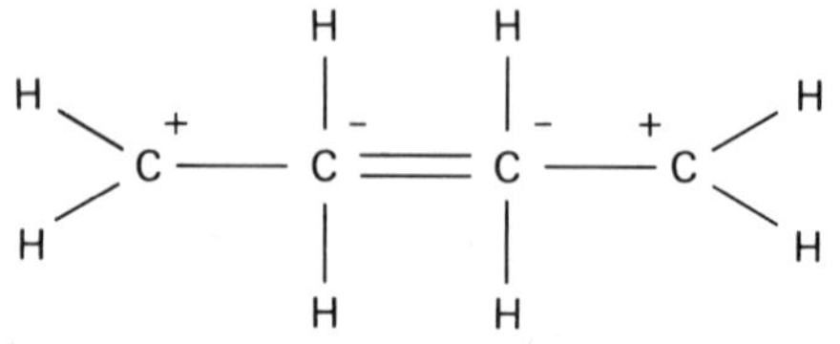

FIGURE 6.16 ● Ionic resonance structure of butadiene.

We may then reduce the Hückel equations for pyridine to the same form as the Hückel equations for benzene in Eq. (124), the only difference being that in the case of pyridine the first equation takes the form

$$(-x + \gamma_N)a_1 + a_2 + a_6 = 0 \tag{157}$$

The only unknown parameter is then γ_N.

It is to be expected that γ_N is positive, because nitrogen is more electronegative than carbon. It is more difficult to decide exactly which value should be assigned to γ_N, because this is often more a matter of taste than a decision based on sound theoretical considerations. In the case of pyridine the values that have been proposed in the literature vary from 0.4 to 1.0, and all these values have been used in calculations.

Once a particular value for γ_N has been chosen, the Hückel equations for the pyridine molecule may be solved. The analytical method that we discussed for solving the Hückel equations (124) for the benzene molecule is no longer applicable, because the presence of the nitrogen atom destroys the sixfold symmetry. It is now customary to solve the Hückel equations for conjugated molecules by using electronic computers. The programs are readily available, and the solutions are obtained as a set of eigenvalues ϵ_k and corresponding eigenvectors $\mathbf{c}_k$ so that the eigenfunctions are given by

$$\phi_k = \sum_j c_{k,j} p_{z,j} \tag{158}$$

The π electron charge density on the jth atom due to the molecular orbital ϕ_k is given by $c_{k,j} c_{k,j}^*$. The wave function of the molecular ground state is obtained by placing a pair of electrons each in the lowest molecular orbitals just as in the case of the Hartree-Fock orbitals. The total π electron charge density on the jth atom is then obtained as

$$P_j = 2 \sum_k c_{k,j} c_{k,j}^* \tag{159}$$

where we must sum over all occupied molecular orbitals. The net charge q_j on the jth atom is then

$$q_j = 1 - P_j \tag{160}$$

The final result of the calculation may be represented in a diagram such as we show in Fig. 6.17, which gives the charge densities in pyridine from a calculation where γ_N has been taken equal to 0.4. It may be seen that there is a net electronic charge on the nitrogen and on the two *meta* carbon atoms, and that the *ortho* and *para* carbons have a positive charge. Clearly, this result can be used to make predictions about the chemical reactivity of the pyridine molecule.

244 THE CHEMICAL BOND

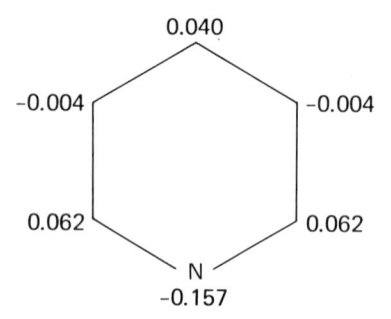

```
           0.040
            /\
  -0.004  /  \  -0.004
         |    |
   0.062  \  /  0.062
           \/
           N
         -0.157
```

FIGURE 6.17 ● Charge densities in pyridine.

The Hückel method has been widely used in organic chemistry to make predictions about charge distributions and chemical reactivities of organic molecules. The method is quite simple, and it is suitable for semiempirical considerations where the parameters are adjusted to make predictions about a series of similar molecules. Its results are surprisingly accurate, considering how approximate the method is.

Appendix

We want to calculate the following three integrals:

$$S = \langle s_a | s_b \rangle = \frac{1}{\pi} \int e^{-r_a} e^{-r_b} \, d\mathbf{r}$$

$$I = \langle s_a | r_b^{-1} | s_a \rangle = \frac{1}{\pi} \int e^{-r_a} \left(\frac{1}{r_b}\right) e^{-r_a} d\mathbf{r} \qquad \text{(A–1)}$$

$$J = \langle s_b | r_b^{-1} | s_a \rangle = \frac{1}{\pi} \int e^{-r_b} \left(\frac{1}{r_b}\right) e^{-r_a} \, d\mathbf{r}$$

The three integrals I, J, and S are calculated by using elliptical coordinates. We feel that it may be useful to discuss these coordinates, because they are often used in quantum mechanical calculations.

We consider the point P in Fig. 6.18, which is described by the vector $\mathbf{r}$. We take two points a and b on the Z axis, each of which is at a distance $\frac{1}{2}R$ from the origin so that the point a has the coordinates $(0, 0, -\frac{1}{2}R)$ and the point b has the coordinates $(0, 0, \frac{1}{2}R)$. We define r_a as the distance from a to P and r_b as the distance from b to P. We know then that

$$r_a + r_b \geq R \geq |r_a - r_b| \qquad \text{(A–2)}$$

APPENDIX 245

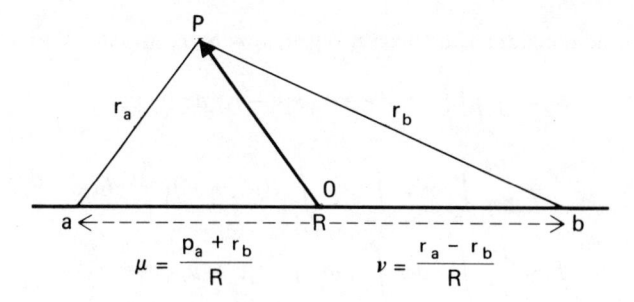

FIGURE 6.18 ● Elliptical coordinates.

Hence

$$\frac{r_a + r_b}{R} \geq 1 \qquad -1 \leq \frac{r_a - r_b}{R} \leq 1 \qquad \text{(A-3)}$$

The point P is determined by r_a, r_b and the angle ϕ between the plane aPb and the X axis. The elliptical coordinates of the point P are now defined as μ, ν, and ϕ, where

$$\mu = \frac{r_a + r_b}{R} \qquad \nu = \frac{r_a - r_b}{R} \qquad \text{(A-4)}$$

The range of values of the coordinates follows from Eq. (A-3),

$$\begin{array}{c} 1 \leq \mu \leq \infty \\ -1 \leq \nu \leq 1 \\ 0 \leq \phi \leq 2\pi \end{array} \qquad \text{(A-5)}$$

It is obvious that

$$r_a = \tfrac{1}{2} R(\mu + \nu) \qquad r_b = \tfrac{1}{2} R(\mu - \nu) \qquad \text{(A-6)}$$

The transformation between the Cartesian coordinates (x, y, z) and the elliptical coordinates (μ, ν, ϕ) is given by

$$\begin{aligned} x &= \tfrac{1}{2} R \cos \phi [(\mu^2 - 1)(1 - \nu^2)]^{1/2} \\ y &= \tfrac{1}{2} R \sin \phi [(\mu^2 - 1)(1 - \nu^2)]^{1/2} \\ z &= \tfrac{1}{2} R \mu \nu \end{aligned} \qquad \text{(A-7)}$$

Three-dimensional integrations are transformed as follows:

$$\int_{-\infty}^{\infty} \int_{-\infty}^{\infty} \int_{-\infty}^{\infty} f(x, y, z) \, dx \, dy \, dz = \frac{R^3}{8} \int_0^{2\pi} d\phi \int_{-1}^{1} d\nu \int_{1}^{\infty} f(\mu, \nu, \phi)(\mu^2 - \nu^2) \, d\mu$$

$$\text{(A-8)}$$

246   THE CHEMICAL BOND

Let us now consider the three integrations S, I, and J. We have

$$S = \frac{1}{\pi} \iiint \exp(-r_a) \exp(-r_b) \, dx \, dy \, dz$$

$$= \frac{R^3}{8\pi} \int_0^{2\pi} d\phi \int_{-1}^{1} d\nu \int_1^{\infty} (\mu^2 - \nu^2) e^{-\mu R} \, d\mu$$

$$I = \frac{R^2}{4\pi} \int_0^{2\pi} d\phi \int_{-1}^{-1} d\nu \int_1^{\infty} (\mu + \nu) e^{-R(\mu+\nu)} \, d\mu \qquad (A-9)$$

$$J = \frac{R^2}{4\pi} \int_0^{2\pi} d\phi \int_{-1}^{1} d\nu \int_1^{\infty} (\mu + \nu) e^{-\mu R} \, d\mu$$

The various integrals can be calculated exactly by standard methods. The results are

$$S = e^{-R}\left(1 + R + \frac{R^2}{3}\right)$$

$$I = \frac{1}{R}[1 - e^{-2R}(1 + R)] \qquad (A-10)$$

$$J = e^{-R}(1 + R)$$

Problems

1. In Eq. (20) we take the variational function for the hydrogen molecular ion as $e^{-\delta\mu}(1 + a\nu^2)$. Why didn't we take the variational function as $e^{-\delta\mu}(1 + a\nu + b\nu^2)$?
2. Show that the probabilities P_{aa}, P_{ab}, and P_{bb} for the hydrogen molecular ion, as defined by Eq. (22), have the values that we report in the text for $R = 2$ atomic units. Calculate the values also for $R = 2.5$ a.u. and compare the two sets of values.
3. Which method gives the best result for the bond energy of the hydrogen molecule, the VB method or the MO method? Can you explain this result?
4. Which method do you expect to give better results for the hydrogen molecule, the VB method with the inclusion of ionic structures or the MO method with the inclusion of configuration interaction?
5. What is the value of the overlap integral $\langle s_a \mid s_b \rangle$ at the internuclear distance $R = 2$ a.u. of the hydrogen molecular ion? What is the value of the same integral for $R = 1$ a.u., $R = 4$ a.u., and $R = 10$ a.u.?
6. Determine the values of the normalized wave functions $s_a + s_b$ for the hydrogen molecular ion at the position of one of the protons and at the center of the molecule. Where does the wave function have a larger value?
7. Why does the wave function of Eq. (36) give a better energy for the hydrogen molecule than the much more complicated Hartree-Fock function?
8. The oxygen molecule O_2 is one of the few molecules whose ground state is a triplet state. Can you explain why this is by making use of Eq. (51)?
9. Until recently, it was believed that the CH_2 radical has a linear configuration

in its ground state. Explain why it may be expected that in this configuration the ground state of the molecule is a triplet state.
10. It has been found recently that the CH_2 radical is slightly bent, and it was also found that its ground state is a triplet state. Can you explain this by means of arguments similar to those used for Problem 9?
11. The electronic dipole moment of the NH_3 molecule is 1.47, and the dipole moment of the NF_3 molecule is 0.21 Debye units. Explain the relative magnitudes of these numbers and, in particular, why the NF_3 moment is so small in spite of the large difference in electronegativity constants between N and F.
12. Why is the H—N—H bond angle of 108° in ammonia smaller than the tetrahedral angle of $109\frac{1}{2}°$?
13. If you have to do some numerical calculations with the set of sp^3 hybridized orbitals of carbon of Eq. (79), what would you substitute for the atomic $(2s)$ and $(2p)$ orbitals? Give the specific form of the orbitals and justify your choice.
14. Calculate the expectation value of the coordinate z with respect to the lone-pair orbital of Eq. (90) of the nitrogen atom. Choose the atomic orbitals yourself and justify your choice. Use your result to calculate the dipole moment in Debye units due to the lone pair of electrons in ammonia.
15. What are the relative magnitudes of the C—H, N—H, and O—H bond moments in methane, ammonia, and water? (Qualitative, not quantitative.) Justify your answer.
16. Why are the C—H bond energies different in ethane, ethylene, and acetylene?
17. Explain the electronic structure of the HCN molecule in terms of σ and π orbitals.
18. The triphenyl methyl radical has a planar structure, consisting of three phenyl rings attached to a carbon atom. How do you explain the relative stability of this radical as compared with the nonplanar triphenyl methane molecule?
19. In the aromatic molecule pyridine, C_5H_5N, the C—N—C bond angle is smaller than 120°. (Pyridine is benzene with one C—H group in the ring replaced by a nitrogen atom.) How do you explain this in terms of the sp^2 hybridized σ orbitals of the nitrogen atom?
20. Derive the Hückel energy eigenvalues for the conjugated molecule hexatriene,

$$\begin{array}{c} \text{H} \quad \text{H} \quad \text{H} \quad \text{H} \\ | \quad | \quad | \quad | \\ H_2C=C-C=C-C=CH_2 \end{array}$$

and derive the resonance energy of this molecule in terms of the Hückel parameter β.
21. Write down the equations for the molecular orbitals in the Hückel calculation of the naphthalene molecule. (In a form similar to the benzene equations).
There is no simple method for solving these equations, and we are not asking you to solve them.
22. The bond order for the central bond in naphthalene is 1.518. Predict the length of this bond from the curve in Fig. 6.15 and compare the result with the experimental value, 1.418 Å.
23. Write down the Hückel equations for cyclopentadiene, C_5H_6. (Consider carefully the electronic structure of the σ electrons before doing this.) Give the solutions of these equations, list the energy eigenvalues and the corresponding molecular orbitals.

248 THE CHEMICAL BOND

Bibliography

In this chapter we have given a brief discussion of the quantum theory of the chemical bond. This subject is also known under the name quantum chemistry, and it is treated in a number of books at different levels and from different points of view. In the following list the books have been divided into different categories. Category A contains some of the older books or, to be candid, the books that I used myself when I was a student. The review article by Van Vleck and Sherman (A1) may be old, but it still gives an excellent brief review of the subject. The book by Hellmann (A2) is no longer available and it is written in German, but it is an excellent book, especially considering its age. Coulson's book (A3) is very well written and gives a very good insight into the subject without using very much mathematics. The book by Eyring, Walter, and Kimball (A4) contains a lot of material, but I have always used it more as a reference book than a text. Kauzmann's book (A5) is just the opposite of Coulson's book; it is highly mathematical and it discusses general theory but not too many chemical applications. In category B I have listed some recent texts on quantum chemistry. These are all good books and I find it difficult to make a comparison of their relative quality, so I have just listed them in alphabetical order. Category C includes books that are concerned mainly with molecular orbital theory and its applications to organic chemistry of conjugated molecules. Finally, category D consists of the book by Pople and Beveridge on the CNDO and INDO methods and their applications which we mentioned in Section 6 of this chapter.

A1. J. H. Van Vleck and A. Sherman. The quantum theory of valence. *Rev. Mod. Phys.* **7,** 167 (1935).

A2. H. Hellmann. *Einführung in die Quantenchemie.* Deuticke, Leipzig (1937).

A3. C. A. Coulson. *Valence,* 2nd ed. Oxford University Press, London (1961).

A4. H. Eyring, J. Walter, and G. E. Kimball. *Quantum Chemistry.* Wiley, New York (1944).

A5. W. Kauzmann. *Quantum Chemistry.* Academic Press, New York (1957).

B1. J. M. Anderson. *Introduction to Quantum Chemistry.* W. A. Benjamin, Menlo Park, Calif. (1969).

B2. M. Karplus and R. N. Porter. *Atoms and Molecules.* W. A. Benjamin, Menlo Park, Calif. (1970).

B3. I. N. Levine. *Quantum Chemistry,* Vols. I and II. Allyn & Bacon, Boston (1970).

B4. P. O'D. Offenhartz. *Atomic and Molecular Orbital Theory.* McGraw-Hill, New York (1970).

B5. R. G. Parr. *The Quantum Theory of Molecular Electronic Structure.* W. A. Benjamin, Menlo Park, Calif. (1964).

B6. F. L. Pilar. *Elementary Quantum Chemistry.* McGraw-Hill, New York (1968).

B7. J. C. Slater. *Quantum Theory of Molecules and Solids,* Vol I. McGraw-Hill, New York (1963).

C1. R. Daudel, R. Lefebvre, and C. Moser. *Quantum Chemistry.* Wiley-Interscience, New York (1959).

C2. A. Streitwieser. *Molecular Orbital Theory for Organic Chemists.* Wiley, New York (1961).

C3. M. J. S. Dewar. *The Molecular Orbital Theory of Organic Chemistry.* McGraw-Hill, New York (1969).

D1. J. A. Pople and D. L. Beveridge. *Approximate Molecular Orbital Theory.* McGraw-Hill, New York (1970).

CHAPTER 7

THE SOLID STATE

1. Crystal Structures

Any solid consists of atoms or molecules, but usually they are arranged in space in an orderly fashion and we then call it a crystal. By definition, a crystal is characterized by translational symmetry. In order to illustrate what this means, we have sketched a one-dimensional crystal in Fig. 7.1, consisting of two kinds of atoms, A and B. The atoms A are all a distance a apart, and so are the atoms B. It may be seen that if we start from the position of an atom A and move a distance na, where n is an integer, then we shall end up on another atom A, that is, in an equivalent position. The same happens when we move a distance na away from an atom B. In order to describe the crystal structure we need consider only the elementary cell that we have drawn in Fig. 7.1, because the whole crystal consists of an infinite succession of identical elementary cells of length a. The elementary cell contains one atom A and one atom B. It may be seen that there are two atoms A at the boundaries of the elementary cell, but each of the two atoms is shared with the adjacent cells so that we count them as half; this gives a total of one atom A in our cell.

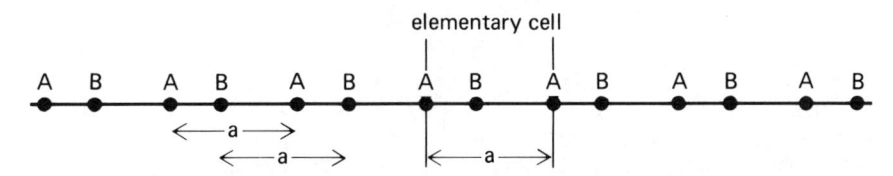

FIGURE 7.1 • One-dimensional crystal.

250 THE SOLID STATE

In a two-dimensional crystal the translational symmetry is characterized by two vectors, **a** and **b**. Here we should move from a particular atom to an equivalent atom if we move over a distance

$$\mathbf{d} = k\mathbf{a} + l\mathbf{b} \tag{1}$$

where k and l are integers. We illustrate this in Fig. 7.2, where we have drawn the structure of graphite. The graphite crystal has a layered structure, and each layer looks like a giant benzene molecule which may be considered a two-dimensional crystal. We have drawn the elementary cell in the middle of Fig. 7.2; it is a parallelogram, determined by the two vectors **a** and **b**. It may be seen that **a** and **b** have the same length and make an angle of 60° with one another.

In three dimensions the crystal structure is determined by three vectors, **a**, **b**, and **c**, and the elementary cell is the space spanned by these three vectors. The crystal is an infinite three-dimensional array of identical elementary cells.

The type of crystal structure is determined by the shape of the elementary cell. There are six types of elementary cells, which we have sketched in Fig. 7.3. In the first three structures, which are known as cubic, tetragonal, and orthorhombic, the three vectors **a**, **b**, and **c** are all perpendicular to one another. In the cubic structure the three vectors also have the same length, and the elementary cell is shaped like a cube. In the tetragonal structure two of the vectors have the same length and the third one is different. In the orthorombic structure the three vectors all have different lengths. In the monoclinic structure two of the vectors, say **a** and **b**, make an arbitrary angle with each other while the third vector **c** is perpendicular to the plane of the other two. In the triclinic structure the three vectors make arbitrary angles (different from 90°) with one another. The sixth crystal structure, hexagonal, stands apart from the above classification. This structure consists of parallel layers of the two-dimensional crystal

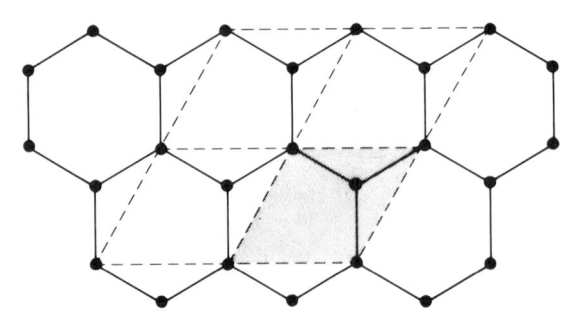

FIGURE 7.2 • Two-dimensional graphite crystal. The elementary cell is the shaded area.

1. cubic   2. tetragonal   3. orthorombic

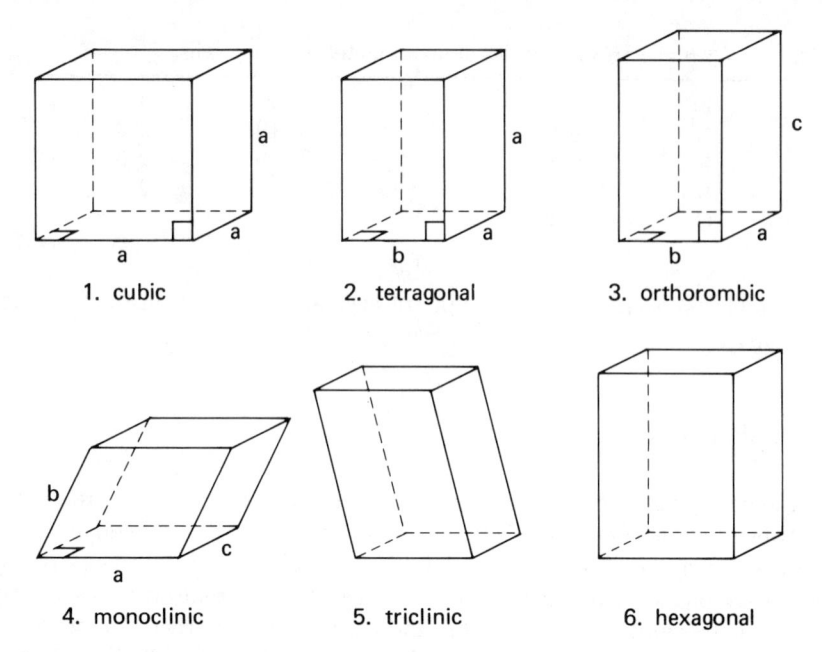

4. monoclinic   5. triclinic   6. hexagonal

FIGURE 7.3 ● The six possible crystal structures, their elementary cells.

structure that we have drawn in Fig. 7.2. The base of the elementary cell is identical with the elementary cell of Fig. 7.2, the two vectors **a** and **b** make an angle of 60° with one another, and the third vector **c** is then perpendicular to the plane of the other two. The length of **c** is usually different from the lengths of the other two vectors.

The above classification of crystals is concerned only with their geometry. A more chemical classification depends on the types of bonds between the atoms in the crystal. According to this point of view, we distinguish four different types of crystals.

1. Ionic crystals, which are composed of positive and negative ions. A typical example is NaCl. Here the forces between the different ions are mainly electrostatic.
2. Metals, where each atom contributes one or more valence electrons that can then move freely through the whole crystal. A metal consists of an array of positive ions and a sea of almost-free electrons.
3. Covalent crystals or semiconductors, such as diamond, graphite, and germanium. These are basically giant molecules in which the atoms are bonded together through covalent bonds.
4. Molecular crystals, such as benzene and anthracene. Here the molecules maintain their separate identity, and the interactions between different molecules is much smaller than the interactions within a specific molecule.

252 THE SOLID STATE

TABLE 7.1 • Electronegativities of alkali metals and halogens.

Li	1.0	F	4.0
Na	0.9	Cl	3.0
K	0.8	Br	2.8
Rb	0.8	I	2.5
Cs	0.7		

We discuss each of the four types of crystals separately in the following sections.

2. Ionic Crystals

An ionic crystal is composed of positive and negative ions, and the forces between the ions are of a purely electrostatic nature. This crystal structure occurs when the atoms involved have very different electronegativities. For example, in Table 7.1 we have listed the electronegativities for the alkali metals and for the halogens, and because the differences between the two groups are quite large, it is understandable that the alkali halides all occur as ionic crystals.

Let us consider two typical structures of alkali halide crystals, namely NaCl and CsCl. The NaCl crystal is composed of cubes of the type of Fig. 7.4a. There are Na$^+$ and Cl$^-$ ions at the corners of the cube in such a way that the nearest neighbors of one species all belong to the other ion species. It is clear that the cube of Fig. 7.4a is not the elementary cell, because the adjacent cubes have a different structure. The elementary cell is obtained by placing eight cubes of the Fig. 7.4a type together; the resulting elementary cell is drawn in Fig. 7.4b. This structure is known as

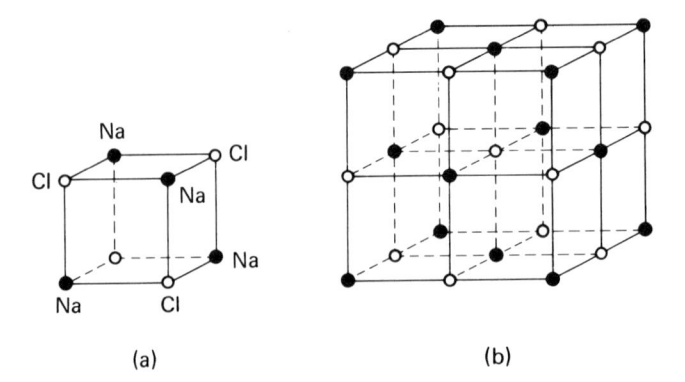

FIGURE 7.4 • Sketch (a) and elementary cell (b) of the NaCl structure.

TABLE 7.2 ● Ionic radii (in Å) of the alkali metal ions and of the halogen ions.

Li$^+$	0.60	F$^-$	1.36
Na$^+$	0.95	Cl$^-$	1.81
K$^+$	1.33	Br$^-$	1.95
Rb$^+$	1.48	I$^-$	2.16
Cs$^+$	1.69		

face-centered cubic, because it contains ions at the centers of the six faces. The length a of the side of the cube is 5.63 Å. The ion in the center of the cube, which we take as Cl$^-$, has six nearest neighbors, which are Na$^+$ ions at a distance of 2.81 Å.

We have seen in Chapter III that in an atom (or an ion) the charge density is a continuous function and that there is no sharply defined boundary for the dimensions of an atom. However, in ionic crystals it is assumed that the charge density of each individual ion drops off sharply at a certain distance and that we may define an ionic radius that represents the effective dimension of the charge cloud. In this model we approximate the charge cloud of the ion as a hard sphere, and the ionic radius is the radius of the sphere. According to this model, the shortest possible distance between two ions A$^+$ and B$^-$ in an ionic crystal must be the sum of the ionic radii $r(A^+)$ and $r(B^-)$. In Table 7.2 we have listed the atomic radii for the alkali metal ions and the halogen ions. It may be seen that the sum of the atomic radii for Na$^+$ and Cl$^-$ is 2.76 Å, which is fairly close to the experimental value of 2.81 Å. A number of ionic crystals have the same lattice structure as NaCl, namely LiH (lattice constant $a = 4.08$ Å), KBr ($a = 6.59$ Å), and RbI ($a = 7.33$ Å).

Another frequently occurring crystal structure is the body-centered cubic lattice, sketched in Fig. 7.5. The CsCl crystal has this structure. Here we have a Cl$^-$ ion in the center of the cube and Cs$^+$ ions at the eight corners. The lattice constant $a = 4.11$ Å. The shortest distance between a Cs$^+$ and a Cl$^-$ ion is $\tfrac{1}{2}a\sqrt{3} = 3.56$ Å. It may be seen from Table 7.2 that the sum of the atomic radii of Cs$^+$ and Cl$^-$ is 3.50 Å, which is again fairly close to the experimental value.

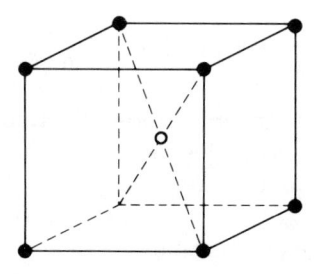

FIGURE 7.5 ● Body-centered crystal structure of CsCl, the elementary cell.

If we assume that every ion in the crystal is a hard sphere with a spherical charge density, then we may calculate the total electrostatic energy of the crystal by assuming that we deal with a set of point charges at the various lattice points. Let us first consider the NaCl structure of Fig. 7.4 and calculate the electrostatic energy of the central Cl⁻ ion. If we take R as the shortest NaCl distance, then there are 6 Na⁺ ions at a distance R, 12 Cl⁻ ions at a distance $R\sqrt{2}$, 8 Na⁺ ions at a distance $R\sqrt{3}$, and so on. The electrostatic energy $E(\text{Cl})$ of the central Cl⁻ ion is therefore given by the series

$$E(\text{Cl}) = -e^2 \left\{ \frac{6}{R} - \frac{12}{R\sqrt{2}} + \frac{8}{R\sqrt{3}} \cdots \right\} \qquad (2)$$

We can write this as

$$E(\text{Cl}) = -\frac{Ae^2}{R} \qquad (3)$$

where the constant A is derived by summing the infinite series in Eq. (2). The constant is known as the Madelung constant, and it depends on the crystal structure. In general, we may write the electrostatic energy for any ion in any crystal structure in the form of Eq. (3), where A is the Madelung constant and R is the closest distance between two ions in the crystal structure. The sum of Eq. (2) is difficult to evaluate, but the result has been derived for a variety of crystal structures. It was found that for the face-centered NaCl structure the Madelung constant is 1.748, and for the body-centered cubic CsCl structure the constant is 1.763. It follows that the body-centered cubic structure has a more favorable electrostatic energy. In general, it may be predicted that an alkali halide crystal has the CsCl structure if the atomic radii of the positive and negative ions are comparable in magnitude. If the radii are widely different, the crystal structure is more likely to be face-centered cubic.

Apart from some small corrections, the lattice energy U_0 is derived by multiplying Eq. (3) by Avogadro's number, N_0,

$$U_0 = -N_0 E(\text{Cl}) = -\frac{NAe^2}{R} \qquad (4)$$

This energy is defined as the energy of formation from 1 mole of Na⁺ ions and 1 mole of Cl⁻ ions. Obviously, this energy cannot be measured directly, because the solid NaCl is obtained experimentally from solid Na metal and gaseous Cl₂ molecules. It can be determined indirectly by means of the Born-Haber cycle, which gives the result:

$$\Delta E = U_0 + \tfrac{1}{2} D_0 + E_s + I(M) - E(X) \qquad (5)$$

Here ΔE is the heat of formation of the crystal MX, starting with solid metal M and gaseous X_2, D_0 is the dissociation energy of X_2 into atoms and $E(X)$ is the electron affinity, E_s is the sublimation energy of the atom and $I(M)$ its ionization energy. It is easily seen that the right-hand side of Eq. (5) represents the formation of ionic M^+ and X^- and if we add the unknown lattice energy U_0, the total should be equal to the experimental heat of formation ΔE. It may be derived in this fashion that the lattice energy of the NaCl crystal is 7.98 eV/NaCl unit. From Eq. (4) we derive that U_0 should be

$$U_0 = \frac{1.748 \times 27.205 \times 0.5292}{2.815} = 8.94 \text{ eV/NaCl unit} \qquad (6)$$

The difference between the theoretical and experimental values is due to a number of secondary effects, such as the polarization of the ions and the repulsive forces between the charge clouds. These effects can all be calculated, and they account almost exactly for the difference of 0.96 eV between the experimental value of U_0 and the value derived in Eq. (6). It may therefore be concluded that the electrostatic model gives an excellent account of the properties of ionic crystals.

3. Metals and Semiconductors

The division of the chemical elements into metals and nonmetals is discussed in most freshman chemistry courses. Some of the characteristic features of metals are their high electrical and thermal conductivity. It is generally assumed that in a solid metal one or more valence electrons can move freely through the whole crystal and that this accounts for most of the characteristic properties of metals. The crystal structures of metals are such that each ion is surrounded by a large number of neighbors; as a result, most metals have either a body-centered cubic, a face-centered cubic, or a hexagonal close-packed structure (the latter is sketched in Fig. 7.6).

Nonmetals usually crystallize in the form of covalent crystals. Typical examples are diamond, graphite, and germanium, in which the crystal may be considered a giant molecule in which the atoms are connected by covalent bonds. In diamond the bonds on each carbon form a tetrahedron, and they are bonded to the set of four nearest neighbors. Graphite is not a typical covalent crystal: It consists of separate layers of giant benzene molecules (without the hydrogens, of course).

In spite of the large differences between metals and covalent crystals, their theoretical descriptions are fairly similar, so we shall present a general

256 THE SOLID STATE

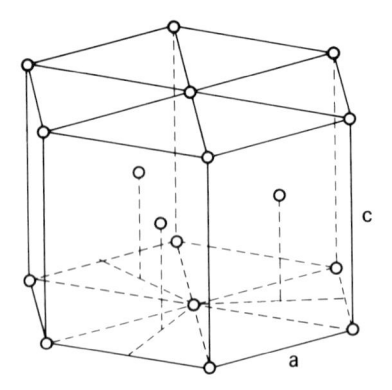

FIGURE 7.6 • Hexagonal close-packed structure, the elementary cell.

quantum mechanical description of both types of crystals. It turns out that the decisive factor that determines whether a crystal is a metal or an insulator is the form of the eigenvalue spectrum and the distribution of the electrons over the energy levels.

In order to understand the structure of a metal we consider Li, which crystallizes in the body-centered cubic structure sketched in Fig. 7.5. A Li atom has the configuration $(1s)^2(2s)$, and in the crystal the $1s$ electrons stay attached to the nucleus while the outer valence electron, the $2s$ electron, moves freely through the whole crystal. The crystal consists of Li$^+$ ions at the lattice points and a set of $2s$ electrons that move freely through the whole crystal. There is a number of different ways to describe the behavior of the $2s$ electrons, but one of the simpler approaches is to represent each electron by means of a one-electron orbital, which may then be written as a linear combination of atomic orbitals,

$$\phi_k = \sum_k a_{k,n}(2s)_n \tag{7}$$

Here $(2s)_n$ is the atomic orbital associated with the nth Li atom, and the summation is to be performed over all Li atoms in the crystal. In this model it is assumed that each electron moves in an effective potential field V_{eff}, so that its behavior is represented by an effective Hamiltonian,

$$\mathcal{H}_{\text{eff}} = \frac{-\hbar^2}{2m}\Delta + V_{\text{eff}} \tag{8}$$

The orbitals ϕ_k should then be solutions of the Schrödinger equation,

$$\mathcal{H}_{\text{eff}}\phi_k = \epsilon_k \phi_k \tag{9}$$

We can obtain some insight into the properties of the eigenvalues and eigenfunctions of this equation by considering a linear array of Li atoms (see Fig. 7.7); in other words, we consider a one-dimensional lattice. The eigenvalue problem (8) then becomes equivalent with the eigenvalue problem of the π electron chain or ring systems discussed in Chapter VI. The matrix elements of the effective Hamiltonian $\mathcal{H}_{\text{eff}}$ depend only on the relative positions of the atomic orbitals, so we have

$$\mathcal{H}_{n,n} = \alpha$$
$$\mathcal{H}_{n,n+1} = \mathcal{H}_{n+1,n} = \beta \tag{10}$$

where

$$H_{n,m} = \langle (2s)_n \mid \mathcal{H}_{\text{eff}} \mid (2s)_m \rangle \tag{11}$$

It is customary to treat a one-dimensional lattice as an infinite ring of N atoms. Its eigenvalues and eigenfunctions may then be obtained from the results of Chapter VI,

$$\epsilon_m = \alpha + 2\beta \cos \frac{2\pi m}{N}$$
$$\phi_m = \sum_n (2s)_n \exp\left(\frac{2\pi i n m}{N}\right) \tag{12}$$

with

$$m = 0, \pm 1, \pm 2, \pm 3, \ldots \tag{13}$$

In Fig. 7.7 we have sketched the distribution of the energy levels for the four cases $N = 2$, $N = 6$, $N = 12$, and $N = 30$. It may be seen that the energy levels must always be between the two limits

$$\alpha + 2\beta \leq \epsilon_m \leq \alpha - 2\beta \tag{14}$$

If N becomes very large then the energy levels are very close together, and in the limit where N tends to infinity we obtain a continuum of energy levels, situated within the limits of Eq. (14). Such a set of energy levels is called an energy band. In the present case the band has a width 4β.

If we deal with crystals, we write the eigenvalues and eigenfunctions of Eq. (9) in a somewhat different form, namely

$$\phi_k = \sum_n (2s)_n \exp(ikr_n) \tag{15}$$

where r_n denotes the position of the nth Li atom. If the distance between two adjacent Li atoms is a, then

$$r_n = na \tag{16}$$

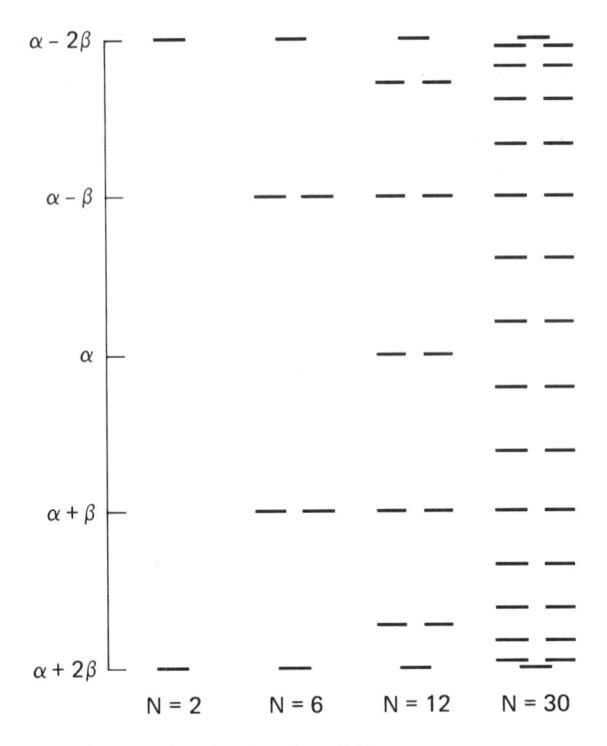

FIGURE 7.7 ● Energy levels of a ring of N atoms.

and by comparing this with Eq. (12) we find that k can assume the values

$$k = \frac{2\pi}{a} \frac{m}{N} \qquad m = 0, \pm 1, \pm 2, \ldots, \pm \tfrac{1}{2}N \tag{17}$$

If N is very large, k can take all values between the limits

$$-\frac{\pi}{a} \leq k \leq \frac{\pi}{a} \tag{18}$$

The energy values may also be rewritten in terms of k,

$$\epsilon(k) = \alpha + 2\beta \cos ak \tag{19}$$

The most important conclusions that we draw from this treatment are that the energy levels form a band and that the energy eigenvalues and the eigenfunctions are both continuous functions of a parameter k. The eigenfunctions are given by Eq. (15) and are known as Bloch functions. It should be noted that the form (15) of the eigenfunctions may be derived

METALS AND SEMICONDUCTORS   259

from the crystal symmetry, because these are the eigenfunctions that describe a system with translational symmetry.

Obviously, the atomic (2p) functions also combine in the form of Bloch functions:

$$\phi_k^\sigma = \sum_n (p_\sigma)_n \exp(ikr_n)$$
$$\phi_k^\pi = \sum_n (p_\pi)_n \exp(ikr_n) \qquad (20)$$

At first sight, it may seem that we have three different energy bands of the type of Eq. (19), namely a band for the (2s) energy levels, one for the $(2p_\sigma)$ levels and one for the $(2p_\pi)$ levels. We have sketched this situation in Fig. 7.8a, but it is obvious that this arrangement of energy bands is not likely to occur. Usually the (2s) and $(2p_\sigma)$ bands overlap, and we have the situations of either Fig. 7.8b or Fig. 7.8c. In Fig. 7.8b we have one broad energy band containing the (2s) and $(2p_\sigma)$ levels. In Fig. 7.8c we have two energy bands separated by an energy gap. Each of the two bands consists of (2s) and (2p) levels, but the lower band has lower energies. In an energy band the one-electron orbitals take the form

$$\psi_k = a_k \phi_k + b_k \phi_k^\sigma \qquad (21)$$

and the values of a_k and b_k depend on the values of k. The corresponding energies are also functions of k.

The above description may be generalized to three dimensions. Imagine that we have a monoclinic structure and that the elementary cell is defined by the three vectors **a**, **b**, and **c** and that we have one atom per elementary cell. The positions of the atoms are then given by

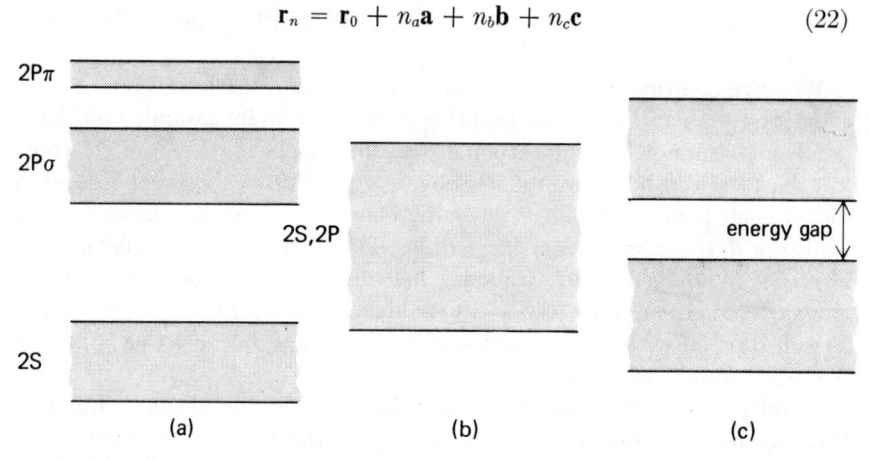

$$\mathbf{r}_n = \mathbf{r}_0 + n_a \mathbf{a} + n_b \mathbf{b} + n_c \mathbf{c} \qquad (22)$$

FIGURE 7.8 ● Possible arrangement of energy bands that may be constructed from (2s) and (2p) atomic orbitals.

where the three numbers n_a, n_b, and n_c denote the elementary cell. The Bloch functions are then defined as

$$\phi(\mathbf{k}) = \sum_n \phi_n \exp(i\mathbf{k} \cdot \mathbf{r}_n)$$
$$= \sum_n \phi_n \exp(ik_x n_a a + ik_y n_b b + ik_z n_c c + i\mathbf{k} \cdot \mathbf{r}_0) \quad (23)$$

where ϕ_n is the atomic orbital belonging to the atom in cell $\mathbf{n}$. The components k_x, k_y, and k_z of the vector $\mathbf{k}$ must satisfy the conditions

$$-\frac{\pi}{a} \le k_x \le \frac{\pi}{a}$$
$$-\frac{\pi}{b} \le k_y \le \frac{\pi}{b} \quad (24)$$
$$-\frac{\pi}{c} \le k_z \le \frac{\pi}{c}$$

by analogy with Eq. (18). Again, the energy levels form an energy band, but the energy is now a function of the three variables k_x, k_y, and k_z.

It is important to note that the theoretical description of a covalent crystal such as diamond is basically the same as the treatment of a metal that we discussed above. Again, we must consider the various (2s) and (2p) functions of each carbon atom in the elementary cell and combine them in the form of Bloch functions. It is then possible to derive the form of the energy bands, that is, the energy as a function of the vector $\mathbf{k}$, from the set of Bloch functions. The fact that diamond is an insulator and that a metal is a conductor is due to the form of the energy bands and the distribution of the electrons over these bands, as we shall discuss in the rest of this section.

We have mentioned before that diamond is a giant molecule, and at first sight it may seem that we can describe such a system by assuming localized bonds between neighboring carbon atoms and by placing a pair of electrons in each localized bond orbital. However, we should realize that in such a model each bond orbital has the same energy, so that this model would lead to a degeneracy of very large order. Due to the interactions between electrons in adjacent bonds, this degenerate energy level broadens into an energy band. If we were to take these effects into account, then we might as well start off with Bloch orbitals in the first place, because this approach is simpler and leads to the same answer.

Usually the form of the function $E(\mathbf{k})$ is fairly complicated, but fortunately we can draw some conclusions from the general behavior of the energy bands without having to know their specific forms.

We note first that the energy levels occur in pairs, corresponding to

METALS AND SEMICONDUCTORS   261

positive and negative values of the vector **k**. The one level represents an electron traveling in the direction **k** and the other level represents an electron traveling in the opposite direction. Such a pair of electrons cannot give rise to any net transport of electrons in any direction, and it does not contribute to the electrical conductivity. It follows that we must have unpaired electrons in order to observe electrical conductivity (We do not consider superconductivity).

Let us now consider how the available electrons are distributed over the available energy levels. At zero temperature all electrons are in the lowest energy levels that are available; that is, we place the first pair of electrons in the lowest energy levels, the next pair in the next lowest, and so on until all electrons are accommodated. In Fig. 7.9 we have sketched the two possible situations that may occur. In Fig. 7.9a the energy band is completely filled up by all the available electrons and the next higher energy band, which is empty at zero temperature, is separated from the filled band by a fairly large energy gap. In Fig. 7.9b the energy band is only partially filled by the electrons.

It may be seen that Fig. 7.9b represents the situation that exists in a metal. At finite temperatures the electrons in the higher occupied levels will move to the lower of the empty levels because of thermal excitation. Since there is no separation between the filled and the empty energy levels (they form a continuum), there will be a substantial fraction of the electrons in singly occupied energy levels even at low temperatures. All the unpaired electrons contribute to the electrical conductivity, and it follows that we have a conductor.

The situation in Fig. 7.9a is quite different. Here the highest occupied level and the lowest empty level are separated by a fairly large energy gap ΔE. The fraction of electrons that is excited into the empty energy band is given by $\exp(-\Delta E/kT)$, and it follows that we have a very small elec-

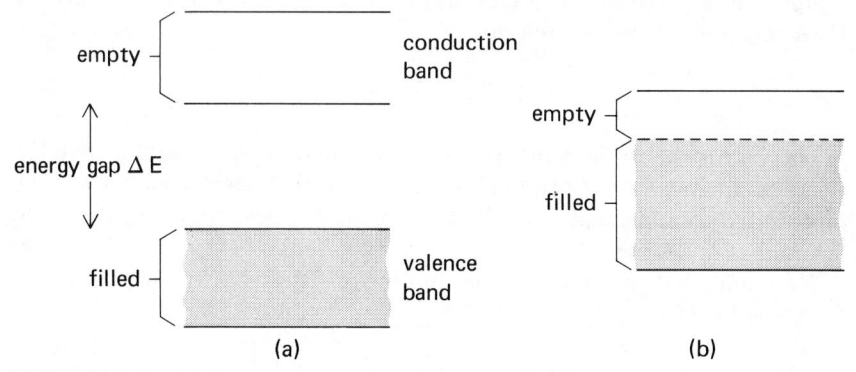

FIGURE 7.9 ● The distribution of electrons in the energy bands in a semiconductor (a) and in a metal (b).

trical conductivity unless the temperature T is high enough so that kT becomes comparable with ΔE. It follows that Fig. 7.9b represents a semiconductor. Its electrical conductivity is quite small, but if the temperature becomes high enough that some of the electrons are thermally excited into the empty band, then we have a measurable electrical conductivity. Of course, if we shine light on a semiconductor, then it may be possible to excite the electrons optically from the filled to the empty energy band, in which case we have photoconductivity.

The electrical conductivity of a semiconductor is determined mostly by the magnitude of the energy gap ΔE. A typical example is germanium, where ΔE is approximately 0.7 eV; this corresponds to a temperature of about 8000°K. It may be seen that at room temperature $\exp(-\Delta E/kT)$ is approximately 10^{-11}, so the conductivity is quite small. In the case of diamond, the energy gap is 6 eV, so for all practical purposes this may be considered an insulator. Gray tin has a fairly small energy gap, namely 0.08 eV, which corresponds to a temperature of 900°K. It follows that gray tin has a sizeable conductivity at room temperature.

We mentioned that the calculation of the energy bands of a solid, that is, the calculation of E as a function of the three variables k_x, k_y, and k_z, is a fairly complicated problem. However, some of the properties of metals and semiconductors may be derived from fairly simple models. We shall illustrate this by deriving the Fermi energy and the electronic heat capacity of a metal.

We have already mentioned that in a metal (or semiconductor) we place the available electrons in the lowest possible energy levels at zero temperature. We start with a pair of electrons in the lowest level, then we place a pair in the next lowest, and we keep doing this until we have accommodated all the available electrons. The highest occupied energy level at zero temperature is known as the Fermi level of the metal, and the corresponding energy is called the Fermi energy $E_F(0)$. The effective Fermi temperature T_F is then defined by the relation

$$kT_F = E_F(0) \qquad (25)$$

The Fermi energy of a metal may be calculated by assuming that the electrons can move freely through the metal, so that their behavior may be represented by the simple model of a particle in a box, which we discussed in Chapter I, Section 4.

We found that in a one-dimensional box of length L, the energy levels are given by

$$E_n = \frac{n^2\hbar^2\pi^2}{2mL^2} = \frac{n^2h^2}{8mL^2} \qquad n = 1, 2, \ldots \qquad (26)$$

where m is the mass of the particle. In a three-dimensional cubic box of length L, the energy levels are given by

$$E_n = \frac{\pi^2 \hbar^2}{2mL^2}(n_x^2 + n_y^2 + n_z^2) = \frac{\pi^2 \hbar^2 n^2}{2mL^2} \qquad (27)$$

with

$$\begin{aligned} n_x &= 1, 2, 3, \ldots \\ n_y &= 1, 2, 3, \ldots \\ n_z &= 1, 2, 3, \ldots \\ n &= (n_x^2 + n_y^2 + n_z^2)^{1/2} \end{aligned} \qquad (28)$$

We have N electrons in the box, and we fill up the energy levels by placing a pair of electrons in each energy level until we have accommodated all N electrons. The highest occupied level is characterized by a value $n_{\max}$, and it follows from the definition that

$$E_F(0) = E(n_{\max}) = \frac{\pi^2 \hbar^2 n_{\max}^2}{2mL^2} \qquad (29)$$

We must determine first how many energy levels there are below the energy value $E(n_{\max})$. If we consider a three-dimensional space and we represent each set of values (n_x, n_y, n_z) by a point in space with the Cartesian coordinates (n_x, n_y, n_z), then a sphere of radius $n_{\max}$ contains

$$S(n_{\max}) = \frac{4\pi (n_{\max})^3}{3} \qquad (30)$$

points because there is one point per unit volume. The values of (n_x, n_y, n_z) must be positive, so only one octant of the sphere is filled. Consequently, the number of energy levels S_0 below $n_{\max}$ is given by

$$S_0 = \frac{\pi}{6}(n_{\max})^3 \qquad (31)$$

The relation between the number of electrons N and $n_{\max}$ is thus given by

$$\frac{1}{2}N = S_0 = \frac{\pi}{6}(n_{\max})^3 \qquad (32)$$

or

$$n_{\max} = \left(\frac{3N}{\pi}\right)^{1/3} \qquad (33)$$

By substituting this into Eq. (29), we find that the Fermi energy is given by

$$E_F(0) = \frac{\pi^2\hbar^2(3N/\pi)^{2/3}}{2mL^2} = \frac{\hbar^2}{2m}\left(\frac{3\pi^2 N}{L^3}\right)^{2/3} \tag{34}$$

If we define

$$N_0 = \frac{N}{L^3} \tag{35}$$

as the number of particles per unit volume, this reduces to

$$E_F(0) = \frac{\hbar^2}{2m}(3\pi^2 N_0)^{2/3} \tag{36}$$

In the case of Li there are approximately 5×10^{22} free electrons/cm³, and the Fermi energy becomes about 5 eV. In general, it may be derived from Eq. (36) that the Fermi energies of most metals range from about 1 to about 6 eV. This is the range of the energy band part that is filled with electrons at zero temperature. The corresponding Fermi temperatures T_F vary from about 10,000°K to about 60,000°K.

Let us now derive an expression for the specific heat of a metal. If we consider a metal at zero temperature and if we raise the temperature to a value T, then only the electrons at the upper part of the filled band will be excited. The fraction of electrons that is excited thermally is roughly equal to T/T_F. Each of these electrons will gain an amount of energy kT. It follows that the increase in electronic energy is then

$$E_{\text{el}}(T) = \frac{NT \cdot kT}{T_F} = \frac{NkT^2}{T_F} = \frac{RT^2}{T_F} \tag{37}$$

where N is the number of electrons. The electronic heat capacity is then

$$C_v(\text{el}) = \frac{2RT}{T_F} \tag{38}$$

This is in agreement with the experimental results, where it is found that the electronic specific heat is indeed proportional to the absolute temperature.

The magnetic susceptibility of a metal may be discussed in a similar fashion. If we consider a free electron with spin magnetic moment μ, then it may be derived from Maxwell's distribution law that its net magnetic moment is given by

$$\mu(T) = \frac{\mu e^{\mu H/kT} - \mu e^{-\mu H/kT}}{e^{\mu H/kT} + e^{-\mu H/kT}} = \frac{\mu^2 H}{kT} \tag{39}$$

If we have N free electrons, then the total magnetization would be

$$\mu_{\text{tot}} = \frac{N\mu H^2}{kT} \tag{40}$$

However, we should remember that in a metal most of the electrons are in doubly occupied orbitals with their spins paired. Only the small fraction NT/T_F that are thermally excited are free, so the magnetic moment of the electrons in a metal is given by

$$\mu_{\text{tot}} = \frac{N\mu H^2}{kT_F} \tag{41}$$

instead of by Eq. (40). This is again in agreement with the experiments, where it was found that the magnetic susceptibility of a metal is small and temperature-independent, as follows from Eq. (41).

Of course, a complete and accurate description of the properties of a metal can be derived only if the exact form of the energy band is known, but the above examples illustrate how some of the properties of a metal may be derived by means of very simple arguments.

4. Molecular Crystals

The metallic and covalent crystals that we have discussed in the previous section are characterized by the fact that the forces between adjacent atoms are all of the same order of magnitude. In a molecular crystal we can recognize the individual molecules because the forces between different molecules are much smaller than the forces between the atoms within one molecule. We may assume that there is no exchange of electrons between different molecules and that the forces between different molecules are the relatively weak van der Waals forces. As a result, molecular crystals should have fairly low melting points. The best-known class of molecular crystals are the aromatic hydrocarbons, such as benzene, naphthalene, and anthracene. Another category are the crystals of gases such as hydrogen, argon, and nitrogen. Obviously, these gases solidify only at very low temperatures, below 100°K, but they have been studied experimentally and they definitely fall in the category of molecular crystals. In the present discussion we are interested mainly in the aromatic hydrocarbons.

There is ample experimental evidence that in molecular crystals the molecules have practically the same structures and configurations as in the gaseous states. The internuclear distances that are measured in crystals are almost identical with the distances for the free molecules, and the molecular configurations in the crystalline and gaseous states are identical in most cases (there are, of course, some exceptions). The infrared and electronic spectra in the two phases are also very similar. Each vibronic transition in the free molecule is easily recognized in the crystal spectrum. To a first approximation, the vapor spectrum is identical with the crystal spectrum. However, there are some small differences between the two spectra, and even though they are very small they are important because they serve to give us some understanding of the type of interactions between the molecules in the crystal.

In Table 7.3 are listed some vibronic transitions in the naphthalene molecule as they occur in the free molecule (vapor) and in a molecular crystal. It should be noted first that the crystal transitions are shifted by an amount of about 5,000 cm^{-1} as compared to the free molecule. If we measure the crystal absorption spectra with polarized light, then the crystal also exhibits some anisotropy; the frequencies are slightly different depending on whether the light is polarized along different crystal axes. At sufficiently low temperatures, this same effect may also be observed as a splitting of the spectral line into two different lines. As may be seen from Table 7.3, the distance between the two lines can be quite small: It varies from 13 cm^{-1} for the one doublet to 146 cm^{-1} for the other. However, such splittings have been clearly observed for the majority of vibronic transitions in molecular crystals. They are known as Davydov splittings, and they play an important role in the description of the intermolecular interactions in molecular crystals.

In order to understand the nature of the intermolecular interactions in a molecular crystal, we consider a typical case where there are two molecules per elementary cell; naphthalene falls into this category. Within the elementary cell the first molecule is located at the position $\mathbf{r}_1$ and the second molecule is located at the position $\mathbf{r}_2$. The position of any molecule in the crystal is then given by

$$\begin{aligned} \mathbf{r}_n^{(1)} &= \mathbf{r}_1 + n_x \mathbf{a} + n_y \mathbf{b} + n_z \mathbf{c} \\ \mathbf{r}_n^{(2)} &= \mathbf{r}_2 + n_x \mathbf{a} + n_y \mathbf{b} + n_z \mathbf{c} \end{aligned} \quad (42)$$

TABLE 7.3 ● Some typical vibronic transition lines for naphthalene (in cm^{-1}). The first column gives the values for naphthalene vapor, and the second and third columns give the values for two different crystal directions.

Vapor	b dir.	a dir.	Splitting	Shift
35,905	31,063	31,050	13	4,849
36,398	31,620	31,474	146	4,851
37,302	32,255	32,227	28	5,061

We can label the molecules by means of $(\mathbf{n}, s)$, where $\mathbf{n}$ denotes the elementary cell and s is either 1 or 2 and denotes either one of the two molecules in the cell.

Each molecule can be in its ground state, characterized by an eigenfunction ψ_0, or in one of its excited states, characterized by eigenfunctions ψ_w. The corresponding energies are ϵ_0 and ϵ_w, respectively. It is clear that in the ground state of the crystal all molecules are in their ground states. The corresponding wave function for the crystal is the product of all molecular ground-state eigenfunctions,

$$\Psi_0 = \Pi \psi_0(\mathbf{n}, s) \tag{43}$$

where $\psi_0(\mathbf{n}, s)$ is the ground-state eigenfunction of molecule $(\mathbf{n}, s)$. Let us now consider the situation where one of the molecules is in an excited state, described by an eigenfunction $\psi'(\mathbf{n}, s)$ and an eigenvalue ϵ'. If the excited molecule is located at the position $(\mathbf{n}, s)$, then the corresponding crystal wave function is given by

$$\Psi(\mathbf{n}, s) = \frac{\Psi_0 \psi'(\mathbf{n}, s)}{\psi_0(\mathbf{n}, s)} \tag{44}$$

In this situation the excitation is localized at the position $(\mathbf{n}, s)$ and we speak of a localized exciton.

We should realize that the situation where one molecule is excited, described by the wave function (44), corresponds to a degenerate eigenvalue with a very large degree of degeneracy. In fact, the degeneracy is N-fold, where N is the total number of molecules in the crystal, since all functions $\Psi(\mathbf{n}, s)$ of Eq. (44) for all possible values of $\mathbf{n}$ and s correspond to the same energy.

As a result of the small interactions between adjacent molecules, the degeneracy is lifted and we obtain an energy band, just as in the case of semiconductors. Again, the localized exciton wave functions of Eq. (44) combine to form Bloch functions of the form

$$\begin{aligned}\Phi_1(\mathbf{k}) &= \sum_{\mathbf{n}} \Psi(\mathbf{n}, 1) \exp(i\mathbf{k} \cdot \mathbf{r}_\mathbf{n}) \\ \Phi_2(\mathbf{k}) &= \sum_{\mathbf{n}} \Psi(\mathbf{n}, 2) \exp(i\mathbf{k} \cdot \mathbf{r}_\mathbf{n}) \end{aligned} \tag{45}$$

The crystal energies form an energy band with the energies given as a function of the three variables k_x, k_y, and k_z. The corresponding eigenfunctions are obtained in the form

$$\Phi(\mathbf{k}) = \alpha(\mathbf{k})\Phi_1(\mathbf{k}) + \beta(\mathbf{k})\Phi_2(\mathbf{k}) \tag{46}$$

where the coefficients α and β are also functions of **k** and must be determined from the intermolecular interactions and from the positions of the molecules in the elementary cell.

Let us now consider first the question of why a crystal spectrum consists of sharp lines—in fact, at sufficiently low temperatures the lines of the crystal spectrum may even be sharper than the lines of the vapor spectrum. It may seem at first sight that the width of a crystal spectrum line should be much larger than the corresponding width in the vapor spectrum because of the presence of an energy band. If there were a finite transition probability for all transitions from the crystal ground state to every energy level in the band, then the width of the crystal spectal line would be approximately equal to the width of the energy band, which is contrary to what we observe experimentally.

If we consider the selection rules for transitions between the ground state and an exciton state characterized by a vector **k**, then it can be shown fairly easily that the only allowed transitions are those for which

$$\mathbf{k} \approx \mathbf{0} \qquad (47)$$

This means that we observe only transitions between the ground state and the part of the exciton band where $\mathbf{k} = \mathbf{0}$, usually at the bottom of the exciton band. This is the reason why spectroscopic transitions in a molecular crystal give lines that are just as sharp as in the vapor.

In order to prove the selection rule of Eq. (47), we imagine that the transition is induced by a light wave of the form $\exp(i\mathbf{\sigma} \cdot \mathbf{r})$. The transition between the ground state and the exciton state **k** is then derived from the matrix element

$$\langle \Phi(\mathbf{k}) \mid \exp(i\mathbf{\sigma} \cdot \mathbf{r}) \mid \Psi_0 \rangle = \int \Phi^*(\mathbf{k}) \exp(i\mathbf{\sigma} \cdot \mathbf{r}) \Psi_0 \qquad (48)$$

It easily follows from the translational symmetry of the lattice that this matrix element is zero unless

$$-\mathbf{k} + \mathbf{\sigma} = \mathbf{0} \qquad (49)$$

It should be noted that the vector $\mathbf{\sigma}$, which characterizes the light wave, is much smaller than the ordinary values of the vector **k**. The wavelength λ of the light, which is $(2\pi/\sigma)$, is of the order of 5,000 Å, whereas k is of the order of a, the dimension of the elementary cell, which is around 10 Å. For that reason we may neglect $\mathbf{\sigma}$ in Eq. (49), which thus reduces to the selection rule of Eq. (47).

In the same way it may be argued that the probability for a transition between two exciton states **k** and **k**' is derived from the matrix element

$$\langle \Phi(\mathbf{k}) \mid \exp(i\mathbf{\sigma} \cdot \mathbf{r}) \mid \Phi(\mathbf{k}') \rangle = \int \Phi^*(\mathbf{k}) \exp(i\mathbf{\sigma} \cdot \mathbf{r}) \Phi(\mathbf{k}') \qquad (50)$$

It follows again from the translational symmetry of the crystal that this matrix element is zero unless

$$\mathbf{k}' + \mathbf{\delta} - \mathbf{k} = 0$$
$$\mathbf{k}' \approx \mathbf{k} \tag{51}$$

Again we may neglect $\mathbf{\delta}$ for the reasons stated above.

Let us now again consider the naphthalene crystal, which has two molecules per unit cell. We have seen in Eq. (46) that for a given value of $\mathbf{k}$ the exciton eigenfunctions are given by

$$\Phi(\mathbf{k}) = \alpha(\mathbf{k})\Phi_1(\mathbf{k}) + \beta(\mathbf{k})\Phi_2(\mathbf{k}) \tag{52}$$

where $\Phi_1(\mathbf{k})$ and $\Phi_2(\mathbf{k})$ are Bloch functions for the first and second molecule in the elementary cell, respectively. It is somewhat difficult to determine the coefficients $\alpha(\mathbf{k})$ and $\beta(\mathbf{k})$ for arbitrary values of the vector $\mathbf{k}$. However, when $\mathbf{k} = \mathbf{0}$, it follows easily from symmetry considerations that there are two exciton eigenstates, characterized by the eigenfunctions

$$\Phi_+(\mathbf{0}) = \Phi_1(\mathbf{0}) + \Phi_2(\mathbf{0})$$
$$\Phi_-(\mathbf{0}) = \Phi_1(\mathbf{0}) - \Phi_2(\mathbf{0}) \tag{53}$$

The corresponding energies are $E_+(\mathbf{0})$ and $E_-(\mathbf{0})$, respectively. Because of the selection rule (51), we observe two spectroscopic transitions with frequencies

$$h\nu_+ = E_+(\mathbf{0}) - E_0$$
$$h\nu_- = E_-(\mathbf{0}) - E_0 \tag{54}$$

where E_0 is the energy of the crystal ground state.

Since the exciton band is derived from an excited molecular eigenstate with ϵ', the transition in the isolated molecule corresponds to a frequency ν', given by

$$h\nu' = \epsilon' - \epsilon_0 \tag{55}$$

We see that this transition in the isolated molecule with frequency ν' gives rise to two transitions with frequencies ν_+ and ν_- in the molecular crystal.

If the transition moment for the transition $(\epsilon_0 \to \epsilon')$ in molecule 1 is the vector $\mathbf{\mu}_1$ and the same transition moment for molecule 2 is $\mathbf{\mu}_2$, then the transition moment for the transition ν_+ is given by

$$\mathbf{\mu}_+ = \mathbf{\mu}_1 + \mathbf{\mu}_2 \tag{56}$$

and the transition moment for the transition $\mathbf{\mu}_-$ is given by

$$\mathbf{\mu}_- = \mathbf{\mu}_1 - \mathbf{\mu}_2 \tag{57}$$

It may be seen that if we use polarized light, there are certain directions in the crystal where only the first transition may be observed and other directions where only the other transition is observed. In many cases these directions coincide with the crystal axes.

It is even possible to calculate the magnitude of the shifts and splittings of the spectral lines, because it turns out that they may be derived from the dipole–dipole interactions between the transition moments on the different molecules. It follows, then, that for very intense transitions these quantities are large, and for weak lines the splittings become very small. The various situations that have been calculated all give excellent agreement with the experimental values.

It has now become customary to talk about excitons as if they were particles, in the same way as photons are often treated as particles. Absorption of light is then considered the absorption of a photon and the simultaneous creation of an exciton. Such excitons may either be totally delocalized as we have described them, or they may be localized because of interactions with the environment. The whole area of molecular crystal phenomena is a fairly active research field at the moment. Even though the main properties of molecular crystals are now well understood, there are many problems that can be studied, and both experimental and theoretical studies of molecular crystals are of interest to many scientists.

Problems

1. How many carbon atoms are there per unit cell in the two-dimensional graphite structure of Fig. 7.2?
2. How many Na^+ and Cl^- ions are there per unit cell in the NaCl crystal?
3. The density of a NaCl crystal in the form of rocksalt is 2.18 g/cm³. Use this number to derive the lattice constant a of the NaCl crystal. (This is the length of the lattice vector **a** that forms the elementary cell. The crystal structure of NaCl is cubic, as may be seen from Fig. 7.4.)
4. The elementary cell of a hexagonal structure is determined by the length a of the side of the hexagon and by the height c of the elementary cell (see Fig. 7.6). Derive the distance between one of the particles on the hexagon points and one of the particles inside the structure of Fig. 7.6 in terms of a and c.
5. According to the chemical classification, what type of crystal is solid argon, what type of crystal is SiC, and what type is LiH? Sometimes a crystal is an intermediate between two categories. For example, how would you classify ice and the SnS crystal?

6. The electrical conductivity of a semiconductor at room temperature depends on the value of the energy gap between the valence and the conduction bands. The values are 6 eV for diamond, 1.10 eV for Si, and 0.7 eV for Ge. Which of these compounds has the highest conductivity and why?
7. The lattice constant of NaCl is 5.63 Å. What is the shortest distance R between a Na$^+$ ion and a Cl$^-$ ion? What is the lattice energy of NaCl as derived from the Madelung constant?
8. The KBr crystal has the same crystal structure as NaCl, and its lattice constant is 6.59 Å. What is the lattice energy of the KBr crystal?
9. The CsCl crystal has a lattice constant of 4.11 Å. What is the shortest distance R between a Cs$^+$ and a Cl$^-$ ion in the crystal? What is the lattice energy of CsCl as derived from purely electrostatic interactions?
10. Derive the energy levels of a ring of eight Li atoms by considering only the $2s$ electrons in MO theory. The energy levels should be expressed in terms of the parameters α and β.
11. If we assume that in metallic Li each Li atom contributes one free electron, how many free electrons are there per cubic centimeter of Li? The density of Li is 0.534 g/cm^3.
12. Metallic Na has 2.5×10^{-22} free electrons/cm^3. Derive the Fermi energy of metallic Na in terms of electron volts.
13. Ordinarily the paramagnetic susceptibility of a paramagnetic substance (such as oxygen) is proportional to the inverse of the absolute temperature. However, the paramagnetic susceptibility of a metal is temperature-independent and much smaller than for nonmetallic paramagnetic substances. Explain why this is so.
14. Explain why a vibronic transition in an isolated naphthalene molecule gives rise to two spectral lines in the molecular crystal of naphthalene.
15. What is the Davydov splitting? Explain it.

Bibliography

The bulk of the material in this chapter falls in the category of solid state physics. We list two books on this subject, the book by Kittel (1), which contains a lot of up-to-date applications, and the book by Peierls (2), which does not contain too many applications but gives an elegant treatment of the general quantum theory of solids. The third book we list deals with the theory of molecular crystals.
1. C. Kittel. *Quantum Theory of Solids*. Wiley, New York (1963).
2. R. E. Peierls. *Quantum Theory of Solids*. Oxford University Press, London (1955).
3. A. S. Davydov. *Theory of Molecular Excitons*. McGraw-Hill, New York (1962).

CHAPTER 8   MAGNETIC RESONANCE

1. Introduction

Magnetic resonance concerns itself with the study of transitions between the various spin states within an atom or a molecule. In chapter III we discussed the properties of the electron spin. We mentioned that the electron spin has a constant magnitude characterized by the quantum number $S = \frac{1}{2}$. Most nuclei also possess a spin angular momentum and magnetic moment, characterized by a spin quantum number I. We shall show that the formal theoretical description of the nuclear spin is very similar to the theory of the electron spin.

For magnetic resonance we can either study the transitions between different electronic spin states or between different nuclear spin states. In the first case we speak of electron spin resonance (ESR), and in the second we speak of nuclear magnetic resonance (NMR). Almost every atom has at least one isotope that has a nonzero nuclear spin, and NMR is in principle applicable to any of the nuclei in a molecule. In practice, though, the bulk of the work in NMR has been on protons. An ESR experiment can be performed only on a molecule with a nonzero total electronic spin. We have seen that practically all stable molecules have singlet ground states, and it follows that they cannot be studied by means of electron spin resonance. Initially, ESR experiments were confined to the study of free radicals, which have an unpaired electron spin. Many free radicals are intermediate products in chemical reactions, and ESR has been used extensively to study chemical reactions. More recently it has also become possible to measure ESR of molecules in their triplet states.

The possible observation of magnetic resonance was predicted as early as 1936, but the early attempts to measure the phenomenon all failed. We shall discuss later why the experiment is not quite as easy to do as it may seem at first sight. Magnetic resonance was discovered in 1945–1946. The discovery may be attributed to three independent and almost simultaneous achievements: (1) the first electron spin resonance measurement by Zavoiski in the USSR, (2) the nuclear resonance absorption measurement by Purcell, Torrey, and Pound at Harvard, and (3) the nuclear induction work by Bloch, Hansen, and Packard at Stanford. We should realize that the electromagnetic radiation that is absorbed in electron spin resonance has a wavelength of the order of 1 cm, which is in the range of radar waves. The advances in radar technology made during World War II were essential in developing the experimental techniques needed for measuring magnetic resonance.

We have seen in Chapter III that a spinning electron has a magnetic moment $\mathbf{\mu}_s$, which is given by

$$\mathbf{\mu}_s = -\frac{e}{mc}\mathbf{S} \tag{1}$$

where $\mathbf{S}$ is the spin operator (see Eq. 47 in Chapter III). If we place a magnetic moment $\mathbf{\mu}$ in a homogeneous magnetic field $\mathbf{H}$, its energy is given by

$$E = -\mathbf{\mu}_s \cdot \mathbf{H} = -(\mu_x H_x + \mu_y H_y + \mu_z H_z) \tag{2}$$

The interaction between the electron spin and a homogeneous magnetic field $\mathbf{H}$ is therefore represented by a Hamiltonian,

$$\mathcal{H} = -\mathbf{\mu}_s \cdot \mathbf{H} = \frac{e}{mc}(\mathbf{H} \cdot \mathbf{S}) \tag{3}$$

It is convenient to choose the Z axis as the direction of the magnetic field $\mathbf{H}$ because the Hamiltonian then reduces to

$$\mathcal{H} = \frac{e}{mc} H S_z \tag{4}$$

The electron spin can have two possible orientations with respect to the Z axis. The first one, where the spin is parallel to the Z axis and where it points in the positive Z direction, is represented by a spin function α. In the second one the spin points in the negative Z direction and is represented by a spin function β. The spin functions α and β are eigenfunctions of the operator S_z, as we have seen in Eq. 48 of Chapter III. We thus have

INTRODUCTION

$$S_z\alpha = \frac{\hbar}{2}\alpha \qquad S_z\beta = -\frac{\hbar}{2}\beta \qquad (5)$$

Obviously, the functions α and β are also eigenfunctions of the Hamiltonian (4), because

$$\mathcal{H}\alpha = \frac{e\hbar H}{2mc}\alpha \qquad \mathcal{H}\beta = -\frac{e\hbar H}{2mc}\beta \qquad (6)$$

It is convenient to make use of the constant μ_0, the Bohr magneton, which is given by

$$\mu_0 = \frac{e\hbar}{2mc} = 0.9273 \times 10^{-20} \text{ erg/gauss} = 4.670 \times 10^{-5} \text{ cm}^{-1}/\text{gauss} \qquad (7)$$

It follows easily from Eq. (6) that the energy difference ΔE between the two eigenstates of the electron spin is given by

$$\Delta E = 2\mu_0 H \qquad (8)$$

The magnetic fields customarily used for magnetic resonance experiments vary between 3,000 and 10,000 gauss; in electron spin resonance the field is usually around 3,400 gauss. In that case the energy difference is

$$\Delta E = 2\mu_0 H = 2 \times 4.670 \times 10^{-5} \times 3.4 \times 10^3 \text{ cm}^{-1} = 0.318 \text{ cm}^{-1} \qquad (9)$$

The resonance radiation that corresponds to ΔE has a wavelength of about 3.15 cm, which is in the radar region.

In magnetic resonance theory the spin operators are usually defined somewhat different from the above treatment: The spin operators are defined by

$$S_z\alpha = \tfrac{1}{2}\alpha \qquad S_z\beta = \tfrac{1}{2}\beta \qquad (10)$$

Instead of by Eq. (5). This means that the factor $\hbar$ is included in the Hamiltonian, which takes the form

$$\mathcal{H} = \frac{e\hbar}{mc}(\mathbf{H} \cdot \mathbf{S}) = 2\mu_0(\mathbf{H} \cdot \mathbf{S}) \qquad (11)$$

instead of Eq. (4). Customarily, this is written as

$$\mathcal{H} = g\mu_0(\mathbf{H} \cdot \mathbf{S}) \qquad (12)$$

By comparing Eqs. (11) and (12), it would follow that $g = 2$, but it has been found experimentally that g is slightly different from this value if we take Eq. (12) as the definition of g. The experimental value for a free electron spin is $g = 2.002292$. This discrepancy in the g factor is due to a number of small corrections that are not considered in the simple classical theory we have discussed above.

The quantum theory of nuclear spins is very similar to the theory of the electron spin, especially if we use the formulation of Eqs. (10), (11), and (12). The interaction between a nuclear spin **I** and a homogeneous magnetic field **H** is represented by a Hamiltonian

$$\mathcal{H} = -g_N \mu_N (\mathbf{H} \cdot \mathbf{I}) \tag{13}$$

Here μ_N is the nuclear magneton, which is defined as

$$\mu_N = \frac{e\hbar}{2Mc} \tag{14}$$

where M is the mass of the proton. The spin operator **I** plays the same role as the spin operator **S** for electron spin, but there is one important difference between the two cases. We have seen that the total electron spin has a value $\frac{1}{2}$ and, consequently, the electron spin can have two possible projections, $\pm\frac{1}{2}$ in a given direction. The spin functions α and β are defined with respect to the Z direction, consequently they are eigenfunctions of the operator S_z, with the eigenvalues given by

$$S_z \alpha = \tfrac{1}{2}\alpha \qquad S_z \beta = -\tfrac{1}{2}\beta \tag{15}$$

Certain nuclei (we have listed some of them in Table 8.1) also have spin $\frac{1}{2}$. Here the situation is identical with the electron, the nuclear spin has two possible projections with respect to the Z axis, and the two situations may

TABLE 8.1 ● Nuclei with spin $\frac{1}{2}$ and their g values.

Nucleus	g
^{1}H	5.5854
^{3}H	5.9576
^{13}C	1.4043
^{15}N	−0.5661
^{19}F	5.2546
^{29}Si	−1.1095
^{31}P	2.2610
^{195}Pt	1.2008
Hg	0.9986

INTRODUCTION

TABLE 8.2 ● Nuclei with spin 1 and their g values.

Nucleus	g
D	0.85738
^{14}N	0.40357

be represented by the spin functions $\zeta_{1/2}$ and $\zeta_{-1/2}$. The eigenfunctions are given by

$$I_z\zeta_{1/2} = \tfrac{1}{2}\zeta_{1/2} \qquad I_z\zeta_{-1/2} = -\tfrac{1}{2}\zeta_{-1/2} \tag{16}$$

There is no simple way to calculate the gyromagnetic ratio g_N for a nucleus, and the g-factors have to be determined experimentally. We have listed some of them in Table 8.1.

An important difference between the nuclear and the electronic spins is that the spin quantum number I for a nucleus can have a value different from one half. For example, the nuclear spins of deuterium and of ^{15}N both have a magnitude $\hbar$ (see Table 8.2). In this case the projections of the spin in a given direction can have the values 1, 0, and -1. The operator I_z has three eigenfunctions, which we denote by ζ_1, ζ_0, and ζ_{-1}, and the eigenvalues are given by

$$I_z\zeta_1 = \zeta_1 \qquad I_z\zeta_0 = 0 \qquad I_z\zeta_{-1} = -\zeta_{-1} \tag{17}$$

In general, if the spin quantum number of a nucleus X has the value I_X, the projection in the Z direction can have the values I_X, $I_X - 1$, $I_X - 2$, ..., $-I_X + 1$, $-I_X$ (see Fig. 8.1). The corresponding eigenfunctions and

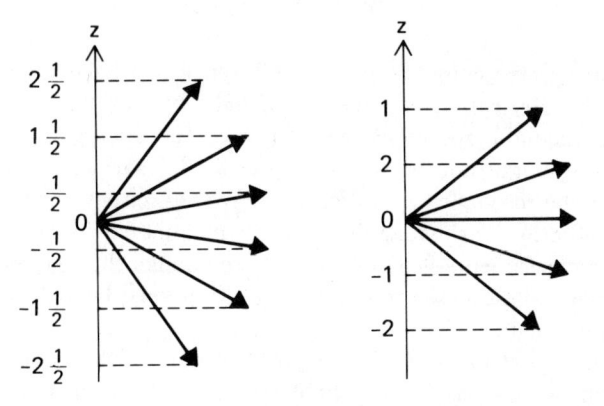

FIGURE 8.1 ● Quantization of the nuclear spin along the Z axis for the two cases $I = \tfrac{5}{2}$ and $I = 2$.

eigenvalues are given by

$$I_z \zeta_m = m \zeta_m \qquad m = I_X, I_X - 1, \ldots, -I_X + 1, -I_X \qquad (18)$$

The nucleus with the largest spin is ^{50}V, which has spin 6. Other values are $\frac{3}{2}$ for ^{7}Li, ^{9}Be, and ^{11}B, $\frac{5}{2}$ for ^{17}O, ^{25}Mg, and ^{127}I, $\frac{7}{2}$ for ^{123}Cs, and so on. Some nuclei, such as ^{12}C and ^{16}O have no spin, and their spin quantum number is zero.

The rest of this chapter is concerned mainly with the magnetic resonance of proton spins, and it may be useful to calculate some of the values that are relevant for proton magnetic resonance. The nuclear magneton μ_N has the value

$$\mu_N = \frac{e\hbar}{2Mc} = 2.5428 \times 10^{-8} \text{ cm}^{-1}/\text{gauss} \qquad (19)$$

In a homogeneous magnetic field H, the energy difference between the two levels of the proton spin is

$$\Delta E = g_N \mu_N H = 1.4202 \times 10^{-7} H \text{ cm}^{-1} \qquad (20)$$

Customarily, the magnetic field in nuclear magnetic resonance is 10,000 gauss. The resonance radiation in such a field has a wave number of 1.4202×10^7 cycles/sec.

It is obvious from our discussion in Chapter IV that radiation will be absorbed by a spin system only if its frequency ν satisfies the resonance condition

$$h\nu = \Delta E = g_N \mu_N H \qquad (21)$$

In optical spectroscopy the energy difference is a fixed quantity, and we can observe absorption or emission of light by varying the frequency of light. In magnetic resonance we can vary the energy difference ΔE by varying the magnitude of the magnetic field H, so we have two ways of satisfying the resonance condition (21). We can either vary the frequency ν of the radiation or the magnitude of the magnetic field. In most magnetic resonance experiments it is not possible to change the frequency beyond a very narrow range, so the resonance condition must be satisfied by varying the magnetic field.

The value of H in Eq. (21) is the magnitude of the magnetic field at the position of the nucleus. This is slightly different from the exterior field H_ext that we apply, because it is the sum of H_ext and the small magnetic fields that are due to the motion of the electrons in the vicinity and to the interaction with the other nuclei. Magnetic resonance experiments are accurate enough so that these small magnetic fields can be measured, and this is the reason magnetic resonance is of such general interest to chemists. The small magnetic fields H_{ind} that are due to the motion of the electrons in the vicinity of a proton are proportional to the exterior field H_{ext}; the proportionality factor is of the order of 10^{-5}. The magnitude of H_{ind} is slightly different for different protons; for example, in the ethyl alcohol molecule there are three different types of protons, and each of them will have a slightly different resonance line, because of differences in the induced field H_{ind}. The interactions between the magnetic moments of the different protons, the spin-spin interactions, also affect the resonance frequencies. As a result of these various effects, the proton magnetic resonance spectrum of the ethyl alcohol molecule contains a number of different lines. In this chapter we first discuss the experimental conditions under which magnetic resonances may be observed, and then we discuss the various types of induced magnetic fields and how they affect the magnetic resonance spectra.

2. Relaxation Phenomena

If it were not for relaxation, it would never be possible to measure magnetic resonance. We should realize that in magnetic resonance we want to observe transitions between energy levels that are very close together—the separation is 0.3 cm^{-1} in electron spin resonance and 1.4×10^{-3} cm^{-1} in proton magnetic resonance. As a result, there are two important differences between magnetic resonance spectroscopy and optical spectroscopy. First, in magnetic resonance the probability of spontaneous emission is negligible because of the low frequency of the radiation. This means that the probability of emission is due to induced emission only, and it is therefore exactly equal to the probability of absorption. Second, all the energy levels are about equally populated at ordinary temperatures, which means that if we have a large number N of spins there will be equal numbers of spins N_l and N_u in the lower and the upper states. Since N_l and N_u are equal to one another and since the two transition probabilities W_{abs} and W_{em} are also equal, the total number of absorption transitions is equal to the total number of emission transitions. Consequently, there is no net transfer of energy between the spin system and the radiation field, and we do not observe any magnetic resonance for the above-described system.

It may be helpful to use a situation in everyday life as a model for the behavior of the spin system. Imagine that I start out with a number of cigarettes N_1 and that I have a friend who has a number N_2 of the same brand of cigarettes. We make a deal where I give him every day 10% of my cigarettes and he gives me at the same time 10% of his cigarettes. Clearly, after a certain number of days each of us ends up with the same number,

$$n = \tfrac{1}{2}(N_1 + N_2) \qquad (22)$$

of cigarettes. Any future exchange of cigarettes is somewhat futile, because we just pass the same number of cigarettes back and forth and the net effect of all this trading is zero. We have assumed that neither I or my friend smokes any of the cigarettes we have. Obviously, the situation changes if my friend starts smoking, either a few cigarettes a day or a few packs. In that case I shall begin to lose cigarettes also as a result of the exchange, since I shall get fewer cigarettes back each day than I am giving away. In fact, through the continual exchange I am partially financing his smoking.

A magnetic resonance experiment works in a very similar way as the above cigarette exchange. If we have just a spin system, then we have equal amounts of upward and downward transitions and we do not observe any transfer of energy between the radiation field and the spin system. What we need is some kind of an energy leak through which the energy can leak away from the system. In practice this is easily accomplished, because in most cases the spin system is in thermal equilibrium with its surroundings. In a solid this is the set of lattice vibrations, and in a gas this may be the translational motion of the molecules. Formally, we can say that the spin system is in thermal contact with a reservoir and that it will try to maintain thermal equilibrium with this reservoir. This causes an energy transfer between the spin system and the reservoir and, consequently, an energy transfer out of the radiation field.

Let us try and give a mathematical description of this situation. We consider a system of N spins of spin quantum number $\tfrac{1}{2}$. At a given time t there are $N_+(t)$ spins in the lower state $\zeta_{1/2}$ and $N_-(t)$ spins in the upper state $\zeta_{-1/2}$. The probability for a downward transition is W_- and for an upward transition it is W_+ (see Fig. 8.2). The changes in time of $N_+(t)$ and

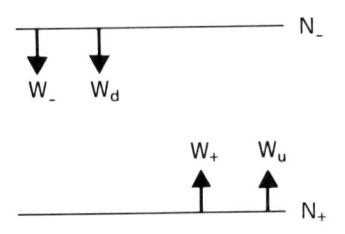

FIGURE 8.2 • Transitions between a two-level spin system that are due to the radiation field (W_- and W_+) and due to thermal transitions (W_d and W_u).

$N_-(t)$ are then given by

$$\frac{dN_+(t)}{dt} = W_- N_-(t) - W_+ N_+(t)$$
$$\frac{dN_-(t)}{dt} = W_+ N_+(t) - W_- N_-(t) \quad (23)$$

The two transition probabilities W_- and W_+ due to the radiation field are equal to one another, since we can neglect the spontaneous emission,

$$W_+ = W_- = W \quad (24)$$

If we define the difference in population $n(t)$ as

$$n(t) = N_+(t) - N_-(t) \quad (25)$$

it follows from Eq. (23) that

$$\frac{dn(t)}{dt} = -2Wn(t) \quad (26)$$

This is a simple differential equation, which has the solution

$$n(t) = n(0)e^{-2Wt} \quad (27)$$

We see that in the stationary state (for $t \to \infty$), the population difference is zero and the total number of upward transitions is equal to the total number of downward transitions.

Let us now consider the same spin system without the radiation field, but try and analyze the interactions with the reservoir. We know that in the equilibrium situation the spin system is in thermal equilibrium with the reservoir. The ratio between N_+ and N_- is then determined by Maxwell's equation,

$$\frac{N_+(\infty)}{N_-(\infty)} = \exp\left(\frac{\hbar\omega}{kT}\right) \quad (28)$$

where $\hbar\omega$ is the energy difference between the two levels and T is the temperature of the reservoir. If at an earlier time t the spin system is not in equilibrium with the reservoir, then there must be a mechanism through which thermal equilibrium is attained. We may represent this mechanism by means of a downward transition probability W_d and an upward transition probability W_u. Both these transition probabilities cause changes in

MAGNETIC RESONANCE

the spin level population that approach the thermal equilibrium. We have again

$$\frac{dN_+}{dt} = W_d N_-(t) - W_u N_+(t)$$
$$\frac{dN_-}{dt} = W_u N_+(t) - W_d N_-(t) \qquad (29)$$

It is easily shown that the two transition probabilities W_d and W_u must be different, because in the stationary state we have

$$\frac{dN_+}{dt} = 0 = W_d N_-(\infty) - W_u N_+(\infty) \qquad (30)$$

or

$$\frac{W_d}{W_u} = \frac{N_+(\infty)}{N_-(\infty)} = \exp\left(\frac{\hbar\omega}{kT}\right) \qquad (31)$$

We solve the differential equations (29) by introducing the population difference $n(t)$,

$$n(t) = N_+(t) - N_-(t) \qquad (32)$$

and the equilibrium population $n(\infty)$,

$$n(\infty) = \frac{W_d - W_u}{W_d + W_u} [N_+(t) + N_-(t)] \qquad (33)$$

Equation (29) may then be written as

$$\frac{dn(t)}{dt} = 2W_d N_-(t) - 2W_u N_+(t) = (W_d - W_u)N - (W_d + W_u)n(t)$$
$$= (W_d + W_u)[n(\infty) - n(t)] = \frac{n(\infty) - n(t)}{T_1} \qquad (34)$$

Here we have substituted

$$T_1 = (W_d + W_u)^{-1} \qquad (35)$$

and we call T_1 the relaxation time. The solution of Eq. (34) is

$$n(t) = n(\infty) + \alpha \exp(-t/T_1) \qquad (36)$$

where α is an integration constant. Clearly, the population difference $n(t)$ approaches its stationary state value $n(\infty)$ exponentially.

In the above treatment we have considered the two mechanisms through which transitions may occur. First we considered the transitions due to the radiation field only and then we considered the transitions that represent the approach to thermal equilibrium. In reality, of course, the two mechanisms occur at the same time. It is assumed now that this situation may be described theoretically by adding the two separate transition probabilities (see Fig. 8.2). We obtain, then, from Eqs. (26) and (34),

$$\frac{dn(t)}{dt} = -2Wn(t) + \frac{n' - n(t)}{T_1} \tag{37}$$

Here we have written n' rather than $n(\infty)$ because n' is no longer the stationary state value.

There is no need to solve the differential equation (37) because we only wish to know the value $n(\infty)$ in the stationary state. This is obtained by setting

$$\left(\frac{dn}{dt}\right)_\infty = 0 = -2Wn(\infty) + \frac{n' - n(\infty)}{T_1} \tag{38}$$

or

$$n(\infty) = \frac{n'}{1 + 2WT_1} \tag{39}$$

It is helpful to compare this result with Eq. (27), where we found that without the thermal transition probabilities $n(\infty)$ would be zero. Because of the thermal relaxation, the population difference $n(t)$ has now become different from zero in the stationary state.

Let us now return to our original question, namely, what is the energy transfer from the radiation field to the spin system? Every time we have a transition from a lower spin state $(+)$ to a higher spin state $(-)$, the field must supply an amount of energy $\hbar\omega$ and, vice versa, for every transition from a state $(-)$ to a state $(+)$, the field receives an amount of energy $\hbar\omega$. Per second there are WN_+ transitions from a lower to a higher state, and WN_- transitions from a higher to a lower state. It is easily seen that the field supplies per unit of time an amount of energy

$$\frac{dE}{dt} = N_+ W\hbar\omega - N_- W\hbar\omega = nW\hbar\omega \tag{40}$$

284 MAGNETIC RESONANCE

In the stationary state this becomes

$$\left(\frac{dE}{dt}\right)_{\infty} = n(\infty)W\hbar\omega = \frac{n'W\hbar\omega}{1 + 2WT_1} \qquad (41)$$

according to Eq. (39).

We see that it is possible to have a net transfer of energy from the radiation field to the spin system if there is thermal relaxation, in which case magnetic resonance is in principle observable. It follows from Eq. (41) that the energy transfer becomes very small if T_1 becomes large. This is understandable because a large T_1 value means that the energy transfer from the spin system to the reservoir is a very slow process, so that it becomes negligible. We again approach the situation where there is no reservoir, and we have seen that in that case magnetic resonance cannot be observed.

Magnetic resonance of a given sample (either a solid, a liquid, or a gas) is formally described by the Bloch equations. These were obtained from the classical equations of motion for the total magnetic moment **M** of the sample by adding the thermal relaxation effects. The classical equations of motion are

$$\frac{dM_x(t)}{dt} = \gamma_n[H_z M_y(t) - H_y M_z(t)]$$

$$\frac{dM_y(t)}{dt} = \gamma_n[H_x M_z(t) - H_z M_x(t)] \qquad (42)$$

$$\frac{dM_z(t)}{dt} = \gamma_n[H_y M_x(t) - H_x M_y(t)]$$

where γ_n is the ratio between the magnetic moment **μ** of a spin and its angular momentum **J**.

In a magnetic resonance experiment the magnetic field **H**(t) is always the sum of a constant homogeneous magnetic field **H**₀, which causes the splitting of the spin energy levels, and a small, time-dependent field **H**'(t), which acts as the radiation field with which we measure the absorption (Fig. 8.3). If we take the homogeneous field **H** along the Z axis, then the Bloch equations are

$$\frac{dM_z(t)}{dt} = \gamma_n[H_y M_x(t) - H_x M_y(t)] + \frac{M_z' - M_z(t)}{T_1}$$

$$\frac{dM_x(t)}{dt} = \gamma_n[H_z M_y(t) - H_y M_z(t)] - \frac{M_x(t)}{T_2} \qquad (43)$$

$$\frac{dM_y(t)}{dt} = \gamma_n[H_x M_z(t) - H_z M_x(t)] - \frac{M_y(t)}{T_2}$$

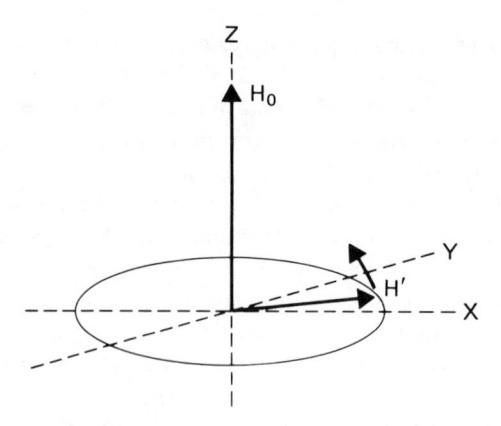

FIGURE 8.3 ● A magnetic resonance experiment is performed by applying a large constant magnetic field $\mathbf{H}_0$ to split the spin levels and a rotating field $\mathbf{H}'$ to cause transitions.

It is important to note that there are two different relaxation times, T_1 and T_2. In solids the relaxation time T_1 corresponds generally to the energy transfer between the spin system and the lattice, and T_1 is then called the spin-lattice relaxation time. The transverse relaxation time T_2 may be due to a variety of mechanisms, but in many cases it may be calculated from the dipole-dipole interactions between different spins, and it is often referred to as the spin-spin relaxation time.

By solving the Bloch equations, it may be found that magnetic resonance may be observed only if the two relaxation times T_1 and T_2 have suitable values. We have seen already in our model calculation that the effect cannot be observed if the two relaxation times are too long, in which case the intensity of the signal is too small. It may be shown that magnetic resonance cannot be observed either if T_2 is too small, because the resonance line becomes too broad in that case. The line shape of the absorption line has the general form

$$I(\omega) = \frac{A}{1 + T_2^2(\omega - \omega_0)^2 + B} \tag{44}$$

This is known as a Lorentzian function with a sharp maximum for $\omega = \omega_0$. The resonance frequency ω_0 is given by the resonance condition of Eq. (21),

$$\hbar\omega_0 = h\nu = g_N\mu_N H \tag{45}$$

for nuclear magnetic resonance. It is clear now why the early attempts at measuring magnetic resonance failed: The samples that had been chosen for the experiments had unsuitable relaxation times, so the resonance signals could not be observed.

In the majority of routine magnetic resonance experiments, we just wish to determine the resonance frequency ω_0 and we are interested only in the relaxation times to the extent that the resonance signal should be observable. In other experiments, though, the main purpose is to measure the relaxation times in order to obtain some information about the various relaxation mechanisms. In the latter situation we would measure the intensities and the line widths of the resonance signals as a function of the temperature and of the concentrations of the various components in the sample. The relaxation mechanisms relate mostly to intermolecular phenomena, such as interactions between different molecules, diffusion, and molecular rotations. However, in the rest of this chapter we shall confine our discussion to intramolecular phenomena, that is, to the magnetic interactions within a single isolated molecule at rest. For that reason we shall not discuss the relaxation mechanisms any further.

3. Chemical Shifts

If we place a molecule in a homogeneous magnetic field **H**, then the field causes small changes in the motion of the electrons. These changes give rise to a small magnetic field called the induced field. A general law of physics, Lenz's law, states that the induced magnetic field must always be in the opposite direction to the original field **H**.

The induced magnetic fields can be measured in magnetic resonance. We should realize that in the resonance equation (45) we ought to substitute for the magnetic field H the field H_{loc} at the position of the nucleus. For a particular nucleus a, we have

$$H^a_{\text{loc}} = H_{\text{ext}} - H^a_{\text{ind}} \tag{46}$$

where H_{ext} is the homogeneous field that we apply and H^a_{ind} is the value of the induced magnetic field at the positions of nucleus a. It may be shown that the induced field H_{ind} is proportional to the exterior field H_{ext}, and we may write

$$H^a_{\text{ind}} = \sigma_a H_{\text{ext}} \tag{47}$$

The constant σ_a depends on the electronic environment of nucleus a. By substituting Eqs. (46) and (47) into the resonance equation (45), we find that the resonance frequency for nucleus a is given by

$$h\nu_a = g_N \mu_N H^a_{\text{loc}} = g_N \mu_N H_{\text{ext}}(1 - \sigma_a) \tag{48}$$

The constant σ_a depends on the chemical environment of the nucleus that we measure, and it follows that the resonance frequency of the NMR signal depends on the type of molecule that the nucleus is in and on the way it is bonded to the rest of the molecule. For example, in the ethyl alcohol molecule, CH_3—CH_2—OH, we have three different kinds of protons, the three protons on the methyl group, the two protons on the central carbon atom, and the proton on the hydroxyl group. These different types of protons have different values for the constant σ_a, and we therefore expect that the NMR spectrum of C_2H_6O consists of three lines at different frequencies and that the intensities of these lines have a ratio $3:2:1$.

We have already mentioned that the majority of nuclear magnetic resonance experiments are concerned with protons. For protons the value of σ varies between 2×10^{-5} and 4×10^{-5}, depending on the electronic environment of the proton. A resonance frequency in proton magnetic resonance can be measured with an accuracy of $1:10^8$ to $1:10^9$, so we can derive the value of the constants σ_a with an accuracy of $1:10^3$ to $1:10^4$.

In the initial magnetic resonance experiments the physicists were interested mainly in determining the magnetic moment $g_N \mu_N$ of the proton. It was an unpleasant surprise for them to find that the magnetic moment depends on the chemical environment when the accuracy of their measurements surpassed $1:10^5$. The effect was called "chemical shift," and the constants σ_a are sometimes known as chemical shift constants. A more precise name for σ_a is the proton magnetic shielding constant.

The quantum mechanical theory of the chemical shift is fairly complicated, and the numerical results for chemical shifts that have been derived from calculations are not particularly accurate. It may be useful to outline the theoretical problem.

We consider the situation where we have a very small magnetic dipole $\mathbf{\mu}_a$ at the position of one of the nuclei a in a molecule and where we have placed the molecule in a homogeneous magnetic field $\mathbf{H}$. We have seen already that in zeroth approximation the interaction between the dipole and the field is given by

$$E_{int} = -\mathbf{\mu}_a \cdot \mathbf{H} \qquad (49)$$

The electrons in the molecule interact both with the dipole and the magnetic field, and the total Hamiltonian $\mathcal{H}$ for the electrons in the molecule contains both $\mathbf{H}$ and $\mathbf{\mu}$:

$$\mathcal{H} = \mathcal{H}_{0,0} + \mathcal{H}_{1,0}\mathbf{H} + \mathcal{H}_{0,1}\mathbf{\mu} + \mathcal{H}_{1,1}\mathbf{\mu} \cdot \mathbf{H} + \cdots \qquad (50)$$

Here $\mathcal{H}_{0,0}$ is the Hamiltonian of the molecule in the absence of $\mathbf{\mu}$ and $\mathbf{H}$, and its lowest eigenvalue is E_0. The lowest eigenvalue ϵ_0 of the full Hamiltonian $\mathcal{H}$ may be derived by means of perturbation theory. It is again obtained as a power series in μ and H,

$$\epsilon_0 = E_0 + \mathbf{H}E_{1,0} + \mathbf{\mu}H_{0,1} + (\mathbf{\mu} \cdot \mathbf{H})E_{1,1} + \cdots \tag{51}$$

It is easily seen that the perturbation term $E_{1,1}$ relates to the shielding constant σ_a for nucleus a. The value of $E_{1,1}$ can be calculated by using second-order perturbation theory, but the expression that is obtained in this way contains an infinite sum of terms that depend on the eigenfunctions of all excited states of the molecule. In practice, it is difficult to obtain reliable numbers from this expression.

The only systems for which it is relatively easy to calculate the shielding constants are atomic S states (states with zero angular momentum). Here the theoretical expression for σ reduces to

$$\sigma = \frac{e^2}{3mc^2 a_0} \left\langle \Psi_0 \left| \sum_j \frac{1}{r_j} \right| \Psi_0 \right\rangle \tag{52}$$

In the case of atomic hydrogen, $\sigma = 1.8 \times 10^{-5}$. Also, a number of fairly accurate calculations have been performed for molecular hydrogen. The best calculations give results that are in the vicinity of $\sigma = 2.70 \times 10^{-5}$ ($\pm 0.05 \times 10^{-5}$), and we take this as the best value for H_2.

All these theoretical results are obtained with the bare proton as the reference point. Unfortunately, it has not been possible to measure the proton magnetic resonance frequency for the bare proton. This means that we have no experimental values for the absolute magnitudes of the proton shielding constants; instead we have results for the differences in proton shielding constants for different molecules. Also, we cannot measure the magnetic moment of the proton with a greater accuracy than $1:10^5$. The most accurate value of the proton magnetic moment is obtained by measuring the proton resonance frequency for the hydrogen molecule and by correcting this with the theoretical value of the proton shielding constant for molecular hydrogen.

The relative values of the proton magnetic shielding constants in different molecules can be measured fairly accurately. It is not practical to report the absolute values of the chemical shift constants, because we cannot measure the resonance frequency of the bare proton. We must therefore choose a reference compound, which we take as a "zero point" for reporting the σ values. In Table 8.3 we have reported the chemical shifts of some molecules with respect to methane; that is, we have chosen the relative constant in σ in such a way that the methane value is equal to zero.

The chemical shifts are easy to measure, and if we could find some simple

CHEMICAL SHIFTS   289

TABLE 8.3 ● Values for proton shielding constants σ in some selected molecules. In the first column σ_1 represents the experimental values, taken with respect to CH_4. In the second column we have listed the absolute values σ_2, which were obtained by taking $\sigma = 2.70 \times 10^{-5}$ for the hydrogen molecule.

Molecule	$\sigma_1 \times 10^5$	$\sigma_2 \times 10^5$
H_2	-0.422	2.70
H_2O	-0.060	3.062
HF	-0.251	2.871
HCl	0.045	3.167
HBr	0.435	3.557
HI	1.325	4.447
NH_3	-0.008	3.114
CH_4	0	3.122
C_2H_6	-0.075	3.047
C_2H_4	-0.518	2.604
C_2H_2	-0.135	2.987
C_6H_6	-0.713	2.409
C_6H_{12}	-0.129	2.993

rules to correlate the value of σ with the molecular structure, then this would be very useful. Numerous attempts have been made to find such correlations, but the situation is not so simple. The theoretical expression for σ consists of a term that is equal to Eq. (52) and, in addition, of an infinite sum over all excited states of the molecule. If the infinite sum could be neglected, then we could try and relate the value of σ_a to the electronic charge density in the vicinity of atom a. It may be seen in Table 8.3 that such a correlation exists for the series H_2, HF, HCl, HBr, and HI. However, there are no general simple rules to relate the constant σ to the electronegativity constants of the neighboring atoms. In the series CH_4, NH_3, H_2O, the proton shielding constants are almost the same in spite of the differences in structure.

Organic chemists are particularly interested in possible applications of NMR spectroscopy, because the structure of an organic molecule can be quite complicated and any information that may be derived from an NMR spectrum is welcome. It may be seen from Table 8.3 that the σ values for the protons in saturated hydrocarbons are all fairly close together. The protons in benzene have a completely different σ value, and it is usually possible to recognize the resonance frequencies of protons in saturated CH bonds and the signals from protons attached to aromatic rings. For example, in toluene the σ value for the CH_3 protons is 0.480×10^{-5}, higher than the σ value for the other five protons.

We cannot even begin to describe the many semiempirical theories for the values of proton shielding constants in organic molecules, because there are just too many of them. The only straightforward rule links the intensity of the signal to the number of equivalent protons. For example, in toluene there are three protons in the CH_3 group and five protons in the phenyl group and we should observe two absorption lines with an intensity ratio of 5:3. None of the other theories on proton shielding constants is too reliable, but we should realize that any semiempirical rule that works can save organic chemists a very large amount of work. Therefore, it is well worth while to look for such rules, even if they cannot be fully justified from theoretical arguments.

4. Spin-Spin Coupling

We have mentioned several times that the theoretical description of an NMR spectrum may be derived from the resonance equation (45) as long as we realize that we must take the magnetic field **H** in the equation as the magnetic field at the position of the nucleus. This field is the sum of the field $\mathbf{H}_{\text{ext}}$ that we apply and the magnetic fields due to the interactions between the nuclear spin $\mathbf{I}_a$ and its environment. In the previous section we have discussed the chemical shifts, which represent the interaction between the nuclear magnetic dipole and the motion of the electrons in the molecule. In this section we discuss the spin-spin interactions, which represent the interactions between the magnetic moment $\mathbf{\mu}_a$ and the magnetic moments due to the other spins.

In classical electromagnetic theory the interaction energy of two magnetic dipoles $\mathbf{\mu}_a$ and $\mathbf{\mu}_b$ is given by

$$E_{\text{int}} = \frac{R_{a,b}^2(\mathbf{\mu}_a \cdot \mathbf{\mu}_b) - 3(\mathbf{R}_{a,b} \cdot \mathbf{\mu}_a)(\mathbf{R}_{a,b} \cdot \mathbf{\mu}_b)}{R_{a,b}^5} \tag{53}$$

where $\mathbf{R}_{a,b}$ is the distance from a to b.

If we have two nuclei with spin $\frac{1}{2}$ and if we place them in a homogeneous magnetic field, then each of the two spins can be either parallel or antiparallel to the field. The magnetic moment of the second spin gives rise to a magnetic field at the position of the first spin, and this magnetic field depends on the orientation of the second spin. In principle, this magnetic field can be in the same direction as $\mathbf{H}_{\text{ext}}$ or it can have the opposite direction, depending on the orientation of the second spin. This effect causes a splitting of the energy levels of the first spin, as shown in Fig. 8.4. This argument is somewhat of an oversimplification of the situation, but in general it may be said that the interactions between the spins cause splittings in the energy levels, which give rise to a fine structure in the magnetic resonance spectrum.

SPIN-SPIN COUPLING 291

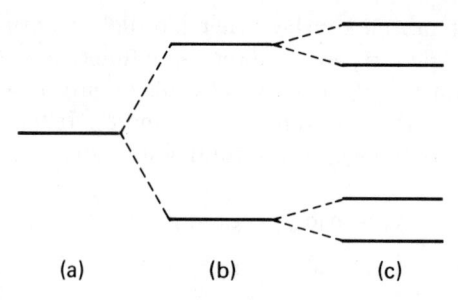

(a) (b) (c)

FIGURE 8.4 ● A simple picture of the effect of spin-spin coupling for two protons. In (b) we show the splitting of the first proton due to the magnetic field **H**. The second proton is parallel or antiparallel to **H**, and it either helps or opposes **H**. As a result, each level in (b) is split further in (c) into two different levels.

The magnitude of the spin-spin interactions may be determined from the fine structure in the magnetic resonance spectrum. However, this is not such a simple procedure. We must start with the Hamiltonian for the spin system, which contains as parameters the various magnetic shielding constants and the terms that represent the spin-spin interactions. Then we must find the eigenvalues of this Hamiltonian, and we must determine the selection rules for the transitions between the energy levels. By comparing this calculated spectrum with the experimental absorption lines, we then obtain the values of the parameters. It was found that in this method the spin-spin interaction between two nuclei i and j may be represented by a term

$$\mathcal{H}_{i,j} = J_{i,j}(\mathbf{I}_i \cdot \mathbf{I}_j) \tag{54}$$

Here $\mathbf{I}_i$ and $\mathbf{I}_j$ are the spin operators for nuclei i and j, and the parameter $J_{i,j}$ represents the magnitude of the spin-spin interaction between the two nuclei. The total Hamiltonian $\mathcal{H}$ for the nuclear spin system is then obtained by combining Eq. (54) with Eq. (48),

$$\mathcal{H} = \sum_i (1 - \sigma_i) g_i \mu_i (\mathbf{I}_i \cdot \mathbf{H}) + \sum_{j>i} J_{i,j}(\mathbf{I}_i \cdot \mathbf{I}_j) \tag{55}$$

It may be useful to illustrate how the eigenvalues and eigenfunctions of $\mathcal{H}$ are obtained for a simple system. We take a system of two protons and we take **H** along the Z axis. The Hamiltonian then reduces to

$$\mathcal{H} = (1 - \sigma_1) g_p \mu_p I_{1z} H + (1 - \sigma_2) g_p \mu_p I_{2z} H + J(\mathbf{I}_1 \cdot \mathbf{I}_2) \tag{56}$$

We write this as

$$\mathcal{H} = \gamma_1 H I_{1z} + \gamma_2 H I_{2z} + J(I_{1x}I_{2x} + I_{1y}I_{2y} + I_{1z}I_{2z}) \tag{57}$$

The spin of the first proton can be either parallel or antiparallel to the Z axis. The first situation is represented by a spin function α_1 and the second situation by a spin function β_1. The two possible orientations of the second spin are represented by the spin functions α_2 and β_2. It follows that there are four possible spin functions for the total spin system, namely

$$\zeta_1 = \alpha_1 \alpha_2 \qquad \zeta_3 = \beta_1 \alpha_2$$
$$\zeta_2 = \alpha_1 \beta_2 \qquad \zeta_4 = \beta_1 \beta_2 \tag{58}$$

The effect of the Hamiltonian $\mathcal{H}$ on this spin function may be derived from the properties of the spin functions that we discussed in Chapter III, Eq. (51). We have

$$I_x \alpha = \tfrac{1}{2}\beta \qquad I_x \beta = \tfrac{1}{2}\alpha$$
$$I_y \alpha = \tfrac{1}{2}i\beta \qquad I_y \beta = -\tfrac{1}{2}i\alpha \tag{59}$$
$$I_z \alpha = \tfrac{1}{2}\alpha \qquad I_z \beta = -\tfrac{1}{2}\beta$$

and

$$(\mathbf{I}_1 \cdot \mathbf{I}_2)\zeta_1 = (I_{1x}I_{2x} + I_{1y}I_{2y} + I_{1z}I_{2z})(\alpha_1\alpha_2) = \tfrac{1}{4}(\alpha_1\alpha_2) = \tfrac{1}{4}\zeta_1$$
$$(\mathbf{I}_1 \cdot \mathbf{I}_2)\zeta_2 = (I_{1x}I_{2x} + I_{1y}I_{2y} + I_{1z}I_{2z})(\alpha_1\beta_2)$$
$$= \tfrac{1}{4}\beta_1\alpha_2 + \tfrac{1}{4}\beta_1\alpha_2 - \tfrac{1}{4}\alpha_1\beta_2 = -\tfrac{1}{4}\zeta_2 + \tfrac{1}{2}\zeta_3 \tag{60}$$
$$(\mathbf{I}_1 \cdot \mathbf{I}_2)\zeta_3 = (I_{1x}I_{2x} + I_{1y}I_{2y} + I_{1z}I_{2z})(\beta_1\alpha_2)$$
$$= \tfrac{1}{4}\alpha_1\beta_2 + \tfrac{1}{4}\alpha_1\beta_2 - \tfrac{1}{4}\beta_1\alpha_2 = -\tfrac{1}{4}\zeta_3 + \tfrac{1}{2}\zeta_2$$
$$(\mathbf{I}_1 \cdot \mathbf{I}_2)\zeta_3 = (I_{1x}I_{2x} + I_{1y}I_{2y} + I_{1z}I_{2z})(\beta_1\beta_2) = \tfrac{1}{4}(\beta_1\beta_2) = \tfrac{1}{4}\zeta_4$$

Also,

$$(\gamma_1 I_{1z} + \gamma_2 I_{2z})\zeta_1 = (\tfrac{1}{2}\gamma_1 + \tfrac{1}{2}\gamma_2)\zeta_1$$
$$(\gamma_1 I_{1z} + \gamma_2 I_{2z})\zeta_2 = (\tfrac{1}{2}\gamma_1 - \tfrac{1}{2}\gamma_2)\zeta_2$$
$$(\gamma_1 I_{1z} + \gamma_2 I_{2z})\zeta_3 = (-\tfrac{1}{2}\gamma_1 + \tfrac{1}{2}\gamma_2)\zeta_3 \tag{61}$$
$$(\gamma_1 I_{1z} + \gamma_2 I_{2z})\zeta_4 = (-\tfrac{1}{2}\gamma_1 - \tfrac{1}{2}\gamma_2)\zeta_4$$

It follows easily that the functions ζ_1 and ζ_4 are eigenfunctions of the Hamiltonian (57), because

$$\mathcal{H}\zeta_1 = [\tfrac{1}{2}H(\gamma_1 + \gamma_2) + \tfrac{1}{4}J]\zeta_1$$
$$\mathcal{H}\zeta_4 = [-\tfrac{1}{2}H(\gamma_1 + \gamma_2) + \tfrac{1}{4}J]\zeta_4 \tag{62}$$

SPIN-SPIN COUPLING   293

Since there exist only four eigenfunctions of the operator $\mathcal{H}$ and since two of the eigenfunctions are given by Eq. (62), the remaining two eigenfunctions must be linear combinations of the spin functions ζ_2 and ζ_3. It follows from Eqs. (60) and (61) that

$$\mathcal{H}\zeta_2 = [-\tfrac{1}{4}J + \tfrac{1}{2}H(\gamma_1 - \gamma_2)]\zeta_2 + \tfrac{1}{2}J\zeta_3$$
$$\mathcal{H}\zeta_3 = \tfrac{1}{2}J\zeta_2 + [-\tfrac{1}{4}J - \tfrac{1}{2}H(\gamma_1 - \gamma_2)]\zeta_3 \tag{63}$$

or

$$\mathcal{H}(\zeta_2 + \zeta_3) = \tfrac{1}{4}J(\zeta_2 + \zeta_3) + \tfrac{1}{2}H(\gamma_1 - \gamma_2)(\zeta_2 - \zeta_3)$$
$$\mathcal{H}(\zeta_2 - \zeta_3) = \tfrac{1}{2}H(\gamma_1 - \gamma_2)(\zeta_2 + \zeta_3) - \tfrac{3}{4}J(\zeta_2 - \zeta_3) \tag{64}$$

We define the eigenfunctions by means of the equation

$$(\mathcal{H} - \lambda)\chi = 0$$
$$\chi = a\chi_1 + b\chi_2 = a(\zeta_2 + \zeta_3) + b(\zeta_2 - \zeta_3) \tag{65}$$

It follows then easily that the eigenvalue λ and the coefficients a and b must satisfy

$$(\tfrac{1}{4}J - \lambda)a + \tfrac{1}{2}H(\gamma_1 - \gamma_2)b = 0$$
$$\tfrac{1}{2}H(\gamma_1 - \gamma_2)a + (-\tfrac{3}{4}J - \lambda)b = 0 \tag{66}$$

The equation for the eigenvalue λ is then

$$(-\lambda + \tfrac{1}{4}J)(-\lambda - \tfrac{3}{4}J) - \tfrac{1}{4}H^2(\gamma_1 - \gamma_2)^2 = 0 \tag{67}$$

which has two solutions,

$$\lambda = -\tfrac{1}{4}J \pm \tfrac{1}{2}\sqrt{J^2 + H^2(\gamma_1 - \gamma_2)^2} \tag{68}$$

We have assumed here that $\gamma_1 \neq \gamma_2$.

It may be useful to summarize the four eigenvalues that we have listed in Eqs. (62) and (68):

$$\begin{aligned}\lambda_1 &= \tfrac{1}{4}J + \tfrac{1}{2}H(\gamma_1 + \gamma_2)\\ \lambda_2 &= -\tfrac{1}{4}J + \tfrac{1}{2}\sqrt{J^2 + H^2(\gamma_1 - \gamma_2)^2}\\ \lambda_3 &= -\tfrac{1}{4}J - \tfrac{1}{2}\sqrt{J^2 + H^2(\gamma_1 - \gamma_2)^2}\\ \lambda_4 &= \tfrac{1}{4}J - \tfrac{1}{2}H(\gamma_1 + \gamma_2)\end{aligned} \tag{69}$$

It may be derived from the selection rules that the transitions between the states 1 and 4 and between the states 2 and 3 are forbidden, so we observe only the four transitions $1 \to 2$, $1 \to 3$, $4 \to 2$, and $4 \to 3$. The frequencies of these four transitions are derived by taking the corresponding energy differences in Eq. (69),

$$\begin{aligned}\nu_{12} &= \tfrac{1}{2}J + \tfrac{1}{2}H(\gamma_1 + \gamma_2) - \tfrac{1}{2}\sqrt{J^2 + H^2(\gamma_1 - \gamma_2)^2} \\ \nu_{13} &= \tfrac{1}{2}J + \tfrac{1}{2}H(\gamma_1 + \gamma_2) + \tfrac{1}{2}\sqrt{J^2 + H^2(\gamma_1 - \gamma_2)^2} \\ \nu_{24} &= \tfrac{1}{2}J - \tfrac{1}{2}H(\gamma_1 + \gamma_2) - \tfrac{1}{2}\sqrt{J^2 + H^2(\gamma_1 - \gamma_2)^2} \\ \nu_{34} &= \tfrac{1}{2}J - \tfrac{1}{2}H(\gamma_1 + \gamma_2) + \tfrac{1}{2}\sqrt{J^2 + H^2(\gamma_1 - \gamma_2)^2}\end{aligned} \qquad (70)$$

We observe four lines in the proton magnetic resonance spectrum, and we can derive the values of J and the chemical shift constants from the spectrum.

It should be pointed out that we have assumed the two constants σ_1 and σ_2 to be different in the above discussion. This means that the discussion applies only to a pair of nonequivalent protons. If the two protons are equivalent, the two constants γ_1 and γ_2 are equal to each other and Eq. (64) reduces to

$$\begin{aligned}\mathcal{H}\chi_1 &= \tfrac{1}{4}J\chi_1 & \chi_1 &= \zeta_2 + \zeta_3 \\ \mathcal{H}\chi_2 &= -\tfrac{3}{4}J\chi_2 & \chi_2 &= \zeta_2 - \zeta_3\end{aligned} \qquad (71)$$

This means that the spin functions χ_1 and χ_2 are eigenfunctions of the Hamiltonian. In the present case the eigenvalues and corresponding eigenfunctions of the Hamiltonian $\mathcal{H}$ are listed in Table 8.4. The only allowed transitions are between the levels 1 and 2 and between levels 2 and 3 of Table 8.4. Both of these transitions have the same frequency, $h\nu = \gamma H$, which is independent of J. We must therefore conclude that it is not possible to measure the coupling constant between two equivalent protons. As a result, the H–H coupling constant in the H_2 molecule is not known, and the simplest system for which an experimental value is available is HD.

TABLE 8.4 ● Eigenvalues and eigenfunctions for a system of two equivalent spins.

Eigenvalue	Eigenfunction
$\tfrac{1}{4}J + \gamma H$	$\alpha_1 \alpha_2$
$\tfrac{1}{4}J$	$2^{-1/2}(\alpha_1\beta_2 + \beta_1\alpha_2)$
$\tfrac{1}{4}J - \gamma H$	$\beta_1\beta_2$
$-\tfrac{3}{4}J$	$2^{-1/2}(\alpha_1\beta_2 - \beta_1\alpha_2)$

It may be seen from the above illustration that the relation between the experimental NMR frequencies and the various parameters may be fairly complicated. However, a number of typical situations (such as three or four equivalent spins, three equivalent protons coupled to another spin, etc.), have been studied theoretically, and the results are mentioned in most textbooks on the interpretation of NMR spectra. If we want to interpret an unknown NMR spectrum we can often tell from the number of lines and their relative positions and intensities what the nature of the spin system is, and it is then relatively easy to derive the values of the chemical shift constants and of the coupling constants from the NMR spectrum. It should also be noted that the effects of magnetic shielding are proportional to the applied magnetic field $\mathbf{H}$, whereas the splittings due to the spin-spin coupling are field independent. In complex situations we can derive the values of the chemical shift constants by performing an NMR experiment at a high magnetic field, and we can determine the coupling constants by repeating the experiment at low magnetic field strengths.

In the customary definition of the coupling constants by means of Eqs. (54) and (55), the constants J have the dimension of energy and their values are usually listed in terms of cycles per second. The coupling constant for H_2 has been calculated to be about 250 cycles/sec, the proton-proton coupling constant in methane is 12.4 cycles/sec, and the coupling constants in the ethylene molecule are 19.1, 11.6, and 2.5 cycles/sec.

Up to this point we have discussed only how the values of the coupling constants J_{ij} are obtained experimentally from the NMR spectra. Let us now consider how they can be evaluated theoretically from the molecular eigenfunctions. Again, the complete theory of coupling constants is fairly complicated, and we do not intend to discuss it in detail. We just want to outline the various mechanisms that are responsible for the coupling between the nuclear spins.

At first glance it seems that the spin-spin coupling is easily calculated from Eq. (53), which represents the classical electromagnetic interaction between two magnetic dipoles. In this calculation we do not even have to know the molecular eigenfunctions; all we have to do is to substitute the distance between the two nuclei into the equation. However, it turns out that the values obtained in this way are much smaller than the experimental coupling constants and it follows that the direct dipole-dipole interaction between the nuclear spins cannot be responsible for the experimentally observed spin-spin coupling.

It is generally believed now that the main contribution to the nuclear spin interactions is the indirect coupling between the nuclei by means of the electron-nuclear interactions. In order to understand this mechanism, let us look at the situation sketched in Fig. 8.5. Here we have two protons, a and b, and two electrons, 1 and 2. At a given moment one of the electrons, electron 1, is in the vicinity of nucleus a and the other electron, electron 2, is close to nucleus b. Since the distance between proton a and electron 1 is so small, there is a fairly large interaction between the proton spin and the electron spin and the two spins will have a tendency to be parallel.* This means that if the proton spin a points upward as in Fig. 8.5, then the spin of electron 1 also has a tendency to point upward. According to the Pauli principle, the two electron spins must be antiparallel if they are in the same orbital. Consequently, if the two electrons form a chemical bond, then the spin of electron 2 must point down if the spin of electron 1 points upward. Now, electron 2 is quite close to proton b, and as a result there is a fairly strong interaction between the spins of electron 2 and proton b. This means that the two spins have a tendency to point in the same direction and that the spin of proton b is likely to point downward.

It may be seen that in the above model the interaction between the two nuclear spins a and b is due to the two interactions between spins a and 1 and between spins b and 2. These two interactions are much larger than the direct interaction between the spins a and b because the distances $a \rightarrow 1$ and $b \rightarrow 2$ are so much smaller than the distance $a \rightarrow b$. It does not matter that the distance between the two electrons is fairly large, because their coupling is due to the Pauli principle and they must be antiparallel at any distance.

The formal quantum mechanical theory of electron-coupled spin-spin interactions is fairly complicated, but we feel that it may be useful if we at least outline the theory. First we must know the interaction between an electron spin i and a nuclear spin N. This interaction is represented by a Hamiltonian,

---

* Two magnetic dipoles have the lowest energy if they are antiparallel. Since the proton spin has the same direction as the proton magnetic moment and since the electron spin has the opposite direction as the electron magnetic moment, the two spins have the lowest energy if they are parallel.

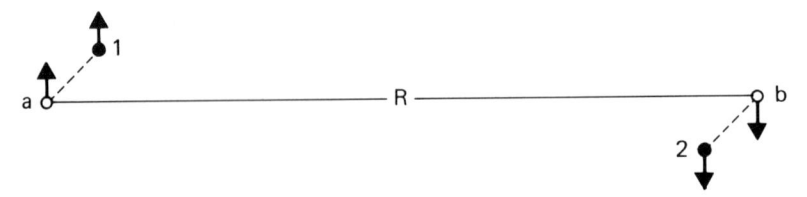

FIGURE 8.5 ● Representation of electron-coupled proton-proton interaction. The two electron spins must be antiparallel because of the Pauli exclusion principle.

SPIN-SPIN COUPLING

$$\mathcal{H}_{i,N} = -2\mu_0 g_N \mu_N \left[\frac{(\mathbf{S}_i \cdot \mathbf{I}_N)(\mathbf{r}_{N,i} \cdot \mathbf{r}_{N,i}) - 3(\mathbf{S}_i \cdot \mathbf{r}_{N,1})(\mathbf{I}_N \cdot \mathbf{r}_{N,1})}{r_{N,i}^5} \right.$$
$$\left. - \frac{8\pi}{3} (\mathbf{S}_i \cdot \mathbf{I}_N) \delta(\mathbf{r}_{N,i}) \right] \quad (72)$$

The first term in this Hamiltonian is the classical dipole-dipole interaction between the magnetic moments of the electron and the nucleus. It has the same form as the nuclear spin-spin interaction of Eq. (53). The second term in Eq. (72) cannot be visualized so easily. It is known as the Fermi contact potential, and its existence can be shown from relativistic arguments or, in an easier way, by taking the finite dimensions of the nucleus into account. The Fermi contact potential is quite important in the theory of coupling constants, because it has been shown in some model calculations that the contribution of the Fermi potential to the electron-nuclear spin interaction is usually much larger than the contribution from the classical dipole-dipole interaction.

In Eq. (72) we have formulated the Fermi contact potential in terms of the three-dimensional δ function. We have sketched the one-dimensional δ function $\delta(x - x_0)$ in Fig. 8.6. It is defined as a function that is very large in the vicinity of the point $x = x_0$ and practically zero when x differs from x_0. The function also satisfies the condition

$$\int_{-\infty}^{\infty} \delta(x - x_0) \, dx = 1 \quad (73)$$

Clearly, if we have a function $f(x)$ that is finite and continuous at the point $x = x_0$, we must have

$$\int_{-\infty}^{\infty} f(x) \delta(x - x_0) \, dx = f(x_0) \quad (74)$$

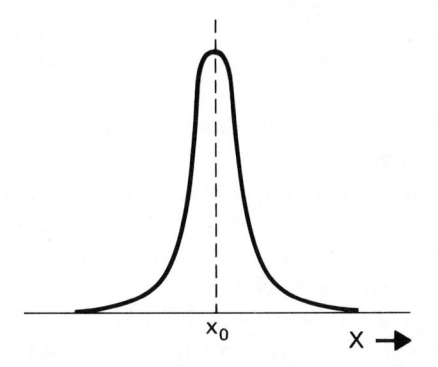

FIGURE 8.6 ● Representation of the one-dimensional δ function $\delta(x - x_0)$.

A three-dimensional δ function is a product of three one-dimensional δ functions,

$$\delta(\mathbf{r} - \mathbf{r}_0) = \delta(x - x_0)\delta(y - y_0)\delta(z - z_0) \tag{75}$$

It follows from Eq. (74) that it has the property

$$\begin{aligned}\int f(\mathbf{r})\delta(\mathbf{r} - \mathbf{r}_0)\, d\mathbf{r} &= \iiint f(x, y, z)\delta(x - x_0)\delta(y - y_0)\delta(z - z_0)\, dx\, dy\, dz \\ &= f(x_0, y_0, z_0) = f(\mathbf{r}_0)\end{aligned} \tag{76}$$

It may be seen that it is always convenient to have a δ function occur in an integral, because then the integration becomes easy to do and the result is usually quite simple.

The quantum mechanical theory of the nuclear spin-spin interactions is fairly complicated, and it is not easy to evaluate the spin-spin coupling constants with high accuracy. In order to do the calculation, we must treat the Hamiltonian

$$\mathcal{H}' = \sum_i \sum_N \mathcal{H}_{i,N} \tag{77}$$

as a perturbation and we must calculate the various perturbations E_0', E_0'', ... to the unperturbed ground-state energy E_0. In Eq. (77) the Hamiltonian $\mathcal{H}'$ is derived from the Hamiltonian (72), which represents the interaction between the spin of electron i and the spin of nucleus N by summing over all electrons and all nuclei in the molecule. It follows from the definition (54) of the coupling constants that they are obtained as the coefficients of terms of the type $(\mathbf{I}_N \cdot \mathbf{I}_M)$ in the energy expression, and that they are derived from the second-order perturbation E_0'' to the energy. We have seen in our discussion on perturbation theory that this energy perturbation is given by

$$E_0'' = \sum_k \frac{\langle \Psi_0 | \mathcal{H}' | \Psi_k \rangle \langle \Psi_k | \mathcal{H}' | \Psi_0 \rangle}{E_k - E_0} \tag{78}$$

where E_k and Ψ_k are the eigenvalues and eigenfunctions of the unperturbed molecular Hamiltonian.

In general, the eigenfunctions of the excited molecular eigenstates are not known too accurately, so it is difficult to perform reliable calculations of the nuclear coupling constants. However, some calculations were performed on the hydrogen molecule in order to find out the relative magnitudes of the contributions due to the different terms in the Hamiltonian (72). It turned out that the main contribution is due to the Fermi contact potential

$$\mathcal{H}'_{i,N} = \frac{16\pi\mu_0 g_N \mu_N}{3}(\mathbf{S}_i \cdot \mathbf{I}_N)\delta(\mathbf{r}_{N,i}) \tag{79}$$

If it is assumed that this is true in general, and not just for the hydrogen molecule, then it may be attempted to relate the coupling constants between two nuclear spins to the electron densities on the two nuclei. In this way it may be attempted to construct semiempirical theories of the values of the various spin-spin coupling constants. However, none of these semiempirical theories are very reliable, because they rest on too many approximations.

It may be seen from the definition that the coupling constants $J_{i,j}$ have the dimension of energy, and they are usually expressed in terms of cycles per second. We derived in Eq. (20) that the proton resonance occurs at 14.2×10^6 cycles/sec, the various proton-proton coupling constants range from 1 to 100 cycles/sec, and their effect is comparable to the effect of the chemical shift. The largest proton-proton coupling occurs in the H_2 molecule, where it is about 300 cycles/sec. It should be noted that the H_2 coupling constant cannot be measured, as we have shown in Table 8.4. However, it can be calculated, and it can also be estimated from the coupling constant of HD, which has been measured.

It should be noted that appreciable coupling can also occur between protons that are not directly bonded to one another. For example, the H—C—H proton-proton coupling constant is around 12 cycles/sec; the exact value varies for different molecules. Coupling has been observed between protons that are separated by several bonds; the record seems to be a coupling constant of the order of 1 cycle/sec between two protons separated by eight bonds.

An interesting situation arises in the ethylene molecule. Here the three coupling constants are $J_{12} = 2.5$, $J_{1,3} = 11.6$, and $J_{1,4} = 19.1$ cycles/sec (the numbering of the protons is illustrated in Fig. 8.7). Here the largest coupling occurs for the two protons that are farthest apart, whereas the H—C—H coupling constant is only 2.5 cycles/sec. The reader may notice that this number is much smaller than the 12 cycles/sec that we quoted in the previous section, but this difference is due to the presence of π electrons; in the previous section we were talking about saturated molecules.

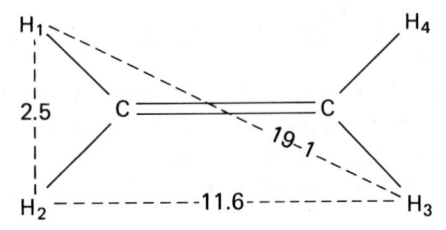

FIGURE 8.7 ● Proton-proton coupling constants in ethylene.

From the various numbers that we quoted, it may be seen that the coupling constants can be quite sensitive to small changes in chemical structure. For that reason the experimental hyperfine splittings are often used by organic chemists to help them in determining the structure of a molecule. If nothing else, it is often possible to find out how many groups of equivalent protons there are by just counting the number of lines in the NMR spectrum. Additional information may be derived from the magnitudes of the various experimental coupling constants. Someone who knows the values of the many experimental coupling constants that have been measured and who has experience in interpreting NMR spectra can make fairly good predictions about the structure of a molecule. As a result, NMR techniques have become very important in determining molecular structures.

5. Electron Spin Resonance

Electron spin resonance can be used only for molecules that have a nonzero net electronic spin. Most molecules have singlet ground states (one of the few exceptions is the oxygen molecule), so ESR cannot be applied to them. In general, ESR may be applied only to free radicals, which have an odd number of electrons so that there is one unpaired electron spin, and to triplet molecules. In this section we shall limit ourselves to a discussion of free radicals.

A free radical has the configuration $(\chi_1)^2(\chi_2)^2(\chi_3)^2 \cdots (\chi_N)^2(\chi_{N+1})$ if we use simple molecular orbital theory to represent its wave function. Here the functions χ_i are the molecular orbitals. Each of the first N molecular orbitals $\chi_1, \chi_2, \ldots, \chi_N$ contains a pair of electrons with antiparallel spins. The next highest molecular orbital, χ_{N+1}, contains a single electron, which can have either spin α or spin β. The complete molecular eigenfunctions $\Psi_{1/2}$ and $\Psi_{-1/2}$ are given by

$$\Psi_{1/2} = [(2N+1)!]^{-1/2} \sum_p P\delta_p \{\chi_1(1)\alpha(1)\chi_1(2)\beta(2)\chi_2(3)\alpha(3)\chi_2(4)\beta(4) \cdots$$
$$\chi_N(2N-1)\alpha(2N-1)\chi_N(2N)\beta(2N)\chi_{N+1}(2N+1)\alpha(2N+1)\}$$
$$\Psi_{-1/2} = [(2N+1)!]^{-1/2} \sum_p P\delta_p \{\chi_1(1)\alpha(1)\chi_1(2)\beta(2)\chi_2(3)\alpha(3)\chi_2(4)\beta(4) \cdots$$
$$\chi_N(2N-1)\alpha(2N-1)\chi_N(2N)\beta(2N)\chi_{N+1}(2N+1)\beta(2N+1)\} \quad (80)$$

We should realize that in ESR of a free radical we measure the resonance radiation between the two spin states of the unpaired electron, that is, the electron in the molecular orbital χ_{N+1}. We have seen in Section 1, Eqs. (8) and (12), that the resonance radiation of a free electron is given by

$$h\nu = g\mu_0 H \quad (81)$$

where the g factor is very close to 2 and where H is the applied homogeneous magnetic field that splits the two spin states. In a free radical we must substitute the magnetic field H_{loc} at the position of the unpaired electron. Just as in nuclear magnetic resonance the field H_{loc} consists of the applied field H_{ext} and of the small fields due to the environment of the electron. These small fields are the sum of the induced fields due to the motion of the other electrons and of the magnetic fields due to the nuclear spins. This means that the spin Hamiltonian for the electron should be written as

$$\mathcal{H} = g\mu_0 H(1 - \sigma)S_z + \sum_n a_n \mathbf{S} \cdot \mathbf{I}_n \tag{82}$$

instead of the Hamiltonian (12) for the free electron.

It may be seen that the parameter σ, which is the analog of the chemical shift constant in NMR, causes a displacement of the resonance line and that the coupling constants a_n cause a splitting of the resonance line into a number of separate lines. It is customary to incorporate the effect of σ into the g factor. Also, since we consider the coupling between the electron and the nuclear spins, we must also consider the interaction between the nuclear spins and the magnetic field H. Consequently, the ESR spectrum should be described by using the Hamiltonian

$$\mathcal{H} = g\mu_0 H S_z - \sum_n g_n \mu_n H I_{nz} + \sum_n a_n \mathbf{S} \cdot \mathbf{I}_n \tag{83}$$

If we neglect higher-order effects, we may approximate this Hamiltonian as

$$\mathcal{H} = g\mu_0 H S_z - \sum_n g_n \mu_n H I_{nz} + \sum_n a_n I_{nz} S_z \tag{84}$$

It may be recalled that the nuclear spin-spin coupling we discussed in the previous section was a second-order effect, and as a result it was relatively difficult to interpret the spectra and to calculate the coupling constants. The electron-nuclear spin-spin coupling is a first-order effect, and its description is therefore much simpler.

In order to illustrate the situation, we consider a system consisting of one electron and two protons. This system has eight possible spin functions, which we write as

$$\begin{aligned}
\zeta_{1/2,1} &= \alpha_{\text{el}}\alpha_1\alpha_2 & \zeta_{-1/2,1} &= \beta_{\text{el}}\alpha_1\alpha_2 \\
\zeta_{1/2,2} &= \alpha_{\text{el}}\alpha_1\beta_2 & \zeta_{-1/2,2} &= \beta_{\text{el}}\alpha_1\beta_2 \\
\zeta_{1/2,3} &= \alpha_{\text{el}}\beta_1\alpha_2 & \zeta_{-1/2,3} &= \beta_{\text{el}}\beta_1\alpha_2 \\
\zeta_{1/2,4} &= \alpha_{\text{el}}\beta_1\beta_2 & \zeta_{-1/2,4} &= \beta_{\text{el}}\beta_1\beta_2
\end{aligned} \tag{85}$$

302 MAGNETIC RESONANCE

where the subscripts 1 and 2 refer to the two protons. These functions are all eigenfunctions of the spin operator (84). It is easily verified that

$$\begin{aligned}
\mathcal{H}\zeta_{1/2,1} &= [(\tfrac{1}{2}g\mu_0 - \tfrac{1}{2}g_1\mu_1 - \tfrac{1}{2}g_2\mu_2)H + \tfrac{1}{4}(a_1 + a_2)]\zeta_{1/2,1} \\
\mathcal{H}\zeta_{-1/2,1} &= [(-\tfrac{1}{2}g\mu_0 - \tfrac{1}{2}g_1\mu_1 - \tfrac{1}{2}g_2\mu_2)H - \tfrac{1}{4}(a_1 + a_2)]\zeta_{-1/2,1} \\
\mathcal{H}\zeta_{1/2,2} &= [(\tfrac{1}{2}g\mu_0 - \tfrac{1}{2}g_1\mu_1 + \tfrac{1}{2}g_2\mu_2)H + \tfrac{1}{4}(a_1 - a_2)]\zeta_{1/2,2} \\
\mathcal{H}\zeta_{-1/2,2} &= [(-\tfrac{1}{2}g\mu_0 - \tfrac{1}{2}g_1\mu_1 + \tfrac{1}{2}g_2\mu_2)H - \tfrac{1}{4}(a_1 - a_2)]\zeta_{-1/2,2} \\
\mathcal{H}\zeta_{1/2,3} &= [(\tfrac{1}{2}g\mu_0 + \tfrac{1}{2}g_1\mu_1 - \tfrac{1}{2}g_2\mu_2)H + \tfrac{1}{4}(-a_1 + a_2)]\zeta_{1/2,3} \\
\mathcal{H}\zeta_{-1/2,3} &= [(-\tfrac{1}{2}g\mu_0 + \tfrac{1}{2}g_1\mu_1 - \tfrac{1}{2}g_2\mu_2)H - \tfrac{1}{4}(-a_1 + a_2)]\zeta_{-1/2,3} \\
\mathcal{H}\zeta_{1/2,4} &= [(\tfrac{1}{2}g\mu_0 + \tfrac{1}{2}g_1\mu_1 + \tfrac{1}{2}g_2\mu_2)H - \tfrac{1}{4}(a_1 + a_2)]\zeta_{1/2,4} \\
\mathcal{H}\zeta_{-1/2,4} &= [(-\tfrac{1}{2}g\mu_0 + \tfrac{1}{2}g_1\mu_1 + \tfrac{1}{2}g_2\mu_2)H + \tfrac{1}{4}(a_1 + a_2)]\zeta_{-1/2,4}
\end{aligned} \quad (86)$$

The selection rules for ESR transitions are

$$\Delta S_{el} = 1 \qquad \Delta S_{nucl} = 0 \tag{87}$$

This means that we observe transitions only between states with identical nuclear spin functions. In the above case these are the transitions $(\tfrac{1}{2}, i) \to (-\tfrac{1}{2}, i)$ with the frequencies

$$h\nu_i = E(\tfrac{1}{2}, i) - E(-\tfrac{1}{2}, i) \tag{88}$$

It follows from Eq. (86) that these frequencies are given by

$$\begin{aligned}
h\nu_1 &= g\mu_0 H + \tfrac{1}{2}(a_1 + a_2) \\
h\nu_2 &= g\mu_0 H + \tfrac{1}{2}(a_1 - a_2) \\
h\nu_3 &= g\mu_0 H - \tfrac{1}{2}(a_1 - a_2) \\
h\nu_4 &= g\mu_0 H - \tfrac{1}{2}(a_1 + a_2)
\end{aligned} \tag{89}$$

We have drawn a sketch of the energy levels of Eq. (86) in Fig. 8.8. It follows that the relation between the experimental splittings and the electron-nuclear spin coupling constants is quite simple. The splittings are even symmetric with respect to the central line. The four frequencies in Eq. (89) may be written as

$$h\nu = g\mu_0 H + \tfrac{1}{2}[(\pm)a_1 (\pm) a_2] \tag{90}$$

This expression is easily generalized to a system of one electron and N protons, where it becomes

$$h\nu = g\mu_0 H + \tfrac{1}{2}(\pm a_1 \pm a_2 \pm a_3 \pm \cdots \pm a_N) \tag{91}$$

The maximum number of lines is 2^N. It is clear that some of the lines will coincide if some of the coupling constants are equal. In the extreme case

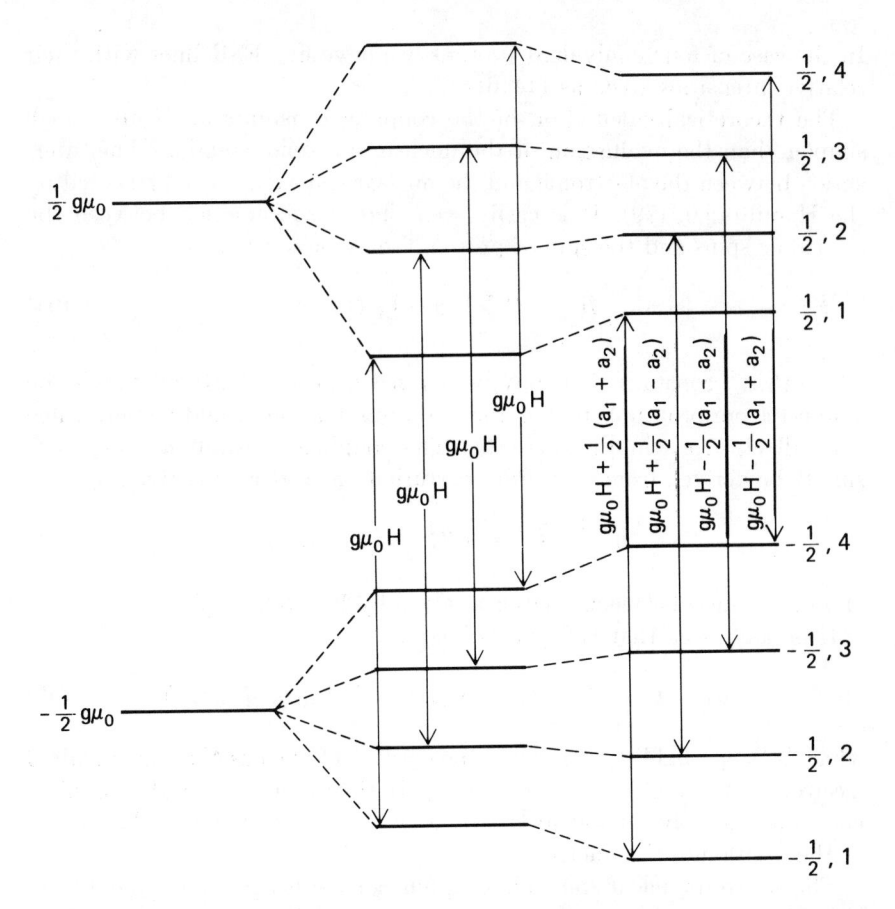

FIGURE 8.8 ● Energy level diagram for a system of one electron and two protons.

where all coupling constants are equal to one another, the lines are given by

$$h\nu = g\mu_0 H + \tfrac{1}{2}a(\pm 1 \pm 1 \pm 1 \pm 1 \pm \cdots \pm 1) \qquad (92)$$

The distribution of the lines and their intensities can be derived from Eq. (92). It may be seen that the relative intensities of the lines are equal to Newton's binomial coefficients, which occur in the expression $(1 + x)^n$. These coefficients are represented in the Pascal triangle,

$$\begin{array}{ccccccccccc}
 & & & & & 1 & & 1 & & & \\
 & & & & 1 & & 2 & & 1 & & \\
 & & & 1 & & 3 & & 3 & & 1 & \\
 & & 1 & & 4 & & 6 & & 4 & & 1 \\
 & 1 & & 5 & & 10 & & 10 & & 5 & & 1 \\
 & & & & & \cdot \cdot \cdot \cdot \cdot & & & & &
\end{array} \qquad (93)$$

In the case of four equivalent protons, we have five ESR lines with their relative intensities given as 1:4:6:4:1.

The theoretical calculation of the coupling constants a_N is also much simpler than the evaluation of the nuclear spin-spin coupling. The interaction between the electronic and the nuclear spins may be represented by the Hamiltonian (79). It is easily seen that the interaction between the electronic spins and the spin of proton N may be written as

$$H_N = C \sum_i (\mathbf{S}_i \cdot \mathbf{I}_N) \delta(\mathbf{r}_{N,i}) \quad (94)$$

where C is a constant that may be derived from Eq. (79) and $\mathbf{r}_{N,i}$ is the distance from proton N to electron i; the summation should be performed over all electrons in the molecule. The coupling constant a_N may now simply be derived from first-order perturbation theory; it is given by

$$a_N(\mathbf{S} \cdot \mathbf{I}_N) = \langle \Psi_{1/2} | H_N | \Psi_{-1/2} \rangle \quad (95)$$

Here the molecular eigenfunction is given by Eq. (80).

It is easily seen that Eq. (95) reduces to

$$a_N = C \langle \chi_{N+1} | \delta(\mathbf{r}_{N,i}) | \chi_{N+1} \rangle = C \chi^*_{N+1}(\mathbf{r}_N) \chi_{N+1}(\mathbf{r}_N) \quad (96)$$

which is the probability density at the position of nucleus N of the unpaired electron in the molecular orbital χ_{N+1}. It thus follows that the coupling constant is simply related to the value of the electronic molecular orbital at the position of the nucleus.

The electron-nuclear spin-spin coupling constants again have the dimension of energy, but they are usually expressed in terms of gauss. It is customary to perform an ESR experiment for a fixed wavelength of the resonance radiation, namely 3.15 cm (see Section 1); the various absorption lines are then observed by varying the value of the homogeneous magnetic field H. In this way we can vary the differences between the energy levels, and we observe resonance absorption whenever the difference between two energy levels corresponds to the radiation of 3.15 cm. For a free electron spin we observe resonance for a field of 3400 gauss, as we have shown in Section 1.

The coupling between the electron and the proton in a hydrogen atom is easily calculated from Eq. (96): It is around 500 gauss. This is about the order of magnitude for the first-order electron-nuclear spin-spin coupling.

Unfortunately, the simple theory that we have outlined above does not cover most of the situations with which we must deal in practice. Most of the molecules that are measured in ESR spectroscopy are fairly large aromatic free radicals; a typical example is diphenyl picryl hydrazyl (see

Fig. 8.9), which is sometimes used for testing the equipment. In these aromatic free radicals the unpaired electron is usually in a π molecular orbital, which is zero in the plane of the molecule. According to Eq. (96), the coupling constants between the electron spin and the various protons should all be zero, because the protons are all located in the molecular plane where the molecular orbital is zero. In practice, we measure finite values for the electron-proton coupling constants in these aromatic molecules, even though their values are much smaller than the effect we calculated for the hydrogen atom. It is interesting that some of the experimental electron-proton coupling constants are negative; this, of course, is not consistent with Eq. (96).

It is clear that we need a more precise theory in order to describe these small coupling constants in aromatic free radicals. In Eq. (80) we wrote the molecular eigenfunction as a product of one-electron functions, and this is too crude an approximation if we wish to account for the aromatic coupling constants. A model calculation was performed on the C—H free radical, and it was shown here that we obtain a finite coupling constant between the π electron and the proton if we introduce configuration interaction. It was shown that in the C—H radical the π electron has a coupling constant of about -20 gauss with the proton spin. This number agrees quite satisfactorily with the experimental coupling value of -23 gauss for the methyl radical. It may be argued that in larger free radicals, where the π electron is in a delocalized molecular orbital, the coupling constant with a proton should be roughly proportional to the charge density of the π electron on the carbon atom to which the proton is bonded. This π-electron atom charge density on the ith carbon atom, ρ_i is easily derived from simple Hückel molecular orbital theory. Accordingly, it might be expected that the coupling constant $a_{\mathrm{H},i}$ for the corresponding proton can be written as

$$a_{\mathrm{H},i} = \phi \rho_i \qquad (97)$$

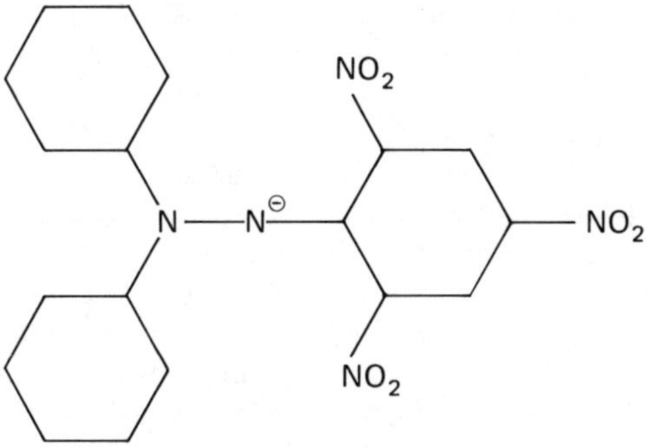

FIGURE 8.9 ● Diphenyl picryl hydrazyl (DPPH).

306 MAGNETIC RESONANCE

This rule seems to work reasonably well, and the empirical values of ϕ range between -22 and -28 gauss.

Most of the applications of electron spin resonance have been in organic chemistry. ESR has been used for structure determinations, for identifying intermediate products in reactions, and for explaining reaction mechanisms. Especially in the area of reaction mechanisms, ESR has been widely used in organic chemistry.

Problems

1. Why did the early attempts in 1936 to measure magnetic resonance fail?
2. Electron spin resonance is performed with radiation of 3.15 cm wavelength. At which magnetic field does resonance occur for a free electron?
3. Calculate the wavelength of the resonance radiation for ^{13}C in a magnetic field of 10,000 gauss.
4. Calculate the magnetic field that produces resonance radiation with a wavelength of 50 m for ^{19}F.
5. If we have 3×10^{22} protons in a homogeneous magnetic field of 10,000 gauss, what is the difference in population between the two spin levels at a temperature of 27°C (300°K) and at a temperature of 20°K?
6. Why are the upward and downward transition probabilities W_d and W_u for establishing thermal equilibrium in a spin system different from each other?
7. Find the solution of the population n as a function of the time t by solving the differential equation (37). How does your result relate to the equilibrium population difference $n(\infty)$?
8. What are the relaxation times T_1 and T_2 that occur in the Bloch equations called? Why are they different, and to which physical processes are they related?
9. How is the chemical shift constant σ defined? Explain what it is caused by and why it shifts the resonance to lower frequencies (as compared with the bare nucleus).
10. Are the absolute values of proton shielding constants σ available from experiments? If not, how are the values of Table 8.3 obtained?
11. If we perform proton magnetic resonance with a fixed homogeneous magnetic field of 10,000 gauss, how far are the resonance lines in HF and HI apart in terms of megacycles per second?
12. The value of the proton shielding constant σ increases in the sequence of molecules HF, HCl, HBr, HI. Can you explain this in terms of general chemical considerations, guided by Eq. (52) for atoms?
13. Is it possible to explain nuclear spin-spin coupling constants in terms of the direct dipole-dipole interactions between the spin magnetic moments? If not, how should the magnitude of the coupling constants be interpreted?
14. Is it possible to measure the spin-spin coupling constant of the H_2 molecule? We mentioned in Section 4 that the value of the coupling constant is 300 cycles/sec; how can this number be obtained?

15. Do the relative magnitudes of the splittings due to chemical shifts and to spin-spin coupling constants depend on the magnitude of the magnetic field **H** in which the NMR experiment is performed? If so, which of the two effects is predominant at high fields, and which at low fields?
16. Find out what the magnitude of the earth magnetic field is where you are and calculate the wavelength of electron resonance radiation in that magnetic field.
17. Calculate the ESR spectrum of the negative benzene ion, assuming that the six protons are equivalent and that the six coupling constants between the unpaired electron and the six spins are all the same. Determine the number of lines in the spectrum and their relative intensities.
18. Which quantum mechanical quantity do we derive by measuring the ESR spectrum of a free radical with one unpaired electron?
19. Why is the spin-spin coupling between the unpaired electron and the protons in a planar, aromatic free radical nonzero, in spite of the result we derived in Eq. (80)?
20. In ammonia, NH_3, there are three equivalent protons, which have the same chemical shift constants. The three proton-proton coupling constants are also equal to one another. Write the spin Hamiltonian for the three-photon system and express it in terms of the spin operators I_1^2, I_2^2, and I_3^2 of the three protons and of the operator I, where I is the total proton spin ($\mathbf{I} = \mathbf{I}_1 + \mathbf{I}_2 + \mathbf{I}_3$).

Bibliography

We have listed first a few general textbooks on magnetic resonance from a physicist's point of view (1, 2) and then a few books that deal with applications of magnetic resonance to chemical problems. Memory's book (5) lists the various theories for calculating the parameters that occur in NMR. As an example, we also list a table of available experimental NMR data to show how they can be used for the purpose of identification (6).

1. A. Abragam. *The Principles of Nuclear Magnetism.* Oxford University Press, London (1961).
2. C. P. Slichter. *Principles of Magnetic Resonance.* Harper & Row, New York (1963).
3. J. A. Pople, W. G. Schneider, and H. J. Bernstein. *High-Resolution Nuclear Magnetic Resonance.* McGraw-Hill, New York (1959).
4. A. Carrington and A. D. McLachlan. *Introduction to Magnetic Resonance.* Harper & Row, New York (1967).
5. J. D. Memory. *Quantum Theory of Magnetic Resonance Parameters.* McGraw-Hill, New York (1968).
6. F. A. Bovey. *NMR Data Tables for Organic Compounds.* Wiley-Interscience, New York (1967).

INDEX

A

Ab initio calculations, 203
Acetylene, 227, 230
Additivity rule, 217
Allowed transitions, 139
Ammonia, 221
Ammonia maser, 140
Ångstrom unit, 12
Angular momentum, 55, 60, 65
Angular velocity, 49
Anti-bonding orbitals, 193, 209
Aromatic molecules, 232
Asymmetric top, 188
Atomic configurations, 94
Atomic orbitals, 92, 114
Atomic units, 69

B

Band spectrum, 148
Benzene, 190, 232
Beveridge, 227
Black-body radiation, 128
Bloch, 274
Bloch equations, 284
Bloch functions, 258, 267
Body-centered cubic structure, 253, 255
Bohr, 131
Bohr mageton, 275
Bohr model, 11, 13
Bond angle, 224, 225
Bond energy, 187, 216, 227
Bond length, 187, 227, 241
Bond order, 240, 241
Bonding orbital, 193, 209
Born-Haber cycle, 254
Born-Oppenheimer approximation, 150, 153, 188
Boundary conditions, 23
Butadiene, 20, 238

C

Carbon monoxide, 215
Center of gravity, 48, 152
Chemical shift, 287
Closed-shell state, 113
CNDO method, 227
Complete set, 30
Configuration, 32
Configuration interaction, 202, 305
Conjugated molecules, 232
Constant of motion, 10, 58
Continuity, 15
Coolidge, 201
Corkscrew rule, 56
Correlation energy, 201
Coulomb integral, 197, 204
Coulson, 240
Coupling constants, 295
Covalent crystals, 251
Cross product, 56
Crystal structures, 249
Cubic cell, 250
Cylindrical symmetry, 166, 205

D

Davydov splitting, 266

INDEX

Debeye, 131
Debeye unit, 215
Degeneracy, 23
Delta function, 297
Dewar structures, 190
Diamond, 255, 260
Diatomic molecules, 45, 48, 145
Dirac constant, 10
Displacement coordinate, 159
Dissociation energy, 217
Doppler effect, 149
Double bonds, 227
DPPH, 305

E

Effective nuclear charge, 91, 115, 118, 194, 200
Effective potential, 115
Eigenfunction, 23
Eigenvalue, 23
Einstein, 128
Einstein coefficients, 132, 135, 138
Electric dipole moment, 182, 187, 215, 224, 226
Electric field strength, 124
Electromagnetic field, 124
Electron spin, 102, 274
Electronegativity, 212
Electronegativity values, 214, 252
Electronic transition, 169
Elementary cell, 249, 250
Elliptical coordinates, 194, 201, 244
Energy band, 257
Energy gap, 261
Energy leak, 280
ESR, 273, 300
Ethane, 227, 228
Ethyl alcohol, 279, 287
Ethylene, 227, 228
Euler polynomial, 52, 53, 67, 156
Even permutations, 111
Exchange integral, 198, 204
Exciton, 270
Expectation value, 24

F

Face-centered cubic, 253, 255
Fermi contact potential, 297, 298
Fermi energy, 262, 264
Fermi temperature, 262
Fine structure, 105
Fluorine molecule, 211
Forbidden transitions, 139
Force constant, 162, 187
Fortrat parabola, 179
Franck-Condon approximation, 176
Franck-Condon factors, 175
Free radicals, 300
Frequency, 125

G

g Factor, 276, 301
g Values, 276
Gamma rays, 125
Gas laser, 142
Gaussian function, 176
Gerade, 168, 205
Germanium, 255, 262
Goudsmit, 102
Graphite, 250, 255

H

Hamiltonian operator, 23, 25
Hansen, 274
Harmonic approximation, 161, 163
Harmonic oscillator, 37, 38, 146, 162
Hartree-Fock limit, 201
Hartree-Fock method, 117, 203
Hartree-Fock orbitals, 115
HCl molecule, 163
Heisenberg, 4
Heitler-London function, 198
Helium atom, 88, 93, 96
Helium-neon laser, 142
Hermite polynomials, 42
Hermitian operators, 25
Heterocyclic molecules, 242
Heteronuclear molecules, 211
Hexagonal cell, 250
  close-packed, 255
Homogeneous polynomials, 52
Hückel MO method, 233
Hund's case (a), 181
  case (b), 181
Huygens, 123
Hybridization, 216
Hybridized orbitals, 208
Hydrocarbons, 227
Hydrogen atom, 37, 63
Hydrogen eigenfunctions, 72, 78
Hydrogen s, p, d, f states, 73
Hydrogen $2p$ functions, 76
Hydrogen $3d$ functions, 77, 81
Hydrogen quantum numbers, 73
Hydrogen molecular ion, 189
Hydrogen molecule, 196
Hypergeometric series, 70, 71

I

INDO method, 227
Infrared spectrum, 149, 182

INDEX

Inversion center, 167
Ionic crystals, 251, 252
  integral, 204
  radius of, 253
Ionic structures, 199, 242

J

James, 201

K

Kekulé structures, 189
Kinetic energy, 9
Kolos, 201
Kuhn, 20

L

Laplace operator, 22
Laser, 139
LCAO MO method, 198
Lenz's law, 286
Lewis, 130
Line width, 149
Lithium, 256, 264
Lone-pair orbitals, 209, 223, 225
Lorentzian function, 285

M

Madelung constant, 254
Magnetic field strength, 124
Magnetic resonance, 273
Magnetic shielding, 287
Magnetic susceptibility, 264
Maser, 139
Maxwell, 281
Metals, 251, 255
Methane, 217
Microwave spectrum, 149, 183
MO method, 202
MO theory, 198
Molecular crystals, 251, 265
Molecular orbitals, 193, 233
Moment of inertia, 52, 188
Momentum, 5
Monoclinic, 250, 259
Morse potential, 163
Mulliken, 203, 212, 214

N

Naphthalene, 266
Natural line width, 149
Neumann expansion, 198
Newton, 123
Newtonian mechanics, 1

Nitrogen molecule, 205, 209
NMR, 273
Normal coordinates, 188
Normal modes, 188
Normalizable wave functions, 15
Nuclear magneton, 276
Nuclear spin, 276

O

Odd permutations, 111
Operators, 22
Orbit, 2
Orbital, 92, 193
Orbital exponent, 204, 205
Orthogonality, 26, 218
Orthonormal functions, 30
Orthorhombic, 250
Overlap charge, 195
Overlap integrals, 174, 175, 217, 234
Oxygen molecule, 210

P

P branch, 179
π electrons, 20
π electron density, 243
π orbitals, 205, 230
Packard, 274
Particle in a box, 14, 16, 262
Pauli exclusion principle, 20, 91, 109, 113
Pauling, 212, 214
Permutations, 109
Perturbation theory, 28, 32
Photoelectric effect, 128, 130
Photon, 130
Planck, 128
Planck's constant, 5
Plane wave, 124
Polar coordinates, 66, 97
Polarization, 126, 194
Pople, 227
Population inversion, 141
Potential curve, 155, 158, 187, 192
Potential energy, 9
Pound, 274
Probability density, 8
Probability distribution, 6
Probability pattern, 3, 5
Promotion energy, 221
Proton shielding constant, 287, 289
Purcell, 274
Pyridine, 242

Q

Q branch, 179
Quantized radiation, 130

INDEX

R

R branch, 179
Radar waves, 124
Radial distribution function, 78
Radial eigenfunctions, 70
Ransil, 205
Reduced mass, 45, 64, 146
Relaxation, 279
Relaxation time, 282
Reservoir, 280
Resonance energy, 237, 240
Resonance principle, 190
Rigid rotor, 37, 48, 146, 156
Ritz, 12
Roothaan, 201, 203
  SCF method, 203, 226
Rotational constant, 51, 53, 146, 147
Rotational motion, 48, 188
Rotational selection rules, 178
Rotational transitions, 183
Ruby laser, 141
Ruedenberg, 203
Rutherford model, 13
Rydberg constant, 13

S

σ orbitals, 205, 230
Scaling parameter, 33
SCF orbitals, 115
SCF method, 116
Schrödinger, 37
Schrödinger equation, 8
Schrödinger time-dependent equation, 136
Selection rules, 139, 166, 169, 171, 268, 302
Semiconductors, 255
Separability theorem, 150
Shielding, 90
Shielding constant, 118
Single bonds, 227
Singlet space function, 110
Singlet spin function, 107, 110
Singlet state, 107
Slater orbitals, 117, 119, 204
Slater rules, 118
Sommerfeld, 14
sp hybridization, 232
sp^2 hybridization, 229
sp^3 hybridization, 229
Σ-Π approximation, 232
Specific heat, 264
Spherical top, 188
Spin angular momentum, 103
Spin energy, 274
Spin magnetic moment, 104
Spin selection rule, 139
Spin variable, 106
Spin-lattice relaxation, 285
Spin-orbit coupling, 181
Spin-spin coupling, 290
Spin-spin relaxation, 285
Spontaneous emission, 133
Stationary states, 136
Stimulated emission, 133
Superposition of states, 30
Symmetry, 166, 205
Symmetric top, 188

T

Tetragonal cell, 250
Thermal equilibrium, 281
Tin, 262
Toluene, 289
Torrey, 274
Transition moment, 138, 169
Transition probability, 131, 138
Translational symmetry, 249
Triclinic, 250
Triple bonds, 227
Triplet space function, 110
Triplet spin functions, 108, 110
Triplet state, 107, 210
Two-center integral, 204

U

Uhlenbeck, 102
Uncertainty relations, 5
Ungerade, 168, 205
Unsaturated molecules, 227

V

Valence-bond function, 198
Valence state, 221
Variation principle, 28, 31, 101
Vector model, 60, 103, 105
Vibrational frequencies, 148
Vibrational motion, 45, 158
Vibrational structure, 174
Vibrational transitions, 182
Vibronic, 175

W

Water molecule, 224
Wave function, 8
Wave number, 12
Wave velocity, 125
Wavelength, 12, 125

X

X-rays, 125

Z

Zavoisky, 274